CAMBRIDGE LIBRARY COLLECTION

Books of enduring scholarly value

History of Oceania

This series focuses on Australia, New Zealand and the Pacific region from the arrival of European seafarers and missionaries to the early twentieth century. Contemporary accounts document the gradual development of the European settlements from penal colonies and whaling stations to thriving communities of farmers, miners and traders with fully-fledged administrative and legal systems. Particularly noteworthy are the descriptions of the indigenous peoples of the various islands, their customs, and their differing interactions with the European settlers.

Polynesian Researches during a Residence of Nearly Six Years in the South Sea Islands

From humble origins, and trained by the London Missionary Society in theology, printing and rudimentary medicine, William Ellis (1794–1872) sailed for the Society Islands in 1816. He found himself at the cusp of major cultural change as Western influences affected the indigenous Polynesians. During his time there, Ellis became a skilled linguist and able chronicler of the traditional yet rapidly shifting way of life. He succeeded in capturing vivid stories of a leisured people who, without written language, had developed a rich oral tradition, social structure and belief system. Published in 1829, this two-volume collection proved to be an important reference work, notably for its natural history; it soon accompanied Darwin aboard the *Beagle*. Volume 1 covers the voyage to Tahiti, the development of Tahitian orthography, the conversion of chief Pomare II, the establishment of a printing press on Moorea, and Ellis's first sermon in Tahitian, delivered on Huahine.

Cambridge University Press has long been a pioneer in the reissuing of out-of-print titles from its own backlist, producing digital reprints of books that are still sought after by scholars and students but could not be reprinted economically using traditional technology. The Cambridge Library Collection extends this activity to a wider range of books which are still of importance to researchers and professionals, either for the source material they contain, or as landmarks in the history of their academic discipline.

Drawing from the world-renowned collections in the Cambridge University Library and other partner libraries, and guided by the advice of experts in each subject area, Cambridge University Press is using state-of-the-art scanning machines in its own Printing House to capture the content of each book selected for inclusion. The files are processed to give a consistently clear, crisp image, and the books finished to the high quality standard for which the Press is recognised around the world. The latest print-on-demand technology ensures that the books will remain available indefinitely, and that orders for single or multiple copies can quickly be supplied.

The Cambridge Library Collection brings back to life books of enduring scholarly value (including out-of-copyright works originally issued by other publishers) across a wide range of disciplines in the humanities and social sciences and in science and technology.

Polynesian Researches
during a Residence of Nearly Six Years in the South Sea Islands

VOLUME 1

WILLIAM ELLIS

CAMBRIDGE
UNIVERSITY PRESS

University Printing House, Cambridge, CB2 8BS, United Kingdom

Published in the United States of America by Cambridge University Press, New York

Cambridge University Press is part of the University of Cambridge.
It furthers the University's mission by disseminating knowledge in the pursuit of education, learning and research at the highest international levels of excellence.

www.cambridge.org
Information on this title: www.cambridge.org/9781108065870

This edition first published 1829
This digitally printed version 2014

ISBN 978-1-108-06587-0 Paperback

Titles known to have formed part of Charles Darwin's library during the *Beagle* voyage, available in the CAMBRIDGE LIBRARY COLLECTION

Abel, Clarke: *Narrative of a Journey in the Interior of China, and of a Voyage to and from that Country in the Years 1816 and 1817* (1818) [ISBN 9781108045995]

Aubuisson de Voisins, J.F. d': *Traité de Géognosie* (2 vols., 1819) [ISBN 9781108029728]

Bougainville, L. de, translated by John Reinhold Forster: *A Voyage Round the World, Performed by Order of His Most Christian Majesty, in the Years 1766–1769* (1772) [9781108031875]

Buch, Leopold von, translated by John Black, with notes and illustrations by Robert Jameson: *Travels through Norway and Lapland during the years 1806, 1807, and 1808* (1813) [ISBN 9781108028813]

Byron, John: *The Narrative of the Honourable John Byron, Commodore in a Late Expedition Round the World* (1768) [ISBN 9781108065368]

Caldcleugh, Alexander: *Travels in South America, during the Years, 1819–20–21* (2 vols., 1825) [ISBN 9781108033732]

Callcott, Maria (née Graham): *Voyage of H.M.S. Blonde to the Sandwich Islands, in the Years 1824–1825* (1826) [ISBN 9781108062114]

Candolle, Augustin Pyramus de, and Sprengel, Kurt: *Elements of the Philosophy of Plants* (1821) [ISBN 9781108037464]

Colnett, James: *A Voyage to the South Atlantic and Round Cape Horn into the Pacific Ocean* (1798) [ISBN 9781108048354]

Cuvier, Georges: *Le règne animal distribué d'après son organisation* (4 vols., 1817) [ISBN 9781108058872]

Cuvier, Georges, edited by Edward Griffith: *The Animal Kingdom* (16 vols., 1827–35) [ISBN 9781108049702]

Daniell, J. Frederic: *Meteorological Essays and Observations* (1827) [ISBN 9781108056571]

De la Beche, Henry T.: *A Selection of the Geological Memoirs Contained in the Annales des Mines* (1824) [ISBN 9781108048408]

Earle, Augustus: *A Narrative of a Nine Months' Residence in New Zealand in 1827* (1832) [ISBN 9781108039789]

Ellis, William: *Polynesian Researches during a Residence of Nearly Six Years in the South Sea Islands* (2 vols., 1829) [ISBN 9781108065382]

Falkner, Thomas: *A Description of Patagonia, and the Adjoining Parts of South America* (1774) [ISBN 9781108060547]

Fleming, John: *The Philosophy of Zoology* (2 vols., 1822) [ISBN 9781108001649]

Flinders, Matthew: *A Voyage to Terra Australis* (2 vols., 1814) [ISBN 9781108018203]

Forster, John Reinhold: *Observations Made During a Voyage Round the World* (1778) [ISBN 9781108031882]

Greenough, George Bellas: *Critical Examination of the First Principles of Geology* (1819) [ISBN 9781108035323]

Hawkesworth, John: *An Account of the Voyages Undertaken by the Order of His Present Majesty for Making Discoveries in the Southern Hemisphere* (3 vols., 1773) [ISBN 9781108065528]

Head, Francis Bond: *Rough Notes Taken during some Rapid Journeys across the Pampas and among the Andes* (1826) [ISBN 9781108001618]

Humboldt, Alexander von: *Essai géognostique sur le gisement des roches dans les deux hémisphères* (1826) [ISBN 9781108049481]

Humboldt, Alexander von, translated by J.B.B. Eyriès: *Tableaux de la nature* (1828) [ISBN 9781108052757]

Humboldt, Alexander von, translated by Helen Maria Williams: *Personal Narrative of Travels* (7 vols., 1814–29) [ISBN 9781108028004]

Humboldt, Alexander von: *Fragmens de géologie et de climatologie Asiatiques* (2 vols., 1831) [ISBN 9781108049443]

Jones, Thomas: *A Companion to the Mountain Barometer* (1817) [ISBN 9781108049375]

King, Phillip Parker: *Narrative of a Survey of the Intertropical and Western Coasts of Australia, Performed between the Years 1818 and 1822* (2 vols., 1827) [ISBN 9781108045988]

Kirby, William and Spence, William: *An Introduction to Entomology* (4 vols., 1815–26) [ISBN 9781108065597]

Kotzebue, Otto von, translated by H.E. Lloyd: *A Voyage of Discovery, into the South Sea and Beering's Straits, for the Purpose of Exploring a North-East Passage* (3 vols., 1821) [ISBN 9781108057608]

La Pérouse, Jean-François de Galaup de, edited by L.A. Millet-Mureau: *A Voyage Round the World, Performed in the Years 1785, 1786, 1787, and 1788, by the Boussole and Astrolabe* (2 vols., 1799) [ISBN 9781108031851]

Lamarck, Jean-Baptiste Pierre Antoine de Monet de: *Histoire naturelle des animaux sans vertèbres* (7 vols., 1815–22) [ISBN 9781108059084]

Lyell, Charles: *Principles of Geology* (3 vols., 1830–3) [ISBN 9781108001342]

Macdouall, John: *Narrative of a Voyage to Patagonia and Terra del Fuego* (1833) [ISBN 9781108060981]

Mawe, John: *Travels in the Interior of Brazil* (1821) [ISBN 9781108052788]

Miers, John: *Travels in Chile and La Plata* (2 vols., 1826) [ISBN 9781108072977]

Molina, Giovanni Ignazio: *The Geographical, Natural, and Civil History of Chili* (2 vols., 1782–6, English translation 1809) [ISBN 9781108049474]

Owen, William Fitzwilliam, translated by Heaton Bowstead Robinson: *Narrative of Voyages to Explore the Shores of Africa, Arabia, and Madagascar* (2 vols., 1833) [ISBN 9781108050654]

Pernety, Antoine-Joseph: *The History of a Voyage to the Malouine (or Falkland) Islands* (1770, English translation 1771) [ISBN 9781108064330]

Phillips, William: *An Elementary Introduction to the Knowledge of Mineralogy* (1816) [ISBN 9781108049382]

Playfair, John: *Illustrations of the Huttonian Theory of the Earth* (1802) [ISBN 9781108072311]

Scrope, George Poulett: *Considerations on Volcanos* (1825) [ISBN 9781108072304]

Southey, Robert: *History of Brazil* (3 vols., 1810–19) [ISBN 9781108052870]

Spix, Johann Baptist von, and Martius, C.F.P. von, translated by H.E. Lloyd: *Travels in Brazil, in the Years 1817–1820* (2 vols., 1824) [ISBN 9781108063807]

Turnbull, John: *A Voyage Round the World, in the Years 1800, 1801, 1802, 1803, and 1804* (1805, this edition 1813) [ISBN 9781108053983]

Ulloa, Antonio de, translated and edited by John Adams: *A Voyage to South America* (2 vols., 1806) [ISBN 9781108031707]

Volney, Constantin-François: *Voyage en Syrie et en Égypte pendant les années 1783, 1784 et 1785* (2 vols., 1787) [ISBN 9781108066556]

Webster, William Henry Bayley: *Narrative of a Voyage to the Southern Atlantic Ocean, in the Years 1828, 29, 30, Performed in H.M. Sloop Chanticleer* (2 vols., 1834) [ISBN 9781108041898]

Weddell, James: *A Voyage towards the South Pole: Performed in the Years 1822–24* (1825) [ISBN 9781108041584]

Wood, James: *The Elements of Algebra* (1815) [ISBN 9781108066532]

For a complete list of titles in the Cambridge Library Collection please visit:
http://www.cambridge.org/features/CambridgeLibraryCollection/books.htm

Pomare

Po-ma-re, King of Tahiti, Eimeo, &c.

H. Fisher, Son & Co. London, 1829

POLYNESIAN RESEARCHES,

DURING

A RESIDENCE OF NEARLY SIX YEARS

IN THE

SOUTH SEA ISLANDS;

INCLUDING

DESCRIPTIONS OF THE NATURAL HISTORY AND SCENERY OF THE ISLANDS—WITH REMARKS ON THE HISTORY, MYTHOLOGY, TRADITIONS, GOVERNMENT, ARTS, MANNERS, AND CUSTOMS OF THE INHABITANTS.

BY

WILLIAM ELLIS,

MISSIONARY TO THE SOCIETY AND SANDWICH ISLANDS, AND AUTHOR OF THE "TOUR OF HAWAII."

"In so vast a field, there will be room to acquire fresh knowledge for centuries to come, coasts to survey, countries to explore, inhabitants to describe, and perhaps to render more happy."

COOKE.

IN TWO VOLUMES.

VOL. I.

LONDON:

FISHER, SON, & JACKSON, NEWGATE STREET,

M,DCCC,XXIX.

TO

THE DIRECTORS AND SUPPORTERS

OF

THE LONDON MISSIONARY SOCIETY;

THESE VOLUMES,

DESCRIBING THE SCENES OF THEIR EARLIEST EXERTIONS,

AND THE IMPORTANT RESULTS

OF THEIR OPERATIONS,

AMONG THOSE WHO WERE THE FIRST OBJECTS

OF THEIR BENEVOLENT SOLICITUDE,

ARE RESPECTFULLY INSCRIBED,

BY THEIR OBLIGED,

AND OBEDIENT SERVANT,

THE AUTHOR.

PREFACE.

Accurate information respecting the different parts of the world, is probably possessed in a greater degree, and diffused to a wider extent, at the present day, than it has been at any former period. The mariner has encountered the dangers of untraversed and hitherto impenetrable seas; and the traveller has explored remote and inhospitable countries, in order to increase general knowledge, and add new facilities to the prosecution of enlightened philosophical research.

Without depreciating the pursuits of science, or the advantages of a more enlarged acquaintance with the natural history of our globe, the Christian philanthropist directs his attention to objects still more important, and is led to contemplate, with growing intensity of interest, the moral and spiritual condition of mankind. The dominion and extent of delusive and sanguinary idolatries, with their moral debasement and attendant misery, have excited his liveliest concern; and to the melioration of human wretchedness thus induced, and the extension of true religion, as the only solid basis of virtue and happiness, his energies are directed, and his resources consecrated.—Animated by the predictions of inspiration which refer to the moral renovation of the world, and cheered by "the signs of the times," his anticipations of ultimate success are strengthened by the effects that already reward his exertions.

The results of efforts combined for the accomplishment of these objects, though various, have been such as materially to affect some of the most interesting portions of the human race. Their influence is at the present moment felt among the aborigines of Africa, the victims of colonial slavery, the millions of civilized China and India, the population of the inhospitable regions of Siberia and Greenland, and the inhabitants of the distant islands of the South Sea.

In this latter part of the world the author has spent a number of years, endeavouring to promote the knowledge of Christianity among the natives; and while engaged in this pursuit, he regarded it as perfectly consistent with his office, and compatible with its duties, to collect, as opportunity offered, information on various subjects relative to the country and its inhabitants.

Although circumscribed in geographical extent, and comparatively insignificant in amount of population, the South Sea Islands have been regarded with unusual interest ever since their discovery; and the descriptions already given to the public, of the loveliness of their general appearance, and the peculiar character and engaging manners of their Inhabitants, have excited a strong desire to obtain additional information relative to the varied natural phenomena of the Islands themselves; the early history; the moral, intellectual, and physical character of the people, and the nature of their ancient institutions.

All their usages of antiquity having been so entirely superseded by the new order of things that has followed the subversion of their former system, the knowledge of but few of them is retained by the majority of the inhabitants, while the rising generation is growing up in total ignorance of all that distinguished their ancestors from themselves. The present, therefore, seems to be the only time in which a variety of facts, connected with the former state of the Inhabitants, can be secured; and to furnish, as far as possible, an authentic record of these, and thus preserve them from oblivion, is one design of the following Work.

To those whose attention has been directed to the systems of polytheism that have at different times prevailed among mankind, the account of the ancient religion of the Islanders will not be uninteresting. Although established among a people scarcely above the rudest barbarism, destitute of letters, hieroglyphics, and symbols, and by their isolated situation deprived of all intercourse with the rest of the world; it is, as a system, singularly complete.

The invention displayed in the fabrication and adjustment of its several parts, the varied and imposing imagery under which it was exhibited, and the mysterious and complicated machinery which sustained its operations, were truly remarkable; and, in the standard of virtue which it fixed, in the future destinies it unfolded, and in its adaptation to the untutored but ardent mind, the Polynesian system will not suffer by comparison with any systems which have prevailed among the most polished and celebrated nations of ancient or modern times.

The following work will exhibit numerous facts, which may justly be regarded as illustrating the essential characteristics of idolatry, and its influence on a people, the simplicity of whose institutions affords facilities for observing its nature and tendencies, which could not be obtained in a more advanced state of society.

In some respects, the mythology of Tahiti presents features peculiarly its own: in others it exhibits a striking analogy to that of the nations of antiquity. In each, the light of truth occasionally gleams through a mass of darkness and error. The conviction that man is the subject of supernatural dominion, is recognized in all, and the multiplied objects of divine homage, which distinguished the polytheism of the ancients, marked also that of the rude islanders. Nor was the fabulous religion of the latter deficient in the mummeries of sorcery and witchcraft, the delusion of oracles, and the influence of other varieties of juggling, and oppressive spiritual domination.

The South Sea Islanders appear under circumstances peculiarly favourable to happiness, but their idolatry exhibits them as removed to the farthest extreme from such a state. The baneful effects of their delusion was increased by the vast preponderance of malignant deities, frequently the personifications of cruelty and vice. They had changed the glory of God into the image of corruptible things, and instead of exercising those affections of gratitude, complacency, and love, in the objects of their worship, which the living God supremely requires, they regarded their deities with horrific dread, and worshipped only with enslaving fear.

While the false system of Tahiti shews the distance to which those under its influence departed from the knowledge and service of the true God; it also furnishes additional confirmation of the fact, that polytheism, whether exhibited in the fascinating numbers of classic poetry, the splendid imagery of eastern fable, or the rude traditions of unlettered barbarians, is equally opposed to all just views of the being and perfections of the only proper object of religious homage and obedience; and that, whether invested with the gorgeous trappings of a cumbrous and imposing superstition, or appearing in the naked and repulsive deformity of rude idolatry, it is alike unfriendly to intellectual improvement, moral purity, individual happiness, social order, and national prosperity.

These volumes also contain a brief, but it is hoped satisfactory history of the origin, progress, and results of the Missionary enterprise, which, during the last thirty years, has, under the Divine blessing, transformed the barbarous, cruel, indolent, and idolatrous inhabitants of Tahiti, and the neighbouring Islands, into a comparatively civilized, humane, industrious, and Christian people. They also comprise a record of the measures pursued by the native governments, in changing the social economy of the people, and regulating their commercial intercourse with foreigners, in the promulgation of a new civil code, (a translation of which is given,) the establishment of courts of justice, and the introduction of trial by jury.

Besides information on these points, the present work furnishes an account of the intellectual culture, Christian experience, and general conduct, of the converts; the proceedings of the Missionaries in the several departments of their duty; the administration of the ordinances of Christianity; the establishment of the first churches, with their order and discipline; the advancement of education; the introduction of arts; the improvement in morals; and the progress of civilization.

During an absence of ten years from England, the author made copious notes of much that came under his notice, and, while residing in the South Seas, kept a daily journal. From these papers, from the printed and manuscript documents in the possession of the London Missionary Society, (to which the most ready access has been afforded,) from the very ample communications by the Missionaries in the islands, especially his respected colleagues Messrs. Barff and Williams, and from information derived by daily intercourse for several years with many of the natives, who have been identified with the most important events of the last thirty years in Tahiti, the present volumes have been written. He has studiously and constantly endeavoured to render the accounts accurate, and trusts they will prove not only interesting, but useful.

For the defects that may appear in the execution of the work, he feels it necessary to apologize. It has been prepared amidst incessant public engagements, and some parts have passed through the press during his absence on a distant journey in behalf of the Missionary Society.

To the Rev. JOSEPH FLETCHER, A. M. of London, who amidst his numerous and important engagements, has kindly inspected most of the sheets, and to Captain R. Elliot, R. N. who has favoured the author with the use of his drawings for the embellishment of the Work, he takes this opportunity of tendering his sincere and grateful acknowledgments.

July, 1829.

CONTENTS OF VOL. I.

CHAP. IX.

CHAP. X.

CHAP. XI.

CHAP. XII.

CHAP. XIII.

CHAP. XIV.

CHAP. XV.

CHAP. XVI.

CHAP. XVII.

CHAP. XVIII.

PLATES IN VOL. I.

WOOD ENGRAVINGS

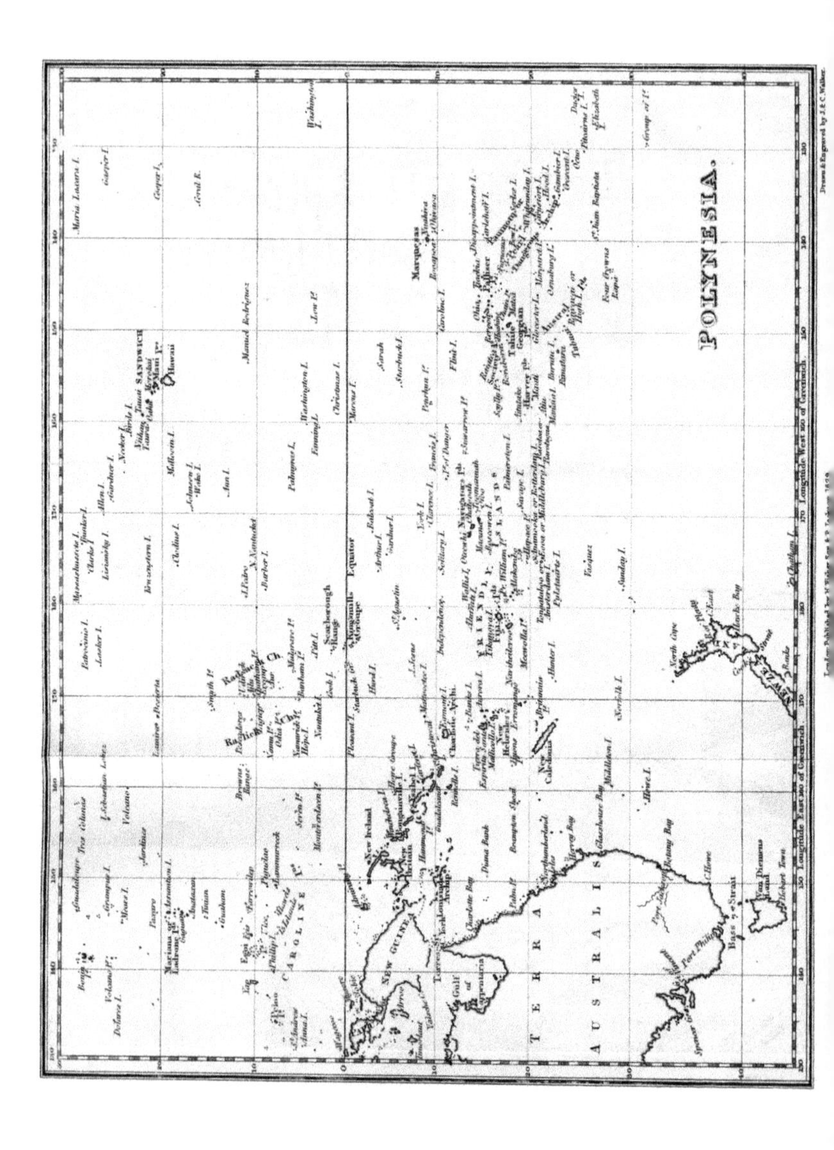

The material originally positioned here is too large for reproduction in this reissue. A PDF can be downloaded from the web address given on page iv of this book, by clicking on 'Resources Available'.

POLYNESIAN RESEARCHES.

CHAP. I.

Historical notice of the discovery of the Pacific—Voyage of Magellan—Discoveries of Cook—Impressions produced by his voyages—Missionary appointment to the South Sea Islands—Embarkation at Portsmouth—Last view of England. Reflections on leaving our native country—View of Madeira—Arrival at Rio de Janeiro—Appearance of the harbour—Slave ship—Incidents on shore—Voyage to New Holland—Tempest off the coast—Residence in New South Wales—Observations on the aborigines.

THE Pacific, the largest ocean in the world, extending over more than one third of the surface of our globe, was discovered in the year 1513, by Vasco Nugnez de Balboa, a courageous and enterprising Spaniard, governor of the Spanish colony of Santa Maria, in the isthmus of Darien.

The desire of finding a more direct communication with the East Indies had prompted Columbus to the daring voyage which resulted in the discovery of the new world. In that immense and unexplored region, his followers pursued their career of enterprise, until Balboa, by discovering the great South Sea, accomplished what Columbus, notwithstanding his most splendid achievements, had in vain attempted. In his march across the isthmus which separates the Atlantic from the Pacific, an enterprise designated by Robertson as the boldest on which the Spaniards had hitherto ventured in the New World, Balboa, having been informed by his Indian guides, that he might view the sea from the next mountain, advanced alone to its summit; and beholding the vast ocean spread out before him in all its majesty, fell on his knees, and rendered thanks to God for having conducted him to so important a discovery. He hastened towards the object he had so laboriously sought, and, on reaching its margin, plunged up to his middle in its waves, with his sword and buckler, and took possession of it in the name of his sovereign, Ferdinand of Spain.

Seven years after this important event, Magellan, a Portuguese, despatched by the court of Spain to ascertain the exact situation of the Molucca Islands, sailed along the eastern coast of South America, discovered the straits that bear his name; and, passing through them, first launched the ships of Europe in the Southern Sea. It is, however, probable, that neither Balboa, while he gazed with transport on its mighty waters, nor Magellan, when he first whitened with his canvass the waves of that ocean whose smooth surface induced him to call it the PACIFIC, had any idea either of its vast extent, of the numerous islands that studded its bosom, the diversified and beautiful structure of those foundations, which myriads of tiny architects had reared from the depths of the ocean to the level of its highest wave, or of the varied tribes of man by whom they were inhabited. Boldly pursuing his way across the untraversed surface of this immense ocean, Magellan discovered the Ladrone, and subsequently the Philippine islands. The object of the voyage was ultimately accomplished; the Victory, the vessel in which Magellan sailed, having performed the first voyage ever made round the world, returned to Europe: but the intrepid commander of the expedition terminated his life without reaching his original destination, having been killed in a quarrel with the natives of one of the Philippine Islands.

Several distinguished Spanish, Dutch, and British navigators followed the adventurous course of Magellan across the waters of the Pacific, and were rewarded by the discoveries they made in that part of the world, which, under the appellation of POLYNESIA, from a Greek term signifying *many islands*, geographers have since denominated the sixth division of the globe.*

But, although many single islands, and extensive groups of diversified forms and structure, some inhabited by isolated families of men, others peopled only by pelicans or aquatic birds, have been visited and explored, fresh discoveries continue to be made by almost every voyager; and it is by no means improbable, that there are still many islands, and even groups of islands, which remain unknown to the inhabitants of the other parts of the globe.

Most of the early voyages of discovery in this ocean attracted unusual attention; but none appear to have excited a livelier interest, or produced a deeper impression, than those performed by Captain Cook, in the latter part of the eighteenth century. These were instrumental, in a great degree, in diverting public attention from the

* According to Pinkerton, Malte Brun, and others, Polynesia includes the various islands found in the Pacific, from the Ladrones to Easter Island. The principal groups are, the Ladrone Islands—the Carolinas—the Pelew Islands—the Sandwich Islands—the Friendly Islands—the Navigators' Islands—the Harvey Islands—the Society Islands—the Georgian Islands, and the Marquesas.

splendid and stupendous discoveries in the New World, and directing it to the clustering islands spread over the Pacific; exhibiting them in all the loveliness of their natural scenery, the interesting simplicity, and novel manners, of their inhabitants. The influence of Cook's discoveries appears to have been felt by voyagers and travellers of other countries, as well as by those of his own. Humboldt, speaking of his laborious researches in South America, remarks, that, "the savages of America inspire less interest, since the celebrated navigators have made known to us the inhabitants of the South Sea, in whose character we find such a mixture of perversity and meekness: the state of half-civilization in which these islanders are found, gives a peculiar charm to the description of their manners. Here, a king, followed by a numerous suite, comes and presents the fruits of his orchard; there, the funeral festival embrowns the shade of the lofty forest. Such pictures, no doubt, have more attraction than those which portray the solemn gravity of the inhabitants of the Missouri or the Maranon."

Since the death of Captain Cook, several intelligent and scientific men from England, France, and Russia, have undertaken voyages of discovery in the South Seas, and have favoured the world with the result of their enterprises. Their accounts are read with interest by the philosopher, who seeks to study human nature under all its diversified forms; and by the naturalist, who investigates the phenomena of our globe, and the varied productions of its surface. Voyages of discovery are also favourite volumes with the juvenile reader. They impart to the youthful mind many delightful and glowing impressions relative to the strange and interesting scenes they exhibit, which in after life are seldom obliterated.—There are few who do not retain the vivid recollections of their first perusal of Prince Leeboo, or Captain Cook's Voyages. Often, when a school-boy, I have found the most gratifying recreation, for a winter's evening, in reading the account of the wreck of the Antelope, the discovery of Tahiti, and other narratives of a similar kind. Little, however, did I suppose, when in imagination I have followed the discoverer from island to island in the Pacific, and have gazed in fancy on the romantic hills and valleys, together with their strange but interesting inhabitants, that I should ever visit any of these scenes, the description of which afforded me so much satisfaction. Yet this, in the providence of God, has since taken place; and I have been led, not indeed on a voyage of discovery, commercial adventure, or naval enterprise, but, as a Christian Missionary, on an errand of instruction; not only to visit, but to reside a number of years among the interesting natives of those isolated regions.

Letters written in 1812 by my esteemed pastor, the Rev. J. Campbell, during a journey in South Africa, undertaken at the request of the London Missionary Society, first directed my attention to Missionary engagements. Subsequent events led me to devote my life to these pursuits, and, under the patronage of the above Society, I was, in the year 1815, appointed a Missionary to the South Seas.

In the month of January, 1816, in company with Mr. and Mrs. Threlkeld, Mrs. Ellis and myself sailed from Portsmouth for the Georgian and Society Islands. It was the morning of the Sabbath when we embarked. Our friends in Gosport were preparing to attend public worship, when we heard the report of a signal-gun. The sound excited a train of feelings, which can be understood only by those who have been placed in similar circumstances. It was a report announcing the arrival of that moment which was to separate, perhaps for ever, from home and all its endearments, and rend asunder every band which friendship and affection had entwined around the heart. The report we had heard might have proceeded from some other vessel; we hastened, therefore, to the windows, which commanded an extensive view of the sea, and, looking towards the anchorage, saw the small cloud of smoke rising up among the rigging, and the signal for sailing flying from the mast of our vessel. Instead of proceeding to the place of worship, we directed our steps towards the sea shore; but, before we left our dwelling, we united in prayer with our friends, and were by them affectionately committed to the guardian care of Him, in obedience to whose sacred injunction, "Go, teach all nations," we were about to embark; and on whose protection and blessing we alone depended for safety and success. A number of kind friends attended us to the beach, where, after waiting a few moments, we bade them farewell, and then raised the last foot from that earth which was our native soil, over which we had often trod under all the varied emotions of our earliest and maturer years, but which we never expected to tread again.

Among those who had walked with us to the shore, several dear brethren, students in the Missionary seminary at Gosport, anxious to defer, as long as possible, the final parting, took their seats beside us in the boat, and accompanied us to the ship. The wind was high, the sea rough, and the snow fell thickly around us. The inclemency of the weather favoured the silence we felt disposed to indulge; and although these were the last moments we were to spend with those whom kindness had prompted to attend us to the ship, the length and nature of the voyage before us, the thoughts that lingered with those, to whom, as we supposed, we had bidden adieu for ever, and the conviction that we must soon part with those who still sat beside us, to meet no more on earth, gave a melancholy solemnity to our thoughts, and predisposed us to silence and reflection, rather than to conversation. When we reached the vessel, a scene was presented very incongenial with the frame of our minds, and unlike the stillness of the Sabbath. All was bustle and confusion. The decks were crowded with live stock, vegetables, &c. the cabins filled with packages and trunks, and the sailors all engaged in the various labours incident to getting ready for sea. The moment had now arrived when we were to separate from our last friends—we took an affectionate, though rather hurried leave of them, and committing each other to the benediction of Heaven, exchanged the parting hand at the vessel's side. As their boat pushed off from the ship, they again bade us farewell by a signal, which we involuntarily returned, while we continued with indescribable emotion to watch their progress, until the intervention of some vessel, or the swelling of the waves, hid them entirely from our view.*

* They shortly afterwards embarked, and commenced their labours in the East nearly as soon as we reached the distant islands in the South: two of them, however, I believe, only remain; the others have died in the Missionary field, and, after a short and laborious course, under a most inhospitable clime, have ended their toil, and entered into rest.

Although we had embarked in the forenoon, the bustle and activity of every one on board, the adjusting and securing different articles in the cabin, brought on the close of the day, before we felt in any degree settled. Towards evening, however, I left the cabin for the deck, and enjoyed an hour of solemn, and, I trust, profitable meditation. Our ship was now under way, and proceeding steadily, though not rapidly, through the water. Every headland we passed on the Isle of Wight, and every point of land on the Hampshire coast, as it receded from my view, awakened the impression that I should never behold it again. I lingered with intensity of feeling on each passing scene, until the shadows of night gathered thickly around, and the only objects visible from the ship were a few distant lights, glimmering amidst the darkness in which every thing besides was concealed. After gazing on these lights until a late hour, I directed, as I supposed, a last glance towards them, and the coasts they illuminated, and retired to rest.

The next morning I hastened on deck, and looking abroad upon the expanse of waters, distinguished with delight a point of land. It was England; my eye rested on it with strong and painful interest; the mighty waters, like those of the deluge, appeared to rise higher and higher; until, at last, the waves of the distant and naked horizon appeared to have rolled over it; and our vessel, like the ark, seemed all that remained to us of the terrestrial world. In every direction there was nothing now to be seen, but one wide waste of water below, and the outstretched heavens above. England, with all its associations and its enjoyments, its tenderest earthly ties, and its distinguished religious privileges, had vanished.

My feelings, though strong, were not discouraging, nor did my choice awaken one emotion of regret; my desire to engage in the work, was as ardent as when my services were first tendered. From many sources of happiness, and sacred Christian privileges long enjoyed, I felt myself, indeed, about to be removed; while dangers and trials, hitherto unknown, could not but be anticipated. The Divine promise, however, "Lo, I am with you always, even unto the end of the world," was my support, and under its cheering influence I could appropriate the language of the poet, and exclaim—

O thou great Arbiter of life and death!
Nature's immortal, immaterial sun!
Whose all-prolific beam late called me forth
From darkness, teeming darkness, where I lay
The worm's inferior—and, in rank, beneath
The dust I tread on—high to bear my brow,
To drink the spirit of the golden day,
And triumph in existence; and couldst know
No motive but my bliss; and hast ordained
A rise in blessing! with the Patriarch's joy
Thy call I follow to the land unknown:
I trust in thee, and know in whom I trust:
Or life or death is equal; neither weighs;
All weighs in this—O LET ME LIVE TO THEE!"

The parting scenes, the embarkation, the last view taken of his native land, when leaving it for a distant clime, in which he expects to end his days, awaken indescribable emotions, and render it a season to which a Missionary is accustomed to look back, during subsequent periods of his life, with no ordinary interest. I have witnessed these emotions in others, as well as experienced them myself, and shall not soon forget the evident feeling with which Mr. Nott, who, after an absence of thirty years, visited England in the summer of 1826, exclaimed, as he a second time left the British shore, to return to the South Sea Islands, not, in the language of the poet, (Camoens,) "Ungrateful country, thou shalt not possess my bones," but, "Io nei oe e tau fenua! eita vau e tahi faahou adu ia oe:" Farewell, my native land, I shall never step on you again.

Out of sight of land, and proceeding every day farther from it, the feelings in immediate connexion therewith gradually began to subside, our thoughts were increasingly occupied with the novel scenes by which we were surrounded; and our attention was engaged by the pursuits which, at sea, we were able to follow. About three weeks after leaving Portsmouth, we touched at Madeira, and, proceeding on our voyage to Rio Janeiro, cast anchor at the mouth of its beautiful harbour in the evening of the 20th of March, 1816.

The light of the next morning presented before us one of the most magnificent and extensive landscapes I ever beheld. The mass of granite rock, surmounted by the fort of Santa Cruz on our right, the towering Sugar-loaf mountain on our left, the picturesque island at the mouth of the harbour, the distant town of St. Sebastian, the turrets of the castle, the convent of St. Antonio, the lofty range of mountains in the interior, whose receding summits were almost lost in aërial perspective, where

"Distance lends enchantment to the view,"

all successively met the eye, together with the widely expanded and beautiful bay, one of the finest in the world, studded with verdant islands, rendered more picturesque by the white cottages with which they were adorned.

The whole scene was enlivened by the numerous boats, with their white and singularly shaped sails, incessantly gliding to and fro on the smooth surface of the water, and the shipping of different nations riding at anchor in the bay, or moored to the shore. Among the vessels, which exhibited almost every variety of size and form, those by no means least interesting to us, were two British frigates; one of which was the Alceste, on her way to China, to join Lord Amherst's embassy. These objects excited in our minds a variety of pleasing sensations, heightened by the circumstance of the country before us being almost the first land we had seen since leaving England.

There is something very exhilarating in approaching land, or entering a friendly port, after a long voyage; and the pleasure we felt on this occasion was so much increased by the novel and delightful landscapes incessantly opening to our view, as we sailed along the bay, that we were unwilling for a moment to leave the deck. Our enjoyment was, however, interrupted by a spectacle adapted to awaken sensations very different indeed from those inspired by the loveliness and peace of the scenery around us.

We had proceeded about half way to the anchorage, when we approached a brig sailing also into the harbour, which, as we came alongside of her, appeared to be a slave ship returning from the coast of Africa. The morning was fine and the air refreshing, and this had probably induced the cruel keepers to bring their wretched captives up from the dungeons of pestilence and death in which they had been confined. The central part of the deck was crowded with almost naked Africans, constituting part of the cargo of the gloomy looking vessel.

Though their ages appeared various, the majority seemed to have just arrived at that period of human life, when the prospects of man are brightest, and the hopes of future happiness more distinct and glowing, than during any other portion of his existence; they were most of them, so far as we could judge, from fourteen to eighteen or twenty years of age; some were younger. We regarded them with a degree of melancholy interest, which for a time rendered us insensible to the beauties of nature every where spread before our eyes. Our passing, however, appeared to affect them but little. The greater part of these unhappy beings stood nearly motionless, though we did not perceive that they were chained: some directed towards us a look of seeming indifference; others, with their arms folded, appeared pensive in sadness; while several, leaning on the ship's side, were gazing on the green islands of the bay, the rocky mountains, and all the wild luxuriance of the smiling landscape; which probably awakened in their bosoms thoughts of "home and all its pleasures," from which they had so recently been torn; and, judging of the future by the past short period of their wretched bondage, their minds were perhaps distressed with painful anticipations of the toils and sufferings that would await them on the foreign shore they were approaching!

Circumstances detained us at Rio Janeiro above six weeks, and although on our arrival we were perfect strangers, we experienced the greatest hospitality and kindness from the English merchants and other residents there. During the whole of our stay, two of these gentlemen accommodated us at their country houses, a few miles distant from the city, where all that friendship could devise for our enjoyment was generously furnished, and every thing provided, when we left, that could make the remaining part of our voyage comfortable.

The heat of the climate was rather oppressive, but the mornings and evenings were pleasant, and, during the forenoon, the sea breezes in general refreshing. The habits of the people, the singularity of the buildings, the narrow streets, projecting balconies, and trellis-work doors and windows, the varied productions of the country, with the sublime grandeur and romantic beauty of the scenery, were all adapted to arrest the attention of those who now, for the first time, found themselves in a foreign land.

To us, the moral and religious state of the people was the subject of greatest interest; and, every observation we made, was adapted to awaken the liveliest gratitude to Him who had cast our lot in a happier land. Ignorance, and disregard of all religious principle, or the substitution of ceremony in its place, appeared every where prevalent. To the freedom of the press, and liberty of conscience, the inhabitants were perfect strangers. No book, we were informed, was allowed to be printed or imported for circulation, without the inspection of individuals appointed for this duty, whose censorship, it appeared, was such as to extinguish every source of light, and perpetuate the darkness of the people. Popery is the religion of the country; and we had an opportunity of beholding it in its own element. The demise of the queen-dowager of Portugal took place about the time of our arrival; and I had an opportunity of witnessing the funeral, which took place by torch-light. Numbers of ecclesiastics, in the habits of their respective orders, appeared in the procession, mounted on mules, which were led by persons bearing large burning tapers or torches; and on the occasion of a ceremony, connected, as we were informed, with the passage of her soul from purgatory to the regions of glory, the royal chapel was most splendidly illuminated. Desiring to see, for myself, their kind of worship, and the appearance of the worshippers, I frequently went to the royal chapel, on our first arrival. The rich gilding and numerous paintings, the images, massy silver candlesticks, and other costly ornaments of the building; the novel habits and sonorous voices of the priests; and, above all, the music mixed with many of their rites, were certainly adapted to produce a powerful impression upon the feelings of the majority of those who resorted thither; the greater part of whom had perhaps never seen a Bible! But notwithstanding there was so much that was imposing in its accompaniments, their worship often appeared a mere heartless attendance on customary ceremonies. Images of the Virgin Mary appeared at the corners of some of the principal streets, in little glass-cases, and in the evening a small lamp was placed before them. In front of these, the poor ignorant Catholic, kneeling in the streets, and offering his prayers to the image, together with other ceremonies performed at this season of the year, presented a most lamentable spectacle. Scenes, the most ludicrous imaginable, sometimes occurred. I was surprised one morning, about the time of Good Friday, to be old what I thought was a man suspended from a tree, on the opposite side of the road: observing my attention attracted, the family informed me that it was the day on hich the Catholics were accustomed to hang Judas. I was surprised to see this representation of the traitor, exhibited in a fashionable coat, waistcoat, and pantaloons, with a pair of Hessian boots, and a cocked hat! The figure hung there till about noon, when it was taken down, and fastened upon the back of a young ox: one end of a rope was tied to each of the animal's horns, and the other end held at a distance of six or eight yards by two young men; who, keeping opposite sides of the road, ran backwards and forwards with the animal, till it became quite furious, and at last, dislodging the image of Judas from its back, the ox tore it to pieces with its horns and its feet. The spectators appeared to derive no small gratification from the exhibition; but such a scene, partaking, according to their opinion, in some degree, of a religious observance, could not be witnessed by a Christian without emotions of pain.

I draw no invidious comparisons between Roman Catholics and Protestants; I desire to cultivate towards the former, as individuals, every feeling of Christian kindness and charity; but I could contemplate Popery with no satisfaction, not because its extension circumscribes the influence of Protestantism, but because it has always appeared to me one of the most absurd and fatal delusions which the powers of darkness ever invented for the destruction of mankind.

Here, for the first time, we came into actual contact with slavery. There are, perhaps, few places where the slaves meet with milder treatment; but it was most distressing, on passing the slave market, to observe the wretched captives there bought and sold like cattle; or to see two or three interesting looking youths, wearing a thin dress, and having a new red cotton handkerchief round their heads, led through the streets by a slave-dealer, who, entering the different houses or workshops as he passed along, offered the young negroes for sale; yet scarcely a day passed while we were in the town, during which we did not meet these heartless traffickers in human beings thus employed. In the English or Portuguese families with which we had any opportunities of becoming acquainted, although the domestic slaves did not appear to be treated with that unkindness which the slaves in the field often experience, yet, even here, the whip was frequently employed in a manner, and under circumstances, revolting to every feeling of humanity.

While we continued in Rio, I had several opportunities of preaching on the Sabbath in the dwelling houses of two of the merchants whom we were visiting. This was shortly after the treaty of peace with Great Britain, which secured to British subjects residing in Brazil, the right of public Protestant worship, but not of proselyting the inhabitants. Several of the English families attended; by whom proposals were made, requesting me to remain as a minister of religion among them. There were at that time fifty-seven British mercantile houses—two hundred and fifty English; and dependent upon them, six hundred servants, including blacks. Having, however, devoted my life to the service of the heathen, I felt it my duty to decline their invitation, and to proceed to my original destination. During the first week of May, we took leave of our friends, thankful for the attentions and kindness we had experienced. Severe domestic affliction detained my colleague, the Rev. L. E. Threlkeld, at Rio, and we were under the necessity of proceeding alone on the remainder of our voyage.

Sailing from Rio, we directed our course across the Atlantic, doubled the Cape of Good Hope, and, traversing the Indian ocean, proceeded towards New South Wales. Our passage was pleasant, and eleven weeks after leaving Brazil, we made the western coast of Van Diemen's Land. We passed through Bass's Straits on the same day, and sailed along the eastern shore of New Holland towards Port Jackson. Soon after day-light the next morning, we perceived a sail some miles before us, which we found on nearer approach to be a small schooner. Our captain on visiting her found only three men on board, who were in the greatest distress. They had been at Kangaroo Island procuring seal-skins, with a quantity of which they were now bound to Sydney. They had remained on the island, catching seals, till their provisions were nearly expended; and during their voyage, they had encountered much heavy weather, had been nearly lost, and were so exhausted by fatigue, want of food, and constant exposure, that they could not even alter the sails, when a change in the wind rendered it necessary. They had been for some time living on seal-skins; pieces of which were found in a saucepan over the fire, when the boat's crew boarded them. The men from our ship trimmed their sails, and our captain offered to take them in tow; but as they were so near their port, which they hoped to reach the next day, they declined accepting his proposal. When he returned to the ship, he sent them some bread and beef, a bottle of wine, and some water; which the poor starving men received with an indescribable degree of eagerness and joy. The seamen who conveyed these supplies returned to the ship, and we kept on our way. We did not, however, hear of their arrival, and as we remained nearly six months in Sydney after this time, and received no tidings of them, it is probable their crazy bark was wrecked, or foundered during a heavy storm that came on in the course of the following day.

The wind from the south continued fresh and favourable, and in the forenoon of the next day we sailed towards the shore, under the influence of exhilarated spirits, and the confident expectation of landing in Port Jackson before sunset. About noon we found ourselves near enough the coast to distinguish different objects along the shore, and soon discovered the flagstaff erected on one of the heads leading to Sydney, our port of destination, about four miles distant from us, but rather to windward. The captain and officers being strangers to the port, some little time was spent in scanning the coast, in the hope of finding an opening still farther northward; but at twelve o'clock our apprehensions of having missed our port were confirmed, as the latitude was then found, by an observation of the sun, to be four miles to the northward of Sydney heads. We had, in fact, sailed with a strong but favourable wind, four miles past the harbour which we ought to have entered. Hope, which had beamed in every eye, and lighted up every countenance with anticipated pleasure, when we first neared the land, had alternated with fear, or given way to most intense anxiety, when we witnessed the uncertainty that prevailed among our companions, as to our actual situation; but disappointment the most distressing, was now strongly marked in every countenance. "About ship," exclaimed the captain; immediately the ship's head was turned from the land, and, steering as near the wind as possible, we proceeded towards the open sea. After sailing in this direction for some time, the ship was again turned towards the shore; but the wind, which during the forenoon had been so favourable, was now against us, and as soon as we could distinguish the flagstaff on the coast, we found ourselves farther from it than before. The wind increased; and as the evening advanced, a heavy storm came on, which raged with fearful violence. The night was unusually dark; the long and heavy waves of the Pacific rolled in foam around our vessel; the stormy wind howled through the rigging; all hands were on deck, and twice or thrice, while in the act of turning the ship from the land, the sails were rent by the tempest; while the hoarse and hollow roaring of the breakers, and the occasional glimmering of lights on the coast, combined to convince us of our situation, and the proximity of our danger. The depression of spirits, resulting from the disappointment, which had been more or less felt by all on board, the noise of the tempest, the vociferations and frequent imprecations of the officers, the hurried steps and almost incessant labours of the seamen on deck, and the heavy and violent motion of the vessel, which detached from their fastenings, and dashed with violence from one side of the ship to the other, chests of drawers, trunks, and barrels, that had remained secure and stationary during the voyage, produced a state of mind peculiarly distressing. The general disorder that prevailed, with the constant apprehension of striking on some fatal rock, that might lie unseen near the craggy and iron-bound shore, and being either ingulfed in the mighty deep, or wrecked on the inhospitable coast, rendered the night altogether one of the most alarming and anxious that we had passed since our departure from England. Amidst the confusion by which we were surrounded, we experienced comparative composure of mind, resting on our God:

"When o'er the fearful depth we hung,
High on the broken wave,
We knew He was not slow to hear,
Nor impotent to save."

In such a season, confidence in Him who holdeth the wind in his fists, and the waters in the hollow of his hand, can alone impart serenity and support.

As the morning advanced, the storm abated; and at sunrise we found ourselves at a considerable distance from the shore. Contrary winds kept us out at sea for nearly a fortnight, which was by far the most irksome part of our voyage. At length we again approached the coast, and were delighted, as we sailed along it on the morning of the eleventh day, to behold a pilot-boat steering towards us. Our vessel had been several times seen from the shore, since the day of our first disappointment; and as soon as we had appeared in sight this morning, the governor of New South Wales, then residing at Sydney, had despatched the pilot, with orders to go out even sixty miles, rather than return without bringing the vessel in. He boarded us about twenty miles from Port Jackson, and conducted us safely within the heads, in the evening of the same day. Early the next morning, we proceeded to Sydney Cove, where we cast anchor on the 22d of July, after a passage, including our stay in Rio Janeiro, of only a few days more than six months.

Five months elapsed before we could meet with a conveyance to the Society Islands. This detention, however, favoured me with an opportunity of visiting the chief settlements of New South Wales, and beholding several of the rare and interesting animals and vegetable productions of that important colony. I was happy also to become acquainted with Mr. Leigh, the

Wesleyan minister, and to experience, during this period, the friendship and kind attentions of the Rev. S. Marsden, senior chaplain of the colony, the steady and indefatigable friend of Missions and Missionaries in the South Seas. He resided at Paramatta, where we passed the greater part of our stay in New South Wales very pleasantly, in the family of the late Mr. Hassel, formerly a Missionary in Tahiti. Mr. and Mrs. Hassel landed at Matavai from the ship Duff, in 1797, but had retired to Port Jackson, in consequence of an attack made by the natives on the Missionaries.

In company with Mr. S. O. Hassel, I made several excursions into the interior of the country, where we frequently saw the inhabitants more completely in a state of nature, than those we met with in the vicinity of the principal towns. The aborigines are but thinly spread over that part of New Holland bordering on the colony; and though the population has been estimated at three millions, I am disposed to think, that, notwithstanding the geographical extent of the country, it does not contain so many inhabitants. Their appearance is generally repulsive, their faces looking more deformed from their wearing a skewer through the cartilage of the nose. Their colour is dark olive, or black, and their hair rather crisped than woolly. In proportion to the body, their limbs are small and weak, while their gait is exceedingly awkward. Excepting in the neighbourhood of the chief towns, they were usually destitute of clothing, though armed with a spear or lance, with which at a great distance they are fatal marksmen. They are represented as indolent, treacherous, and cruel. Agriculture is unknown among them, although the indigenous productions of the country yield them little if any subsistence. Their food is frequently scanty, precarious, and loathsome, sometimes consisting of grubs and reptiles taken in the hollow or decayed trees of the forest. Occasionally, however, they procure excellent fish from the sea, or the lakes, rivers, &c. Their dwellings are low huts of bark, and afford but a mere temporary shelter from the weather.

They are a distinct people from the inhabitants of New Zealand, or the South Sea Islands; altogether inferior to them, and apparently the lowest grade of human kind. Their habits are fugitive and migratory, and this has perhaps greatly contributed to the failure of the benevolent attempts that have been made by the government and others to meliorate their condition, and elevate their character. The school for aboriginal children, under the patronage of the government, was a most interesting institution: I frequently visited it, and was surprised to learn that, though treated with every kindness, the young scholars, when an opportunity occurred, frequently left the school, and fled to their native woods, where every effort to discover the retreat, or to reclaim them, proved ineffectual. Notwithstanding their present abject condition, and all the existing barriers to their improvement, it is most ardently to be hoped, and most confidently to be anticipated, that the period will arrive, when this degraded and wretched people will be raised to the enjoyment of all the blessings of intelligence, civilization, and Christianity.

CHAP. II.

Voyage to New Zealand—Intercourse with the inhabitants—Sabbath on shore—Visit to Waikadie—Instance of parental tenderness—Forest scenery—Sham fight and war-dances—Character of the New Zealanders—Prospects of the Mission—Arrival at the Island of Rapa—Singularity of its structure—Appearance of the natives—Violent proceedings on board—Remarkable interposition of Providence—Visit of the natives to Tahiti—Introduction of Christianity to Rapa—Increased geographical acquaintance with the Pacific.

On the tenth of December, 1816, we sailed from Sydney in the Queen Charlotte, a brig belonging to J. Birnie, Esq., bound for the Society and Marquesan Islands. On the 21st of the same month, we reached New Zealand; and here for the first time saw the rude, untutored inhabitants of the South Sea Islands, in their native state. At daylight, on the morning after our arrival on the coast, we found ourselves off Wangaroa bay, where, six years before, the murderous quarrel took place, in which the crew of the Boyd were cut off by the natives, and near which, subsequently, the Methodist Missionary station at Wesleydale, established in 1823, has been, through the alarming and violent conduct of the inhabitants, abandoned by the Missionaries, and utterly destroyed by the natives. Several canoes, with three or four men in each, approached our vessel at a very early hour, with fish, fishing-lines, hooks, and a few curiosities for sale. Their canoes were all single, generally between twenty and thirty feet long, formed out of one tree, and nearly destitute of every kind of ornament.

The men, almost naked, were rather above the middle stature, of a dark copper colour, their features frequently well formed, their hair black and bushy, and their faces much tataued, and ornamented, or rather disfigured, by the unsparing application of a kind of white clay and red ochre mixed with oil. Their appearance and conduct, during our first interview, was by no means adapted to inspire us with prepossessions in their favour. Our captain refused to admit them into the ship, and after bartering with them for some of their fish, we proceeded on our voyage.

On reaching the Bay of Islands we were cordially welcomed by our Christian brethren, the Missionaries of the Church Missionary Society, who had been about two years engaged in promoting instruction and civilization among the New Zealanders. They were the first Missionaries we had seen on heathen ground, and it afforded us pleasure to become acquainted with those who were in some respects to be our future fellow-labourers. Having been kindly invited to spend on shore the next day, which was the Sabbath, we left the ship soon after breakfast, on the morning of the 22nd. When we reached the landing place, crowds of natives thronged around us, with an idle but by no means ceremonious curiosity, and some time elapsed before we could proceed from the beach to the houses of our friends.

The Missionaries had on the preceding day invited me to officiate for them, and I was happy to have an opportunity of preaching the gospel on the shores of New Zealand. Several of the natives appeared in our little congregation, influenced probably by curiosity, as the service was held in a language unintelligible to them. I could not, however, but indulge the hope that the time was not distant, when, through the influence of the schools already established, and the general instructions given by the Missionaries; my brethren would have the pleasure of preaching, on every returning Sabbath, the unsearchable riches of Christ, to numerous assemblies of attentive Christian hearers. The circumstance of its being exactly two years, this Sabbath day, since Mr. Marsden, who visited New Zealand in 1814—1815, for the purpose of establishing a Christian Mission among the people, preached, not far from this spot, the first sermon that was ever delivered in New Zealand, added to the feelings of interest connected with the engagements of the day.

Circumstances detaining us about a week in the Bay of Islands, afforded me the means of becoming more fully acquainted with the Missionaries, making excursions to different parts of the adjacent country, and witnessing many of the singular manners and customs of the people. I visited, in company with the captain of our ship, and Mr. Hall, one or two of the forests which produce the New Zealand pine, recently discovered to be so valuable as spars for vessels.

In one of these excursions, shortly after leaving the Bay of Islands, we reached Kowakowa, where Mr. Hall proposed to land. As we approached the shore, no trace of inhabitants appeared, but we had scarcely landed when we were somewhat surprised by the appearance of Tetoro and a number of his people. The chief ran to meet us, greeting us in English, with "How do you do?" He perceived I was a stranger, and, on hearing my errand and destination, he offered me his hand, and saluted me, according to the custom of his country, by touching my nose with his. He was a tall, fine-looking man, about six feet high, and proportionably stout; his limbs firm and muscular, and when dressed in his war-cloak, with all his implements of death appended to his person, he must have appeared formidable to his enemies. When acquainted with our business, he prepared to accompany us; but before we set out, an incident occurred that greatly raised my estimation of his character. In front of the hut sat his wife, and around her played two or three little children. In passing from the hut to the boat, Tetoro struck one of the little ones with his foot; the child cried, and though the chief had his mat on, and his gun in his hand, and was in the act of stepping into the boat where we were waiting for him, he no sooner heard its cries, than he turned back, took the child up in his arms, stroked its little head, dried its tears, and giving it to the mother hastened to join us. His conversation in the boat, during the remainder of the voyage, indicated no inferiority of intellect nor deficiency of information, as far as he had possessed the means of obtaining it. On reaching Waikadie, about twenty miles from our ship, we were met by Waivea, Tetoro's brother; but his relationship appeared to be almost all that he possessed in common with him, as he was both in appearance and in conduct entirely a savage.

We accompanied them to the adjacent forests. The earth was completely covered with thick-spreading and forked roots, brambles, and creeping plants, overgrown with moss, and interwoven so as to form a kind of uneven matting, which rendered travelling exceedingly difficult. The underwood was in many parts thick, and the trunks of the lofty trees rose like clusters of pillars supporting the canopy of interwoven boughs and verdant foliage, through which the sun's rays seldom penetrated. There were no trodden paths, and the wild and dreary solitude of the place was only broken by the voice of some lonely bird, which chirped among the branches of the bushes, or, startled by our intrusion on its retirement, darted across our path. A sensation of solemnity and awe involuntarily arose in the mind, while contemplating a scene of such peculiar character, so unlike the ordinary haunts of man, and so adapted, from the silent grandeur of his works, to elevate the soul with the sublimest conceptions of the Almighty. I was remarkably struck with the gigantic size of many of the trees, some of which appeared to rise nearly one hundred feet, without a branch, while two men with extended arms could not clasp their trunks. About three in the afternoon we left Waikadie, but the darkness of night veiled every object from our view, long before we reached our vessel.

Near the settlement at Rangehoo, a small field had been tilled by the Missionaries, in the European manner. I visited it in company with Mr. King, and was pleased to see one of the first crops of wheat that had ever grown, under European culture, in New Zealand, looking green and flourishing. Two years before this, Duatere and 'Honghi had received wheat from Mr. Marsden, which they had carefully sown, and which had arrived at perfection. The introduction of the European methods of culture, and subsequent processes of converting it into bread, may naturally be expected to encourage the natives to facilitate its more extensive growth. In several parts of the low-lands the native flax-plant, *phormium tenax,* was growing remarkably strong. It is by no means like the flax or hemp plants of England, but resembles, in its appearance and manner of growth, the flag or iris; the long broad sword-shaped leaves furnish the fibre so useful in making dresses for the natives, fishing lines, twine, and strong cordage employed as running rigging in most of the vessels that trade with the islanders. It is a most valuable plant, and will probably furnish an important article of commerce with New South Wales, or England.

An unusual noise from the land aroused us early on the morning of the 25th, and, on reaching the deck, a number of war-canoes were seen lying along the shore, while crowds of natives on the beach were engaged in war dances, shouting, and firing their muskets at frequent intervals. On inquiry, we found that on the day we had visited Waikadie, a chief of Rangehoo had committed suicide, by throwing himself from a high rock into the sea. This event had brought the chiefs and warriors of the adjacent country, to investigate the cause of his death; armed and prepared for revenge, in the event of his having been murdered. A council was held for some hours on the beach, when the strangers, being satis fied as to the cause and manner of the chief's death, preparations for war were discontinued, the people of Rangehoo repaired to their fields, to procure potatoes for their entertainment. It was Christmas-day, and about twelve o'clock we went on shore to dine with one of the Mission families. In the afternoon, I walked through the encampment of the strangers, which was spread along the sea-shore. Their long, stately, and in many instances beautifully carved canoes, were drawn up on the pebbly beach, and the chiefs and warriors were sitting in circles, at a small distance from them. Each party occupied the beach opposite their canoes, while the slaves or domestics at some distance further from the shore, were busied round their respective fires, preparing their masters' food. Near his side, each warrior's spear was fixed in the ground, while his *patupatu*, a stone weapon, the tomakawk of the New Zealander, was hanging on his arm. Several chiefs had a large iron hatchet or bill-hook, much resembling those used by woodmen or others, in mending hedges in England. These, which in their hands were rather terrifying weapons, appeared to be highly prized; they were kept clean and polished, and generally fastened round the wrist by a braided cord of native flax. The *patupatu* was sometimes placed in the girdle, in the same manner as a Malay would wear his knife or dagger, or a Turk his pistol. They were generally tall and well-formed men, altogether such as it might be expected the warriors of a savage nation would be. Several of these fighting-men were not less than six feet high; their limbs were muscular and firm, and their bodies stout, but not corpulent. The dress of the chiefs and warriors consisted, in general, of a girdle round the loins, and a short cloak or mantle, worn over the shoulders, and tied with cords of braided flax in front. The rank of the chief appeared to be sometimes indicated by the number of his cloaks fastened one upon the other; that which was smallest, but generally most valuable, being worn on the outside: the whole resembled in this respect the capes of a travelling-coat.

Their physiognomy, indicating any thing but weakness or cowardice, often exhibited great determination. They wore no helmet, or other covering for the head. Their black and shining hair sometimes hung in ringlets on their shoulders, but was frequently tied up on the crown of their heads, and usually ornamented by a tuft of waving feathers. Their dark eyes, though not large, were often fierce and penetrating; their prominent features in general well formed; but their whole countenance was much disfigured by the practice of tatauing. Each chief had thus imprinted on his face, the marks and involutions peculiar to his family or tribe; while the figures tataued on the faces of his dependants or retainers, though fewer in number, were the same in form as those by which the chief was distinguished. The accompanying representation of the head and face of 'Honghi, * the cele-

brated New Zealand warrior, who was among the party that arrived this morning at the settlements, will convey no inaccurate idea of the effect of this singular practice. The tatauing of the face of a New Zealander, answering the

* The bust, from which, by the kindness of the Secretary of the Church Missionary Society, the drawing of the above is taken, was executed with great fidelity by 'Honghi during a visit to Port Jackson.

purpose of the particular stripe or colour of the Highlander's plaid, marks the clan or tribe to which he belongs. It is considered highly ornamental, and, in addition to the distinguishing lines or curves, the intricacy and variety of the pattern, thus permanently fixed on the face, constitutes one principal distinction between the chiefs and common people, and may be regarded as the crest, or coat of arms, of the New Zealand aristocracy. Tatauing is said to be also employed as a means of enabling them to distinguish their enemies in battle. In the present instance, its effect on the countenance, where its marks are more thickly implanted than in any other part of the body, was greatly augmented by a preparation of red ochre and oil, which had been liberally applied to the cheeks and the forehead. Quantities of oil and ochre adhered to my clothes, from close contact with the natives, which I found it impossible to prevent; but this was the only inconvenience I experienced from my visit.

The warriors of New Zealand delight in swaggering and bravado, and while my companion was talking with some of Korokoro's party, one of them came up to me, and several times brandished his patupatu over my head, as if intending to strike, accompanying the action with the fiercest expressions of countenance, and the utterance of words exceedingly harsh, though to me unintelligible. After a few minutes he desisted, but when we walked away, he ran after us, and, assuming the same attitude and gestures, accompanied us till we reached another circle, where he continued for a short time these exhibitions of his skill in terrifying, &c. When he ceased, he inquired, rather significantly, if I was not afraid. I told him I was unconscious of having offended him, and that, notwithstanding his actions, I did not think he intended to injure me. The New Zealanders are fond of endeavouring to alarm strangers, and appear to derive much satisfaction in witnessing the indications of fear they are able to excite.

A number of tribes from different parts of the Bay being now at Rangehoo, the evening was devoted to public sports on the sea beach, which most of the strangers attended. Several of their public dances seemed immoral in their tendency, but in general they were distinguished by the violent gestures and deafening vociferations of the performers. No part of the sports, however, appeared so interesting to the natives, as a sham fight, in which the warriors wore their full dresses, bore their usual weapons, and went through the different movements of actual engagement.

Shungee, or, according to the modern orthography of the Missionaries, 'Honghi, with his numerous dependants and allies, formed one party, and were ranged on the western side of the beach, below the Missionaries' dwelling. The chief wore several mats or short cloaks, of various sizes and texture, exquisitely manufactured with the native flax, one of them ornamented with small shreds of dog's skin, with the hair adhering to it; these were fastened round his neck, while in his girdle he wore a patupatu, and carried a musket in his hand. His party were generally armed with clubs, and spears nine or ten feet long. Their antagonists were ranged at the opposite side of the beach. At a signal given, they ran violently towards each other, halted, and then amidst shouts and clamour, rushed into each others ranks, some brandishing their clubs, others thrusting their spears, which were either parried or carefully avoided by the opposite party. Several were at length thrown down, some prisoners taken, and ultimately both parties retreated to a distance, whence they renewed the combat. As the day closed, these sports were discontinued, and the combatants and spectators retired to their respective encampments.

Having filled our water casks, increased our supply of provender for the cattle and sheep I had on board, procured a number of logs of timber towards the erection of our future dwelling; and having spent a week very pleasantly with our Missionary brethren; we took leave of them, grateful for the assistance of their influence with the natives, and the kindness and hospitality we had experienced at their hands.

New Zealand comprises two large and several smaller islands, extending from 34 degrees to 47 degrees south latitude, and from 166 to about 180 degrees east long. It was discovered by Tasman, a Dutch navigator, in 1642. He sailed from the north point along the eastern shore, which was afterwards called Cook's Straits, where he anchored in a bay, to which, in consequence of an attack from the natives, he gave the name of Murderers' Bay, and finally left the coast without landing. In 1770, New Zealand was visited and explored by Captain Cook, who discovered the straits that are called by his name. The settlement at Rangehoo, and one formed subsequently by the Wesleyan Missionary Society in Wangaroa, and the Church Missionary station at Kerekere, are all on the large northern island. The climate is salubrious, the thermometer ranging between 40 and 80 degrees, avoiding the heat of the tropical climates, yet warmer than most of the temperate latitudes, generally equable, and seldom experiencing those sudden vicissitudes so frequent in the variable climate of England. The soil is in many parts fertile; and though few articles of food are indigenous, or when introduced grow spontaneously, yet it is capable of a high state of cultivation, and would probably favour not only the growth of wheat and other grain, but also of many of the fruits and valuable productions of the temperate and tropical climates. The mountains do not appear so lofty and broken as those of the Society Islands, and consequently the soil may be cultivated with greater facility. In addition to the growth of corn introduced by Mr. Marsden, and the assistants of the Missions at the several stations, the natives have long cultivated the Irish potato with facility and advantage. It is not indigenous, but was left by some of the ships touching here, and not only furnishes a valuable addition to the means of subsistence for the natives, but a very acceptable article of provision for the crews of the vessels by whom they are visited. Other European roots and vegetables have been introduced, but with less success. The kumara, or sweet potato, has been long cultivated, although the fern root furnishes a principal part of the food for the common people at some seasons of the year. The climate is favourable for rearing cattle and sheep, as well as the different kinds of poultry. The pine timber produced in the forests is valuable, not only to the inhabitants, but as an article of export both to New South Wales and to Great Britain. The river Thames to the south-east is a fine and capacious harbour. The coasts are well stocked with fish, which, with potatoes and fern root, constitute the food of the inhabitants. These advantages, together with its local situation in regard to New Holland, render it of importance to that growing colony.

The population of New Zealand has been estimated at

half a million; it may exceed this number. The inhabitants are certainly far more numerous than those of the Society Islands, and appear exempt from many of the diseases which afflict their northern neighbours. They are a hardy industrious race, generally strong and active, not only capable of great physical exertion, but of high moral culture, and are by no means deficient in intellect. Their tatauing and carving frequently display great taste; and when we consider the tools with which the latter is performed, it increases our admiration of their skill and perseverance. They are, nevertheless, addicted to the greatest vices that stain the human character, treachery, cannibalism, infanticide, and murder. Less superstitious than many of the natives of the Pacific, but perhaps as much addicted to war as any of them, if not more so; war appears to be their delight, and the events of their lives are little else than a series of acts of oppression, robbery, and bloodshed. A conquering army, returning from an expedition of murder and devastation, bring home the men, women, and children of the vanquished, as trophies of their victory. These unhappy beings are reduced to perpetual slavery, or sacrificed to satiate the vengeance of their enemies. On these occasions, little children, whose feeble hands could scarcely hold the knife or dagger, have been initiated in the dreadful work of death, and have seemed to feel delight in stabbing captive children, thus imbruing their infant hands in the blood of those who, under other circumstances, they would have hailed as playmates, and have joined in innocent and mirthful pastimes. Their wars are not only sanguinary, but horribly demoralizing and brutal, from the circumstance of the captives, or the slain, furnishing the victors with their triumphal banquet.

The cannibalism of the inhabitants of New Zealand, and other islands of the Pacific, has been doubted by some, and denied by others, and every mind influenced by the common feelings, or exercising the common sympathies of humanity, must naturally resist the conviction of his species ever sinking to a degradation so abject, and a barbarity so horrible, until it be substantiated by the clearest evidence of indisputable facts. But however ardently we may have hoped that the accounts of their anthropophagism were only the result of inferences drawn from their familiarity with and apparent satisfaction in deeds of savage murder; the accounts of the Missionaries who have resided amongst them, no longer admit any doubt to be entertained of the revolting and humiliating fact. The intercourse they have had with the greater part of the foreign shipping visiting their shores, has not been such as to soften the natural ferocity of their character, improve their morals, inspire them with confidence, advance their civilization, or promote peace and harmony among themselves; frequently it has been the reverse, as the affair of the Boyd, and the desolation of the island of Tipahee, affectingly demonstrate.

To the eye of a Missionary, New Zealand is an interesting country, inhabited by a people of no ordinary powers, could they but be brought under the influence of right principles. By the Christian philanthropists of Britain, who are desirous not only to spread the light of revelation and Christian instruction among the ignorant at home, but are also making noble efforts to send its blessings to the remotest nations of the earth, it has not been overlooked.

In 1814, the Church Missionary Society sent their Missionaries to New Zealand; and, under the direction and guardianship of the Rev. S. Marsden, the steady patron of the New Zealand Mission, established their first settlement at Rangehoo in the Bay of Islands. Considerable reinforcements have been sent, and three other stations formed. Since that period, the Wesleyan Missionaries commenced their labours near Wangaroa. The Missionaries and their assistants, who have laboured at these stations ever since their commencement, have not only steadily and diligently applied to the study of the language, which is a dialect of that spoken in all the eastern portion of the Pacific, established schools for the instruction of the natives, and endeavoured to unfold to them the great truths of revelation, but have from the beginning, by the establishment of forges for working iron, saw-pits, carpenters' shops, &c. laboured to introduce among the natives, habits of industry, a taste for the mechanic arts, and a desire to follow the peaceful occupations of husbandry; thereby aiming to promote their advancement in civilization, and improve their present condition, while they were pursuing the more important objects of their mission.

Success indeed has not been according to their desires, but it has not been altogether withheld; the general character of the people, in the neighbourhood of the settlements, is improved, and several pleasing instances of piety among the natives have been afforded. Difficulties attending the introduction of Christianity were from the first, and are still to be expected. The gross ignorance, prejudices, and superstition of the natives, the unbridled influence of their passions, the effects of their intercourse with foreigners inimical to the moral influence of the gospel, combine to resist its establishment. To these may also be added, their habitual treachery and crime, and, above all, their love of war, and their wretched system of government, which is probably one of the greatest barriers to their general reception of Christianity.

It was a favourable circumstance attending the change that has taken place both in the Society and Sandwich Islands, that each island had its chief; and that in some instances several adjacent islands were under the government of a principal chief or king, whose authority was supreme, and whose influence, in uniting the people under one head, predisposed them, as a nation, to receive the instructions imparted by individuals countenanced and protected by their chief or king. Persons of the highest authority not only patronized the Missionaries, but frequently added to their instructions, their commendation, and the influence of their own example in having already received them.

In New Zealand there is no king over the whole, or even over one of the larger islands. The people are generally governed by a number of chieftains, each indeed a king over his narrow territory, supreme among his own tribe or clan, and independent of every other. The same system prevails in the Marquesas, and the Friendly and Figi islands, where no law of right is acknowledged, but that of dominion. A desire to enlarge their territory, increase their power, or satisfy revenge, leads to frequent and destructive war, strengthens jealousy, and cherishes treachery, keeps them without any common bond of union, and prevents any deep or extensive impression being made upon them as a people. This necessarily circumscribes the influence of the Missionaries, and is, in a great degree, the cause which led the Wesleyan Missionaries for a time to suspend altogether their efforts, and has recently so painfully disturbed those of their brethren in connexion with the Church Missionary Society.

The labours of the mechanic and the artisan are valuable accompaniments to those of the Missionary; but Christianity must precede civilization. Little hope is to be entertained of their following to any extent the useful arts, cultivating habits of industry, or realizing the enjoyments of social and domestic life, until they are brought under the influence of those principles inculcated in the word of God. And notwithstanding the discouragements to be encountered, this happy result should be steadily and confidently anticipated by those engaged on the spot, as well as by their friends at home. Their prospect of success is daily becoming more encouraging. They have not yet laboured in hope, so long as their predecessors did in the South Sea Islands; where nearly fifteen years elapsed before they knew of one true convert. The recollection of this circumstance is adapted to inspire those employed in New Zealand with courage, and stimulate to perseverance, as there is every reason to conclude, that when the New Zealanders shall by the blessing of God become a Christian people, they will assume and maintain no secondary rank among the nations of the Pacific.

On the twenty-eighth of December, 1816, we sailed from the Bay of Islands, and proceeded in an easterly direction, with favourable winds, until the 26th of January, when, at daybreak, we discovered an island which we afterwards found to be RAPA, though usually designated Oparo. The first account of this island is given by Vancouver, who discovered it in his passage from New Zealand to Tahiti, on the 22d of December,

1791.* According to the observation made at the time, it was found to be situated in lat. 27. 36. S. and long. 144. 11. W. The mountains are lofty and picturesque, and the summits of those forming the high land in the centre, singularly broken, so as to resemble, in no small degree, a range of irregularly inclined cones, or cylindrical columns, which their discoverer supposed to be towers, or fortifications, manned with natives.

The higher parts of the mountains seemed barren, but the lower hills, with many of the valleys, and the shores, were covered with verdure, and enriched with trees and bushes. The island did not appear to be surrounded by a reef, and, consequently, but little low land was seen. The waves of the ocean dashed against the base of those mountains, which, extending to the sea, divided the valleys that opened upon the eastern shore. As we were not far from the island when the sun withdrew his light, we lay off and on through the night, and, at daybreak, the next morning, found ourselves at some distance from the shore. We sailed towards the island till about 10 A. M.; when, being within two miles of the

* The mingled emotions of astonishment and fear, with which the natives regarded every thing on board Vancouver's ship, prevented their replying very distinctly to the queries he proposed; and he observes, "Their answers to almost every question were in the affirmative, and our inquiries as to the name of their island, &c. were continually interrupted by incessant invitations to go on shore. At length, I had reason to believe the name of the island was *Oparo*, and that of their chief *Korie*. Although I could not positively state that their names were correctly ascertained, yet, as there was a probability of their being so, I distinguished the island by the name of Oparo, until it might be found more properly entitled to another." The explicit declarations of the natives, made under more favourable circumstances, have now determined *Rapa* to be the proper name of this island.

beach, the prow of our vessel was turned to the northward, and we moved slowly along in a direction nearly parallel with the coast. After advancing in this manner for some time, we saw several canoes put off from the land, and not less than thirty were afterwards seen paddling round our vessel. There were neither females nor children in any of the canoes. The men were not tataued, and wore only a girdle of yellow *ti* leaves round their waists. Their bodies, neither spare nor corpulent, were finely shaped; their complexion a dark copper colour; their features regularly formed; and their countenances, often handsome, were shaded by long black straight or curling hair. Notwithstanding all our endeavours to induce them to approach the ship, they continued for a long time at some distance, viewing us with apparent surprise and suspicion. At length, one of the canoes, containing two men and a boy, ventured alongside. Perceiving a lobster lying among a number of spears at the bottom of the canoe, I intimated, by signs, my wish to have it, and the chief readily handed it up. I gave him, in return, two or three middle-sized fish-hooks; which, after examining rather curiously, he gave to the boy, who, being destitute of any pocket, or even article of dress on which he could fasten them, instantly deposited them in his mouth, and continued to hold with both hands the rope hanging from our ship. The principal person in the canoe appearing willing to come on board, I pointed to the rope he was grasping, and put out my hand to assist him up the ship's side. He involuntarily laid hold of it, but could scarcely have felt my hand grasping his, when he instantly drew it back, and, raising it to his nostrils, smelt at it most significantly, as if to ascertain with what kind of a being he had come in contact. After a few moments' pause, he climbed over the ship's side, and as soon as he had reached the deck, our captain led him to a chair on the quarter-deck, and, pointing to the seat, signified his wish that he should be seated. The chief, however, having viewed it for some time, pushed it aside, and sat down on the deck. Our captain had been desirous to have the chief on board, that he might ascertain from him whether the island produced sandal-wood, as he was bound to the Marquesas in search of this article. A piece was therefore procured and shewn to him, with the qualities of which he appeared familiar; for, after smelling it, he called it by some name, and pointed to the shore. While we had been thus engaged, many of the canoes had approached the ship; and when we turned round, a number of the natives appeared on deck, and others were climbing up over the bulwarks. They were certainly the most savage-looking natives I had ever seen, and their behaviour was as unceremonious as their appearance was uninviting. Vancouver found them unusually shy at first, but afterwards remarkably bold, and exceedingly anxious to possess every article of iron they saw: although his ship was surrounded by not fewer than three hundred natives, there were neither young children, women, nor aged persons, in any of their canoes.

A gigantic, fierce-looking fellow, seized a youth as he was standing by the gangway, and endeavoured to lift him from the deck; but the lad, struggling, escaped from his grasp. He then seized our cabin-boy, but the sailors coming to his assistance, and the native finding he could not disengage him from their hold, pulled his woollen shirt over his head, and was preparing to leap out of the ship, when he was arrested by the sailors. We had a large ship-dog chained to his kennel on the deck, and, although this animal was not only fearless but savage, yet the appearance of the natives seemed to terrify him. One of them caught the dog in his arms, and was proceeding over the ship's side with him, but perceiving him fastened to the kennel by his chain, he was obliged to relinquish his prize, evidently disappointed. He then seized the kennel, with the dog in it; when, finding it nailed to the deck, he ceased his attempts to remove it, and gazed round the ship, in search of some object which he could secure. We had brought from Port Jackson two young kittens; one of these now came up from the cabin, but she no sooner made her appearance on the deck, than a native, springing like a tiger upon its prey, caught up the unconscious animal, and instantly leaped over the ship's side into the sea. Hastening to the side of the deck, I looked over the bulwarks, and beheld him swimming rapidly towards a canoe lying about fifty yards from the ship. As soon as he had reached this canoe, holding the cat with both hands, and elevating these above his head, he exhibited her to his companions with evident exultation; while, in every direction, the natives were seen paddling their canoes towards him, to gaze upon the strange creature he had brought from the vessel. When our captain beheld the thief thus exhibiting his prize, he seized his musket, and was in the act of levelling it at the offender, when I arrested his arm, and assured him I had no doubt the little animal would be preserved and well treated. Orders were now given to clear the ship. A general scuffle ensued between the islanders and the seamen, in which many of the former were driven headlong into the sea, where they seemed as much at home as on solid ground, while others clambered over the vessel's side into their canoes. In the midst of the confusion, and the retreat of the natives, the dog, which had hitherto slunk into his kennel, recovered his usual boldness, and not only increased the consternation by his barking, but severely tore the leg of one of the fugitives who was hastening out of the ship, near the spot to which he was chained. The decks were now cleared, but as many of the people still hung upon the shrouds and about the chains; the sailors drew the long knives with which, when among the islands, they were furnished, and by menacing gestures, without wounding any, succeeded in detaching them altogether from the ship. Some of them seemed quite unconscious of the keenness of the knife, and, I believe, had their hands deeply cut by snatching or grasping at the blade. A proposal was now made to entice or admit some on board, and take two of them to Tahiti, that the Missionaries there might become acquainted with their language, gain a knowledge of the productions of their island, impart unto them Christian instruction, and thus prepare the way for the introduction of Christianity among their countrymen, as well as open a channel for commercial intercourse. Our captain offered to bring them to their native island again, on his return from the Marquesas; and, could their consent have been by any means obtained, I should, without hesitation, have acceded to the plan; but, as we had no means of effecting this object, I did not conceive it right to take them by force from their native island.

On a former voyage, about two years before this period, Captain Powel had been becalmed near the shores of this island. Many of the natives came off in their canoes, but did not venture on board; perceiving, however, a hawser hanging out of the stern of the ship, about fifty of them leaped into the sea, and grasping the rope with one hand, began swimming with the other, labouring and shouting with all their might, as they supposed they were drawing the vessel towards the shore. Their clamour attracted the attention of the seamen, and it was found no easy matter, even when all hands were employed, to draw in the rope. While the greater part of the crew were thus engaged, a seaman leaning over the stern with a cutlass in his hand, so terrified the natives, that as they were drawn near the vessel they quitted their hold, and by this means the hawser was secured. A breeze shortly after springing up, they steered away, happy to escape from the savages by whom they had been surrounded.

On the present occasion we experienced a signal deliverance, which, though it did not at the time appear so remarkable, afterwards powerfully affected our minds. As soon as the ship was cleared of the natives, and the wind was wafting us from their shores, I went down to the cabin, where Mrs. Ellis and the nurse had been sitting ever since their first approach to the ship; and when I saw our little daughter, only four months old, sleeping securely in her birth, I was deeply impressed with the merciful providence of God, in the preservation of the child. During the forenoon, the infant had been playing unconsciously in her nurse's lap upon the quarter-deck, under the awning, which was usually spread in fine weather, and she had but recently taken her to the cabin, when the natives came on board. Had the child been on deck, and had my attention been for a moment diverted, even though I had been standing by the side of the nurse, there is every reason to believe that the motives which induced them to seize the boys on the deck, and even the dog in his kennel, would have prompted them to have grasped the child in her nurse's lap or arms, and to have leaped with her into the sea before we could have been aware of their design. Had this been the case, it is impossible to say what the result would have been; bloodshed might have followed, and we might have been obliged to depart from the island, leaving our child in their hands. From the crude food with which they would have fed her, it is probable she would have died; but, from my subsequent acquaintance with the natives of the South Sea Islands, I do not think that during her infancy they would have treated her unkindly. As it was, we felt grateful for the kind Providence which had secured us from all the distressing circumstances which must necessarily have attended such an event.

These brief facts will be sufficient to shew somewhat of the character of the natives of Rapa, in 1791 and 1817. They continued in this state until within the last two or three years, during which a considerable change has taken place.

Towards the close of the summer of 1825, a cutter belonging to TATI, a chief in Tahiti, when on a voyage to the Paumotus, or pearl islands, visited Rapa, and brought two of its inhabitants to Tahiti. On their first arrival they were under evident feelings of apprehension; but the kindness of Mr. Davies the Missionary, and the natives of Papara, removed their suspicions, and inspired them with confidence. They were both delighted and astonished in viewing the strange objects presented to their notice. The European families, the houses, the gardens, the cattle, and other animals, which they saw at Tahiti, filled them with wonder. They also attended the schools and places of public worship, and learned the alphabet. Soon after their arrival, the cutter sailed again for their island, and the two natives of Rapa returned to their countrymen loaded with presents from their new friends, and accompanied by two pious Tahitians, who were sent to gain more accurate information relative to their country, and the disposition of its inhabitants. When the vessel approached their island, and the people saw their countrymen, they appeared highly delighted; and towards the evening, when, accompanied by the two Tahitians, they drew near the beach in the ship's boat, they came out into the sea to meet them, and *carried* the men and the boat altogether to the shore. This to the strangers was rather an unexpected reception; but, though singular, it was not unfriendly, for they were treated with great kindness. The accounts the natives gave their countrymen, of what they had seen in Tahiti, were marvellous to them: the captain of the cutter procured some tons of sandal-wood, and when he left, the Tahitians returned, having received an invitation from the chiefs and people to revisit their island, and reside permanently among them; a request so congenial to their own feelings, that they at once promised to comply.

In the month of January, 1826, the two Tahitian teachers and their wives, accompanied by two others, one a schoolmaster and the other a mechanic, sailed from Tahiti for Rapa. They carried with them not only spelling books, and copies of the Tahitian translations of the Scriptures, but also a variety of useful tools, implements of husbandry, valuable seeds and plants, together with timber for a chapel, and doors, &c. for the teachers' houses. They were conducted to their new station by Mr. Davies, one of the senior Missionaries at Tahiti, who was pleased with his visit, and, upon the whole, with the disposition of the people, although some appeared remarkably superstitious, and, as might be expected, unwilling at once to embrace Christianity. This arose from an apprehension of the anger of their gods, induced by the effects of a most destructive disease, with which they had been recently visited. The gods, they imagined, had thus punished them for their attention to the accounts from Tahiti. The teachers however landed their goods, and the frame-work of the chapel. The chiefs received them with every mark of respect and hospitality, pointed out an eligible spot for their residence, gave them some adjacent plantations of taro, and promised them protection and aid.

The sabbath which Mr. Davies spent there was probably the first ever religiously observed on the shores of Rapa. Several of the natives attended public worship, and appeared impressed with the services. These being performed in the Tahitian language, were not unintelligible to them. The native teachers were members of the church at Papara, although they were but few in number, and were surrounded by a heathen population in a remote and solitary island, and as it was then expected the vessel would sail on that or the following day, they joined with Mr. Davies their pastor in commemorating the death of Christ, under the impression that it was the last time they should ever unite in this hallowed ordinance.

The island of Rapa is about twenty miles in circumference, it is tolerably well wooded and watered, especially on the eastern side, where Aurai, a remarkably fine harbour, extending several miles inland, is situated. The entrance is intricate, but the interior capacious, the beach good, and fresh water convenient. Situated some degrees from the southern tropic, the climate is bracing and salubrious, the soil is fertile, and while it nourishes many of the valuable roots and fruits of the intertropical regions, is probably not less adapted to the more useful productions of temperate climes. Mr. Davies estimates the population at about two thousand. Vancouver supposed that Rapa contained not less than fifteen hundred, merely from those he saw around his ship. In their language, complexion, general character, superstitions, and employments, they resemble the inhabitants of the other islands of the Pacific, though less civilized in their manners, more rude in their arts, and possessed of fewer comforts, than most of their northern neighbours were, when first discovered. Their intercourse with Tahiti will not only increase their knowledge, and their sources of temporal enjoyment, but it is to be hoped will be the means of introducing Christianity among them, and raising them to the participation of its "spiritual blessings."

A fresh avenue is opened for European commerce, and valuable information is likely to result from the visit of the teachers to this solitary abode. The English Missionary from Tahiti was the first foreigner that ever landed on their coasts; but many years before his arrival, an inhabitant of some other island, the only survivor of the party with whom he sailed from his native shores, had been by tempestuous weather drifted to the island, and was found there by the native teachers, who first went from Tahiti. His name was Mapuagua, and that of his country Manganeva, which he stated was much larger than Rapa, and situated in a south-easterly direction. The people he described as numerous, and much tataued; the name of one of their gods the same as that of one formerly worshipped by the Tahitians. An old man, who resided at the same place with the stranger, gave Mr. Davies the name of eleven places, either districts of Manganeva or adjacent islands, which are unknown to the Tahitians. The information thus obtained will be valuable in the search for those islands which has already been commenced; and if no sources of wealth be found, nor important channels of commerce opened, their discovery will increase our geographical knowledge, and extend the range of benevolent operation.

CHAP. III.

Voyage to Tubuai—Notice of the mutineers of the Bounty—Origin of the inhabitants of Tubuai—Visit of Mr. Nott—Prevention of war—Settlement of native Missionaries—Arrival off Tahiti—Beauty of its natural scenery—Anchoring in Matavai Bay—Appearance of the district—Historical notice of its discovery—Of the arrival of the ship Duff—Settlement of the first Mission—Cession of Matavai.—Departure of the Duff—Influence of the mechanic arts on the minds of the people—Comparative estimate of iron and gold—Difficulties attending the acquisition of an unwritten language—Methods adopted by the Missionaries—Propensity to theft among the natives.

On leaving Rapa, we sailed in a northerly direction till the third of February, when we reached the island of Tubuai, situated in lat. 23 degrees 25 minutes S., and long. 149 degrees 23 minutes W. At a distance it appears like two islands, but, on a nearer approach, the high land is found to be united.

Tubuai was discovered by Cook in 1777, and after the mutineers in the Bounty had taken possession of the vessel, and committed to the mercy of the waves, Captain Bligh with eighteen of his officers and men, this was the first island they visited. Hence they sailed to Tahiti, brought away the most serviceable of the live-stock left there by former navigators, and in 1789 attempted a settlement here. Misunderstandings between the mutineers and the natives, and the unbridled passions of the former, led to acts of violence, which the latter resented. A murderous battle ensued, in which nothing but superior skill and fire-arms, together with the advantages of a rising ground, saved the mutineers from destruction. Two were wounded, and numbers of the natives slain. This led them to abandon the island; and after revisiting Tahiti, and leaving a part of their number there, they made their final settlement in Pitcairn's island. Their attempt to settle in this island is celebrated in a poem by the late Lord Byron called, "The Island, or Christian and his Companions," in which are recorded some affecting circumstances connected with the subsequent lives and ultimate apprehension of many of these unhappy men, and several interesting facts relative to the Society and Friendly Islands.

Tubuai was also the first of the South Sea Islands that gladdened the sight of the Missionaries who sailed in the Duff. They saw the land on the morning of the 22d of February, 1797, near thirty miles distant; and as the wind was unfavourable, the darkness of night hid the island from their view before they were near enough distinctly to behold its scenery, or the people by whom it was inhabited. I can enter in some degree into their emotions on this unusually interesting day. All that hope had anticipated in its brightest moments, was no longer to be matter of uncertainty, but was to be realized or rejected. Such feelings I have experienced, and can readily believe theirs were of the same order as those of which I was conscious, when gazing on the first of the isles of the Pacific that we approached. Theirs were probably more intense than mine, as a degree of adventurous enterprise was then thrown around Missionary efforts, which has vanished with their novelty. Our information, also, is now much more circumstantial and explicit than theirs could possibly have been. Tubuai is stated, in the Introduction to the Voyage of the Duff, to have been at that time but recently peopled by some natives of an island to the westward, probably Rimatara, who, when sailing to a spot they were accustomed to visit, were driven by strong and unfavourable winds on Tubuai. A few years after this, a canoe sailing from Raiatea to Tahiti, conveying a chief who was ancestor to Idia, Pomare's mother, was also drifted upon this island, and the chief admitted to the supreme authority; a third canoe was afterwards wafted upon the shores of Tubuai, containing only a human skeleton, which a native of Tahiti, who accompanied the mutineers, supposed belonged to a man he had killed in a battle at sea. The scantiness of the population favoured the opinion that the present race had but recently become inhabitants of this abode; and the subsequent visits of Missionaries from Tahiti, with the residence of native teachers among the people, have furnished additional evidence that the present Tubuaian population is but of modern origin, compared with that inhabiting the island of Raivavai on the east, or Rurutu and Rimatara on the west.

Tubuai is compact, hilly, and verdant; many of the hills appeared brown and sunburnt, while others were partially wooded. It is less picturesque than Rapa, but is surrounded by a reef of coral, which protects the lowland from the violence of the sea. As we approached this natural safeguard to the level shore, a number of natives came out to meet us. Their canoes, resembling those of Rapa, were generally sixteen or twenty feet long; the lower part being hollowed out of the trunk of a tree, and the sides, stem, and stern formed by pieces of thin plank sewn together with cinet made of the fibrous husk of the cocoa nut. The stem projected nearly horizontally, but the stern being considerably elevated, extended obliquely from the seat occupied by the steersman. The sterns were ornamented with rude carving, and, together with the sides, painted with a kind of red ochre, while the seams were covered with the feathers of aquatic birds. A tabu had been recently laid on the island by the priests, which they had supposed would prevent the arrival of any vessel, and they were consequently rather disconcerted by our approach. Among the natives who came on board, was a remarkably fine, tall, well-made man, who appeared, from the respect paid him by the others, to be a chief. His body was but partially tataued, his only dress was a girdle or broad bandage round his loins, and his glossy black and curling hair was tied in a bunch on the crown of his head, while its extremities hung in ringlets on his shoulders. His disposition appeared mild and friendly. His endeavours to induce us to land were unremitted, until it was nearly sunset; when, finding them unavailing, and receiving from the captain an assurance that he would keep near the island till the morrow, he remained on board, although considerably affected by the motion of the vessel.

The next morning we stood in close to the reefs, and a party from the ship accompanied the chief to the shore; the population appeared but small, the people were friendly, and readily bartered fowls, taro, and mountain plantains for articles of cutlery and fish-hooks. Their gardens were unfenced, and the few pigs they had, were kept in holes or wide pits four or five feet deep, and fed with bread-fruit and other vegetables. Only one was brought on board, and very readily purchased. Many of the natives, in addition to the common bandage encircling their bodies, and a light cloth over their shoulders, wore large folds of white or yellow cloth bound round their heads, in some degree resembling a turban, which gave them a remarkably Asiatic appearance. They also wore necklaces of the nuts of the pandanus; the scent of which, though strong, is grateful to most of the islanders of the Pacific. A few weeks before our arrival, a canoe from Tahiti, bound to the Paumotu or pearl islands, had been drifted on Tubuai; and the people on board, although peaceable in their conduct, had incurred the displeasure of the inhabitants by endeavouring to persuade them to renounce idolatry and embrace Christianity. The strangers, though plundered and otherwise ill-treated, forbore to retaliate, from the influence of Christian principles which they had imbibed at Tahiti.

Subsequently, the Tubuaians heard more ample details of the change that had taken place in the adjacent island of Rurutu, as well as in the Society Islands—that the inhabitants had renounced their idolatry, and erected places for the worship of the true God—and determined to follow their example. In the month of March, 1822, they sent a deputation to Tahiti, requesting teachers and books. The messengers from Tubuai were kindly welcomed, and not only hospitably entertained by the Tahitian Christians, but led to their schools and their places of public worship. Two native teachers were selected by the church in Matavai, and publicly designated by the Missionaries to instruct the natives of Tubuai. The churches in Tahiti, so far as their means admitted, furnished them with a supply of articles most likely to be useful in their missionary station; and the 13th of June, 1822, they embarked for the island of Tubuai. Mr. Nott the senior Missionary in Tahiti, embarked in the same vessel, for the purpose of preaching to the people, and affording the native Missionaries every assistance in the commencement of their undertaking.

Finding, on their arrival, the whole of the small population of the island engaged in war, and on the eve of a battle, Mr. Nott and his companions repaired to the encampment of Tamatoa, who was, by hereditary right, the king of the island; acquainted him with the design of their visit, and recommended him to return to his ordinary place of abode. The king expressed his willingness to accede to the proposal, provided his rival, who was encamped but a short distance from him, and whom he expected on the morrow to engage, would also suspend hostilities. Paofai, a chief who accompanied Mr. Nott, went to Tahuhuatama, the chief of the opposite party, with a message to this effect. He was kindly received, his proposal agreed to, and a time appointed for the chiefs to meet midway between the hostile parties, and arrange the conditions of peace.

On the same evening, or early the next morning, the chieftains with their adherents, probably not exceeding one hundred on either side, quitted their encampments, which were about a mile and a half or two miles apart, and proceeded to the appointed place of rendezvous. When they came within fifty yards of each other, they halted. The chiefs then left their respective bands, and met midway between them; they were attended by the Missionaries, and after several propositions had been made by one party, and acceded to by the other, peace was concluded. The chiefs then embraced each other; and the warriors in each little army, wherein the nearest relations were probably arranged against each other, perceiving the reconciliation of their chiefs, dropped their implements of war, and, rushing into each other's arms, presented a scene of gratulation and joy very different from the murderous conflict in which they expected to have been engaged. They repaired in company to the residence of the principal chief, where an entertainment was provided. Here the Missionaries had a second interview with the chiefs, who welcomed them to the island, and expressed their desires to be instructed concerning the true God, and the new religion, as they usually denominated Christianity.

On the following morning, the inhabitants of Tubuai were invited to attend public worship, when Mr. Nott delivered, in a new building erected for the purpose, the first Christian discourse to which they had ever listened. It was truly gratifying to behold those, who had only the day before expected to have been engaged in shedding each other's blood, now mingled in one quiet and attentive assembly, where the warriors of rival chieftains might be seen sitting side by side, and listening to the gospel of peace.

Mr. Nott was unexpectedly detained several weeks at Tubuai; during this time he made the tour of the island, conversed with the people, and preached on every favourable occasion that occurred. The Queen Charlotte at length arrived; when, having introduced the native teachers to the chiefs and people, and recommended them to their protection, he bade them farewell, and prosecuted his voyage to High Island. The chiefs had desired that one teacher might be left with each; and, in order to meet their wishes, two, Hapunia and Samuela, from the church at Papeete, were stationed by Mr. Nott in this island, one with each of the chiefs. The native Missionaries found the productions of Tubuai less various and abundant than those of Tahiti, and the adjacent islands. The habits of the natives were remarkably indolent, and inimical to health, especially the practice of dressing their bread-fruit, &c. only once in five days. Against this the teachers invariably remonstrated, and presented to them, also, a better example, by cooking for themselves fresh food every day. Since that time, a distressing epidemic has, in common with most of the islands, prevailed in Tubuai, and has swept off many of the people. Nevertheless, the native teachers continue their labours, and the condition of the people is improved. In February, 1826, when Mr. Davies visited them, the profession of Christianity was general; 38 adults and four children were baptized. The chiefs and people were assisting the teachers in building comfortable dwellings, and erecting a neat and substantial house for public worship.

In the afternoon of the 4th of February we sailed from Tubuai; but, in consequence of unfavourable winds, did not reach Tahiti till the 10th. As we approached its southern shore, a canoe came off with some natives, who brought a pig and vegetables for sale; but the wind blowing fresh, we soon passed by, and had little more than a glance at the people. About sunset we found ourselves a short distance to the northward of Point Venus, having sailed along the east and northern shores of Tahiti, charmed with the rich and varied scenery of the island, justly denominated the queen of the Pacific, whose landscapes, though circumscribed in extent, are

"So lovely, so adorned
With hill, and dale, and lawn, and winding vale,
Woodland, and stream, and lake, and rolling seas,"

that they are seldom surpassed, even in the fairest portions of the world.

On the morning of the 16th of February, 1817, as the light of the day broke upon us, we discovered that during the preceding night we had drifted to a considerable distance from the island; the canoes of the natives, however, soon surrounded our vessel; numbers of the people were admitted on board, and we had the long desired satisfaction of intercourse with them, through the medium of an interpreter. They were not altogether so prepossessing in person as, from the different accounts I had read, I had been led to anticipate. The impression produced by our first interview was, notwithstanding, far from being unfavourable; we were at once gratified with their vivacity, and soon after with the simple indications of the piety which several exhibited. A good-looking native, about forty years of age, who said his name was Maine, and who came on board as a pilot, was invited to our breakfast. We had nearly finished when he took his seat at the table; yet, before tasting his food, he modestly bent his head, and, shading his brow with his hand, implored the Divine blessing on the provision before him. Several of the officers were much affected at his seriousness; and though one attempted to raise a smile at his expense, it only elicited from him an expression of compassion. To me it was the most pleasing sight I had yet beheld, and imparted a higher zest to the enjoyment I experienced in gazing on the island, as we sailed along its shores.

There is no reason to suppose that Tahiti, or any other island of the group, is altogether volcanic in its origin, as Hawaii and the whole of the Sandwich Islands decidedly are. The entire mass of matter composing the latter, has evidently been in a state of fusion, and in that state has been ejected from the focus of an immense volcano, or volcanoes, originating, probably, at the bottom of the sea, and forming, by their action through successive ages, the whole group of islands; in which, nothing like primitive or secondary rock has yet been found. In Tahiti, and other islands of the southern cluster, there are basalts, whinstone dykes, and homogeneous earthy lava, retaining all the convolutions which cooling lava is known to assume; there are also kinds of hornstone, limestone, silex, breccia, and other substances, which have never, under the action of fire, altered their original form. Some are found in detached fragments, others in large masses. The wild and broken manner, however, in which the rocks now appear, warrants the inference, that since their formation, which was probably of equal antiquity with the bed of the ocean, they have been thrown up by some volcanic explosion, the disruptions of an earthquake, or other violent convulsions of the earth; and have, from this circumstance, assumed their bold, irregular, and romantic forms.

Midday was past before we entered Matavai bay. As we sailed into the harbour, we passed near the coral reef, on which Captain Wallis struck on the 19th of June, 1767, when he first entered the bay. His ship remained stationary nearly an hour; and, in consequence of this circumstance, the reef has received the name of the Dolphin rock. As we passed by it, we felt grateful that the winds were fair and the weather calm, and that we had reached our anchorage in safety. Ma-ta-vai, or Port Royal, as it was called by Captain Wallis, is situated in latitude 17°. 36′. S. and longitude 149°. 35′. W.

It is rather an open bay, and although screened from the prevailing trade winds, is exposed to the southern and westerly gales, and also to a considerable swell from the sea. The long flat neck of land which forms its northern boundary, was the spot on which Captain Cook erected his tents, and fixed his instruments for observing the transit of Venus; on which account, it has ever since been called Point Venus. Excepting those parts enclosed as gardens, or plantations, the land near the shore is covered with long grass, or a species of convolvulus, called by the natives *pohue;* numerous clumps of trees, and waving cocoa-nuts, add much to the beauty of its appearance. A fine stream, rising in the interior mountains, winds through the sinuosities of the head of the valley, and, fertilizing the district of Matavai, flows through the centre of this long neck of land, into the sea.

Such, without much alteration, in all probability, was the appearance of this beautiful bay, when discovered by Captain Wallis, in 1767; and two years after, when first visited by Captain Cook; or when Captain Bligh, in the Bounty, spent six months at anchor here in 1788 and 1789; when Captain Vancouver arrived in 1792; Captain New, of the Dædalus, in 1793; and Captain Wilson, in the Duff, who anchored in the same bay on the 6th of March, 1797.

It was on the northern shores of this bay, that eighteen of the Missionaries, who left England in the Duff, first landed, upwards of thirty years ago. They were

> "——————the messengers
> Of peace, and light and life, whose eye unsealed
> Saw up the path of immortality,
> Far into bliss. Saw men, immortal men,
> Wide wandering from the way, eclipsed in night,

> Dark, moonless, moral night, living like beasts,
> Like beasts descending to the grave, untaught
> Of life to come, unsanctified, unsaved."

To reclaim the inhabitants from error and superstition, to impart to them the truths of revelation, to improve their present condition, and direct them to future blessedness, were the ends at which they aimed; and here they commenced those labours which some of them have continued unto the present time; and which, under the blessing of God, have been productive of the moral change that has since taken place among the inhabitants of this and the adjacent islands. Decisive and extensive as that change has since become, it was long before any salutary effects appeared as the result of their endeavours. And, although the scene before me was now one of loveliness and quietude, cheerful, yet placid as the smooth waters of the bay, that scarcely rippled by the vessel's side, it has often worn a very different aspect. Here the first Missionaries frequently heard the song accompanying the licentious areois dance, the deafening noise of idol worship, and saw the human victim carried by for sacrifice: here, too, they often heard the startling cry of war, and saw their frighted neighbours fly before the murderous spear and plundering hand of lawless power. The invaders' torch reduced the native hut to ashes, while the lurid flame seared the green foliage of the trees, and clouds of smoke, rising up among their groves, darkened for a time surrounding objects. On such occasions, and they were not infrequent, the contrast between the country, and the inhabitants, must have been most affecting, appearing as if the demons of darkness had lighted up infernal fires, even in the bowers of paradise.

Within sight of the spot where our vessel lay, four of the Missionaries were stripped and maltreated by the natives, two of them nearly assassinated, from the anger of the king, and one of them was murdered. Here the first Missionary dwelling was erected, the first temple for the worship of Jehovah reared, and the first Missionary grave opened; and here, after having been obliged to convert their house into a garrison, and watch night and day in constant expectation of attack, the Missionaries were obliged, almost in hopeless despair, to abandon a field, on which they had bestowed the toil and culture of twelve anxious and eventful years.

On the 7th of March, 1797, the first Missionaries went on shore, and were met on the beach by the late Pomare and his queen, then called Otoo and Tetua; by them they were kindly welcomed, as well as by Paitia, an aged chief of the district. They were conducted to a large, oval-shaped native house, which had been but recently finished for Captain Bligh, whom they expected to return. Their dwelling was pleasantly situated on the western side of the river, near the extremity of Point Venus. The natives were delighted to behold foreigners coming to take up their permanent residence among them; as those they had heretofore seen, with the exception of a Spaniard, had been transient visitors. The Spaniard had saved his life by escaping from Langara's ship, while it was lying at anchor in Tairabu, in March 1773, at which time three of his shipmates were executed. The benefit the natives had derived from this individual, and the mutineers of the Bounty, prior to their apprehension by the people of the Pandora, and the residence of several of the crew of the Matilda, which had been wrecked on a reef not far distant, led them to desire the residence of foreigners. The inhabitants of Tahiti had never seen any European females or children, and were consequently filled with amazement and delight, when the wives and children of the Missionaries landed. Several times during the first days of their residence on shore, large parties arrived from different places in front of the house, requesting that the white women and children, would come to the door and shew themselves. The chiefs and people were not satisfied with giving them the large and commodious Fare Beritani (British House,) as they called the one they had built for Bligh, but readily and cheerfully ceded to Captain Wilson and the Missionaries, in an official and formal manner, the whole district of Matavai, in which their habitation was situated. The late Pomare and his queen, with Otoo his father, and Idia his mother, and the most influential persons in the nation, were present, and Haamanemane, an aged chief of Raiatea, and chief priest of Tahiti, was the principal agent for the natives on the occasion. The accompanying plate, representing this singular transaction, is taken from an original painting in the possession of Mrs. Wilson, relict of the late Captain Wilson. It exhibits, not only the rich luxuriance of the scenery, but the complexion, expression, dress, and tatauing of the natives, with remarkable fidelity and spirit. The two figures on men's shoulders are the late king and queen. Near the queen on the right stands Peter the Swede, their interpreter, and behind him stands Idia, the mother of the king. The person seated on the right hand is Paitia, the chief of the district; behind him stand Mr. and Mrs. Henry, Mr. Jefferson, and others. The principal person on this side is Captain Wilson; between him and his nephew Captain W. Wilson, stands a child

Painted by R. Smirke R.A.

Engraved by H. Robinson.

THE HIGH PRIEST OF TAHITI CEDING THE DISTRICT OF MATAVAI TO CAPTN WILSON, of Mr. Hassel; Mrs. Hassel with an infant is before them. On the left, next to the king, stands his father Pomare, the upper part of his body uncovered in homage to his son, and behind him is Hapai, the king's grandfather. Haamanemane, the high-priest, appears in a crouching position, addressing Captain Wilson, and surrendering the district.—Haamanemane was also the *taio*, or friend, of Captain Wilson; and rendered him considerable service, in procuring supplies, facilitating the settlement of the Mission, and accomplishing other objects of his visit.

Presentations of this kind were not uncommon among the islanders, as a compliment, or matter of courtesy, to a visitor; and were regulated by the rank and means of the donors, or the dignity of the guests. Houses, plantations, districts, and even whole islands, were sometimes presented; still, those who thus received them, never thought of appropriating them to their own use, and excluding their original proprietors, any more than a visitor in England, who should be told by his host to make himself perfectly at home, and to do as he would if he were in his own house, would, from this declaration, think of altering the apartments of the house, or removing from it any part of the furniture. It is, however, probable, that such was their estimate of the advantages that would result from the residence of the Mission families among them, that, in order to afford every facility for the accomplishment of an object so desirable, and hold out every inducement to confidence for the Missionaries, as to their future support, they were sincere in thus ceding the district. They might wish them to reside in it, exercise the office of chiefs over the whole, cultivate as much of it as they desired, and receive tribute from those who might occupy the remaining parts; but by no means, perpetually to alienate it from the king, or chief, to whom it originally belonged. This they knew could not be done without their permission, and that permission they could at any time withhold. In 1801, when the Royal Admiral arrived, Pomare was asked, when the Missionaries were introduced to him, if they were still to consider the district theirs; and though he replied in the affirmative, and even asked if they wished the inhabitants to remove, it afterwards appeared that the natives considered them only as tenants at will. All they desired was, the permanent occupation of the ground on which their dwellings and gardens were situated; yet, in writing to the Society, in 1804, they remark, in reference to the district, "The inhabitants do not consider the district, nor any part of it, as belonging to us, except the small sandy spot we occupy with our dwellings and gardens; and even as to that, there are persons who claim the ground as theirs." Whatever advantages the kings or chiefs might expect to derive from this settlement on the island, it must not be supposed that it was from any desire to receive general or religious instructions. This was evident, from a speech once made by Haamanemane, who said that they gave the people plenty of the *parau* (word) talk and prayer, but very few knives, axes, scissors, or cloth. These, however, were soon afterwards amply supplied. A desire to possess such property, and to receive the assistance of the Europeans in the exercise of the mechanic arts, or in their wars, was probably the motive by which the natives were most strongly influenced.

Captain Wilson was, however, happy to find the king,

chiefs, and people so willing to receive the Missionaries, and so friendly towards them; and the latter being now settled comfortably in their new sphere of labour, the Duff sailed for the Friendly Islands on the 26th of March.

Having landed ten Missionaries at Tongatabu, in the Friendly Islands, Captain Wilson visited and surveyed several of the Marquesan Islands, and left Mr. Crook a Missionary there; he then returned to Tahiti, and on the 6th of July, the Duff again anchored in Matavai Bay. The health of the Missionaries had not been affected by the climate. The conduct of the natives had been friendly and respectful; and supplies in abundance had been furnished during his absence. While the ship remained at Tahiti, Mr. W. Wilson made the tour of the island; the iron, tools, and other supplies for the Mission, were landed: the Missionaries, and their friends on board, having spent a month in agreeable intercourse, now affectionately bade each other farewell. Dr. Gilham having intimated to Captain W. his wish to return to England, was taken on board, and the Duff finally sailed from Matavai on the 4th of August, 1797. The Missionaries returning from the ship, as well as those on shore, watched her course as she slowly receded from their view, under no ordinary sensations. They now felt that they were cut off from all but Divine guidance, protection, and support, and had parted with those by whose counsels and presence they had been assisted in entering upon their labours, but whom on earth they did not expect to meet again. Captain Wilson coasted along the south and western shores of Huahine, and then sailed to Tongatabu; where, after spending twenty days with the Missionaries, who appeared comfortably settled, he sailed for Canton, where he received a cargo, with which he returned to England, and arrived safely in the Thames; having completed his perilous voyage, under circumstances adapted to afford the highest satisfaction, and to excite the sincerest gratitude from all who were interested in the success of the important enterprise.

The departure of the Duff did not occasion any diminution in the attention of the natives to the Missionaries in Tahiti. Pomare, Otu, Haamanemane, Paitia, and other chiefs, continued to manifest the truest friendship, and liberally supplied them with such articles as the island afforded. The Missionaries, as soon as they had made the habitation furnished by the people for their accommodation in any degree comfortable, commenced with energy their important work.

Their acquaintance with the most useful of the mechanic arts, not only delighted the natives, but raised the Missionaries in their estimation, and led them to desire their friendship. This was strikingly evinced on several occasions, when they beheld them use their carpenters' tools; cut with a saw a number of boards out of a tree, which they had never thought it possible to split into more than two, and make with these, chests, and articles of furniture. When they beheld a boat, built upwards of twenty feet long, and six tons burden, they were pleased and surprised; but when the blacksmith's shop was erected, and the forge and anvil were first employed on their shores, they were filled with astonishment. They had long been acquainted with the properties and uses of iron, having procured some from the natives of a neighbouring island, where a Dutch ship, belonging to Roggewein's squadron, had been wrecked many years before they were visited by Captain Wallis.

When the heated iron was hammered on the anvil, and the sparks flew among them, they fancied it was spitting at them, and were frightened, as they also were with the hissing occasioned by immersing it in water; yet they were delighted to see the facility with which a bar of iron was thus converted into hatchets, adzes, fish-spears, and fish-hooks, &c. Pomare, entering one day when the blacksmith was employed, after gazing a few minutes at the work, was so transported at what he saw, that he caught up the smith in his arms, and, unmindful of the dirt and perspiration inseparable from his occupation, most cordially embraced him, and saluted him, according to the custom of his country, by touching noses. Iron tools they considered the most valuable articles they could possess; and a circumstance that occurred during the second visit of the Duff, will shew most strikingly the comparative value they placed upon gold and iron. The ship's cook had lost his axe, and Captain Wilson gave him ten guineas to try to purchase one with, supposing that the intercourse the natives had already had with Europeans, would enable them to form some estimate of the value of a guinea, and the number of articles they could procure with it, from any other ship that might visit the island; but, although the cook kept the guineas more than a week, he could meet with no individual among the natives who would part with an axe, or even a hatchet, in exchange for them.

While some of the Missionaries were employed in the exercise of those arts which were adapted to make the most powerful impression upon the minds of the natives, others were equally diligent in exploring the adjacent country, planting the seeds they had brought with them from Europe and Brazil, and studiously endeavouring to gain an acquaintance with the native language, which they justly considered essential to the accomplishment of their objects.

This was a most laborious and tedious undertaking. The language was altogether oral; consequently, neither alphabet, spelling-book, grammar, nor dictionary existed. On their arrival, they found two Swedes, Peter Hagersteine, and Andrew Cornelius Lind; the former had been wrecked in the Matilda, and the latter had been left by Captain New of the Dædalus, only a few years before the Missionaries arrived. Peter had a slight knowledge of the colloquial language of the natives; and in all their early communications with the chiefs and people, the Missionaries were glad to avail themselves of his aid as interpreter. He was a man of low education and bad principles; and if he did not intentionally misrepresent the communications of the Missionaries, his statements must often have conveyed to the natives' minds very erroneous impressions of their sentiments and wishes. From him, as an instructor, they could derive no advantage; as he seldom came near them excepting when he bore some message from the king, or the chief with whom he resided. The remarks of former voyagers, and the specimens of the language they had given, were of little service, as they could only be the names of the principal persons and things that had come under the notice of such individuals, and even in the representation of these, the orthography was as various as the writers had been numerous. In reference to their attempts to acquire the knowledge of Tahitian, they remarked, that they found all Europeans, who had visited Tahiti, had mistaken the language as to spelling, pronunciation, and ease of acquisition. In addition to the printed specimens, they had a small vocabulary, compiled by one of the officers of the mutineers in the Bounty, who had resided some months in Tahiti, prior to the arrival of the Pandora; when he was arrested, and brought a prisoner to England, where he was executed at Portsmouth. This vocabulary he left with the worthy clergyman who attended him in his confinement, and by him it was kindly given to the Missionaries; who found it more useful than every aid besides. On their voyage, they had carefully studied it, but though they were thus put in possession of a number of words, in their proper collocation they discovered they had every thing to learn. They had arranged a number of words in sentences according to the English idiom, which they supposed would be serviceable on landing; but the use of which they soon found it necessary to discontinue. One of these sentences, *Mity po tuaana,* often afterwards amused the king, when he came to know what they intended by it. *Maitai* is good, *po* is night, and *tuaana* brother. Good-night, brother, was the sentiment intended; but if the natives understood the English word *mighty,* it would mean, Mighty night, brother; or, if they understood *mity* as their word *maitai,* the phrase would be an assertion to this effect, Good (is the) night, brother. This circumstance shews the difficulties they had to contend with, even when they had acquired the meaning of many of the substantives and adjectives in the language.

In these embarrassments they had no elementary books to consult, no preceptors to whom they could apply, but were obliged, partly by gestures and signs, to endeavour to obtain the desired information from the natives; who often misunderstood the purport of their questions, and whose answers must, as often, have been quite unintelligible to the Missionaries. A knowledge of the language was, however, indispensable; and many of the Missionaries employed much of their time among the natives, making excursions through the neighbouring districts, spending several days together with the chiefs at their own habitations, for the purpose of observing their customs, and obtaining an acquaintance with the words which they employed in social intercourse among themselves. This was the more necessary, as the natives who reside in those parts visited by shipping, soon pick up a few of the most common English phrases, which they apply almost indiscriminately, supposing they are thereby better understood, than they would be if they used only native words; yet these words are so changed in a native's mouth, who cannot sound any sibilant, or many of our consonants, and who must also introduce a vowel between every double consonant, that no Englishman would recognize them as his own, but would write them down as native words. *Pickaninny* is a specimen of this kind.

It was not in words only, but also in their application, that the most ludicrous mistakes were made by the people. "Oli mani," a corruption of the English words "old man," is the common term for any thing old; hence, a blunt, broken knife, and a threadbare or ragged dress, is called "oli mani." A captain of a ship, at anchor in one of the harbours, was once inquiring of a native something about his wife, who was sitting by The man readily answered his question, and concluded by saying, "Oli mani hoi," she is "also an old man."

Part of each day was by several devoted to the study

of the language, while once a week, the whole met together for conversation and mutual aid in its acquisition. The only means they had of obtaining it, was by observing carefully the native sounds of words, and writing down the characters by which they were expressed. In this they found great difficulty, from what generally proves a source of perplexity to a learner in his first attempt at understanding a foreign tongue, viz. the rapidity with which the natives appeared to speak, and the want of divisions between the distinct words. The singular fact of most of their syllables consisting of a consonant and a vowel, and a vowel always terminating both their syllables and their words, increased their embarrassment in this respect.

It was a circumstance highly advantageous to the Missionaries, that the Tahitians were remarkably loquacious, often spending hours in conversation, however trivial its topics might be, patiently listening to inquiries, and anxious to make themselves intelligible. Although among themselves accustomed to hear critically, and to ridicule with great effect, any of their own countrymen who should use a wrong word, mispronounce or place the accent erroneously on the one they used, yet they seldom laughed at the mistakes of the newly arrived residents. They endeavoured to correct them in the most friendly manner, and were evidently desirous that the foreigners should be able to understand their language, and convey their own ideas to them with distinctness and perspicuity.

When the Missionaries heard the natives make use of a word or sentence with which they were not already acquainted, they wrote it down, and repeated distinctly several times what they had written. If the natives affirmed that the word or sentence was correctly pronounced by the Missionary, it was left for more careful and deliberate investigation. Sometimes they endeavoured to find out words, by presenting to the natives different combinations of the letters of their alphabet: thus they would pronounce the letters a a, and say, "what is that?" The natives would answer by pointing to the fibrous roots of a tree, or the matted fibres round the cocoa-nut stalk, which are called *aa*. They would then pronounce others, as a i, and ask what it meant; the natives, putting their hand to the back of the neck, and repeating *ai*, told them that that part of the body was thus called. By this means they sometimes discovered the meaning of a variety of words, which they did not before know were even parts of the language. In speaking of their progress, shortly after they had commenced this department of labour, they observe, "We have already joined some thousands of words together, and believe some thousands yet remain." Still their progress was but slow, and one of them, who has perhaps made himself most familiar with the native tongue, has frequently assured me, he was ten years on the island, before he knew the meaning of the word *ahiri*, corresponding to the English word *if*, used only in connexion with the past tense of the verb *to have*, as "If I had seen," &c.

While the Missionaries were thus employed, the chiefs continued friendly and attentive; the people, however, began to manifest that propensity to theft, which they evinced even on the first visits they received. This obliged them to watch very narrowly their property. Clothing and iron tools appeared to be most earnestly sought; and, notwithstanding the measures of security which they adopted, their blacksmith's shop was robbed by a native, who dug two or three feet into the ground on the outside, and, burrowing his way under the wall or side of the house, came up through the earthen floor within, and stole several valuable articles.

Their increased acquaintance with the people had awakened their deepest commiseration, when they beheld them, not only wholly given to idolatry, and mad after their idols, but sunk to the lowest state of moral degradation and consequent wretchedness. This furnished a powerful incentive to energetic perseverance in the acquisition of the language, that they might speedily instruct them in the principles of Christianity, and thereby elevate their moral character, diminish their actual suffering, and improve their present condition.

The Tahitian was the first Polynesian language reduced to writing. In acquiring a knowledge of its character and peculiarities, and reducing it to a regular system, the Missionaries had to proceed alone. In adapting letters to its sounds, forming its orthography, and exhibiting the vernacular tongue in writing to the people, presenting to the eye that which had before been applied only to the ear, and thus furnishing a vehicle by which light and knowledge might be conveyed through a new avenue to the mind, they were unaided by the labours of any who had preceded them, and were therefore the pioneers of those who might follow. That their difficulties were great, must be already obvious. They advanced with deliberation and care, and though the Tahitian dialect as written by them is doubtless imperfect, and susceptible of great improvement, the circumstance of its having formed the basis of those subsequently written, the ease with which it is acquired, and the facility with which it is used by the natives themselves, are evidences of its accuracy and its utility.

The Missionaries have been charged with affectation in their orthography, &c. but so far from this, they have studied nothing with more attention than simplicity and perspicuity. The declaration and the pronunciation of the natives formed their only rule in fixing the spelling of proper names, as well as other parts of the language. They aimed at precision, and having adopted the English character, affixed to each letter a distinct and invariable sound. The letters of each word constitute the word, so that a person pronouncing the letters used in spelling a word, would, in fact, pronounce the word itself. Pursuing this plan, they were under the necessity of presenting to the natives a mode of spelling different from that which had been given to Europeans in the narratives of early voyagers. They did this reluctantly. Their early associations and strongest predilections were all in favour of Otaheite, Ulitea, Otahaa, &c., and it was only from the firm conviction that such were not the native designations of these islands, that they adopted others.

As the native names of persons and places will unavoidably occur in the succeeding pages, a brief notice of the sounds of the letters, and the division of some of the principal words, will probably familiarise them to the eye of the reader, and facilitate their pronunciation.

The different Polynesian dialects abound in vowel sounds perhaps above any other language; they have also another striking peculiarity, that of rejecting all double consonants, possessing invariably vowel terminations, both of their syllables and words. Every final vowel is therefore distinctly sounded. Several consonants used in the English language, do not exist in those of the Georgian and Society Islands. There is no sibilant, or hissing sound: s and c, and the corresponding letters, are therefore unnecessary. The consonants that are used retain the sound usually attached to them in English.

The natives sound the vowels with great distinctness; *a* has the sound of a in father, *e* the sound of a in fate, *i* that of i in marine or *e* in me, *o* that of o in no, and *u* that of oo in root. The diphthong *ai* is sounded as i in wine. The following are some of the names most frequently used in the present work.

The first column presents them in the proper syllabic divisions observed by the people. In the second column I have endeavoured to exhibit the native orthoëpy, by employing those letters which, according to their general use in the English language, would secure, as nearly as possible, the accurate pronunciation of the native words. The *h* is placed after the *a* only to secure to that vowel the uniform sound of *a* in father, or *a* in the interjection *ah*, or *aha*. *Y* is also placed after *a*, to secure for the Tahitian vowel *e*, invariably the sound of *a* in *hay* or *day*.

NAMES OF PLACES.

Ta-hi-ti ......pronounced as......Tah-he-te
Ma-ta-vai ....Máh-tah-vye
Pa-re ....Pah-ray
Pa-pe-e-te........................Pah-pay-ây-tay
A-te-hu-ru ....Ah-tay-hoo-roo
Tai-a-ra-bu.......................Tye-ah-rah-boo
Ei-me-o ....Eye-may-o
Mo-o-re-a.........................Mo-o-ray-ah
A-fa-re-ai-tu ....Ah-fah-ray-eye-too

O-pu-no-hu .... O-poo-no-hoo
Hu-a-hi-ne .... Hoo-ah-hé-nay
Fa-re .... Fáh-ray
Rai-a-te-a .... Rye-ah-tay-ah
O-po-a .... O-po-ah
U-tu-mao-ro .... Oo-too-mao-ro
Ta-ha-a .... Tah-ha-ah
Bo-ra-bo-ra .... Bo-rah-bo-rah
Mau-ru-a .... Mou-roo-ah
Ra-pa .... Rah pah
Ai-tu-ta-ke .... Eye-too-tah-kay
Mi-ti-a-ro .... Me-te-ah-ro
Ma-u-te .... Mah-oo-tay
A-tu-i .... Ah-too-e
Ra-ro-to-gna .... Rah-ro-to-na
or .... or
Ra-ro-ton-ga .... Rah-ro-ton-ga
Tu-bu-ai .... Too-boo-eye
Rai-va-vai .... Ry-vah-vye
Ri-ma-ta-ra .... Re-mah-tah-rah

NAMES OF PERSONS.

Po-ma-re .... Po-mah-ray
I-di-a .... E-dee-ah
Ai-ma-ta .... Eye-mah-tah
Te-ri-ta-ri-a .... Tay-ree-tah-re-ah
Ta-ro-a-ri-i .... Tah-ro-ah-ree
Ma-hi-ne .... Mah-he-nay
Te-rai-ma-no .... Tay-rye-mah-no
Tau-a .... Tou-ah
Ta-ma-to-a .... Tah-mah-to-ah
Fe-nu-a-pe-ho .... Fay-noo-ah-pay h
Mai .... Mye
Au-na .... Ou-nah

A-tu-a ....(God)........... Ah-too-ah
Va-ru-a ....(Spirit) .... Vah-roo-ah
Ta-a-ta ....(Man)............ Ta-ah-tah
A-ri-i ....(King) .... Ah-re-e
Ra-a-ti-ra ....(Chief).......... Ra-ah-té-rah.

CHAP. IV.

Character and death of Haamanemane—Efforts to prevent human sacrifices and infant murder—Resolution of the Missionaries, relative to the use of fire-arms—Arrival of the first ship after the Duff's departure—Assault upon the Missionaries—Its disastrous Consequences—Pomare's revenge—Death of Oripaia—Invasion of Matavai—Murder of Mr. Lewis—Pomare's offering for the Mission Chapel—Arrival of a king's ship—Friendly communications from the governor of New South Wales—Government orders—Act of parliament for the protection of the South Sea Islanders—Arrival of the Royal Admiral—Landing of the Missionaries—Departure of Mr. Broomhall—Notice of his subsequent history.

HAAMANEMANE, the old priest, who had been Captain Wilson's *taio*, or friend, was frequently with the Missionaries, and uniformly kind to them. He was evidently a shrewd and enterprising man; yet I should think sometimes rather eccentric. When arrayed in a favourite dress, which was a glazed hat, and a black coat fringed round the edges with red feathers, his appearance must have been somewhat ludicrous; although this was probably his sacerdotal habit, as red feathers were always considered emblematical of their deities. He had formerly been a principal chief in Raiatea, and still possessed great influence over the natives, especially in the adjacent island of Eimeo, where, with a little assistance from the European workmen, he had built a schooner, in which he came over to see his friend Captain Wilson, during the second visit of the Duff to Tahiti. This vessel, considering it as their first effort at ship-building, was an astonishing performance. To him, the Missionaries had frequent opportunities of speaking, though apparently with but little good effect, against many of the sanguinary features of their idolatry, especially the offering of human sacrifices, in which they knew he had been more than once engaged since their arrival. Sometimes, however, he spoke as if he officiated, in these horrid rites, more from necessity than choice.

He was remarkably active and vigorous, and, though far advanced in years and nearly blind, indulged, without restraint, in all the degrading vices of his country. Moral character, and virtuous conduct, were never considered requisite, even in those whose office was most sacred. As a priest, he practised every species of extortion and cruelty; neither was he less familiar with intrigue, nor free from ambition, as a politician. His supposed influence with the gods, his deep skill in the mysteries of their worship, and the constant dread of his displeasure, which would probably have doomed the individual, by whom it was incurred, to immolation on the altar of his idol, favoured, in no small degree, his assumption and exercise of civil power, both in Eimeo and Tahiti. A jealousy appeared to exist between him and Pomare, the father of Otu, who was king of the island; and during the absence of the former, on a visit to a neighbouring island, he formed a league with Otu, to deprive Pomare of all authority in Tahiti. Having offered a human victim to his idol, he invaded the district of the absent chieftain, and brought war to the very doors of the Mission-house, in less than seventeen months after the departure of the Duff. The attack was made at daybreak, in the western border of Matavai: four individuals were killed, and afterwards offered by the priest to his deity. The inhabitants, unable to withstand the young king and his ally, abandoned their plantations and their dwellings, and fled for their lives. The invaders divided the district, and the priest, taking possession of the eastern side, revelled in all the profligacy and insolence of plunder and destruction. His triumph, however, was but short. Pomare sent privately to Idia directions for his assassination. After two or three solicitations from his mother, Otu, though in closest alliance with him, consented to his death, and he was murdered by one of Idia's men, at the foot of One-tree Hill, as he was on his way to Pare, on the 3d of December 1798, ten days after the invasion of Matavai.

The Missionaries sought an early opportunity to unfold to the rulers of the nation the objects of their Mission, and, after several disappointments, held a public interview with Pomare, Otu, and other principal chiefs, in which they stated, as distinctly as possible, through the medium of Peter Hagerstien, as interpreter, their design in coming to reside amongst them; viz. to instruct them in useful arts, teach them reading and writing, and make known to them the only true God, and the way to happiness in a future state; urging the discontinuance of human sacrifices, and the abolition of infanticide. As an inducement to compliance with this last request, they offered to build a house for the accommodation of the children that might be spared, whom they promised to nurse with attention equal to that which they paid to their own. The chiefs and people listened attentively to the proposition, appeared pleased, and said that no more children should be murdered. It was, however, only a promise.

The distressing circumstances under which this unnatural and revolting crime was practised, and the awful extent to which it prevailed, was one of the first of the many horrid cruelties filling these "dark places" of paganism, that deeply affected them. More than once having received intimation of the murderous purpose of the parents, they had, when the period of childbirth drew nigh, used all their influence to dissuade them from its execution, offering as a reward for this act of common humanity, articles highly valued by them When these had failed to move the parents' hearts, and they could obtain no promise from either the father or mother, that they would spare the child, the wives of the Missionaries have, as a last resort, begged that the infant, instead of being destroyed, might be committed to their care. But the people were so much under the slavish influence of cruel custom, that, with one or two exceptions, their efforts were unavailing, and the guilty murderers have in a few days presented themselves at the Missionary dwellings, not only with most affecting insensibility, but apparently with all the impudence of guilty exultation.

The persons and the habitations of the Missionaries had hitherto been secure, excepting from petty thefts; they were, however, occasionally alarmed by rumours of war. Haamanemane had formerly requested their aid in a descent he intended to make upon Raiatea for the recovery of his authority there; but this they had firmly declined. The pilfering habits of the people rendered it necessary for them to watch their property during the night; and the unsettled state of political affairs in the island indicating their exposure to the consequences of actual war, led them to consider the line of conduct it would be their duty under such circumstances to pursue. They were in the possession of fire-arms, which they had brought on shore solely with a view to intimidate the natives, and deter any, who, unrestrained by the influence of those chiefs who had guaranteed their protection, might be disposed to attack them. The propriety of their using fire-arms was, however, questioned by some, and discussed by the whole body; who publicly agreed that it was not their duty even to inflict punishment upon those that might be detected in stealing their property, but to complain to their chiefs; that they could take no part even with their friends in any of their wars. They resolved that their arms should be used for defence, only in the event of an attack being made upon their habitations; and not even then, until every means of avoiding it had been employed. Some of the Missionaries carried their principles of forbearance so far, as to declare that, but for the exposure of the females, even then it would not be right to have recourse to arms. Such were the views of the Missionaries, and the circumstances of the people, when an event transpired which altogether altered the aspect of affairs in reference to the Mission.

On the 6th of March 1798, exactly twelve months from the day on which the Duff first anchored in Matavai bay, a vessel arrived at Tahiti; which, being the first they had seen since the departure of Captain Wilson, awakened considerable interest. She was boarded by three of the Missionaries at the mouth of the harbour, and found to be the Nautilus of Macao, commanded by Captain

Bishop, and originally bound to the north-west coast of America for furs. Being driven by a heavy gale to Kamtschatka, and, unable to pursue her intended voyage, she had altered her course for Massuefero, near the South American coast, but had been compelled by stress of weather to steer for Tahiti. The ship was in great distress, the crew in want of most of the necessaries of life, and the captain had nothing to barter with the natives for supplies, but muskets and powder. These indeed were formerly the only articles of trade, with the exception of ardent spirits, that many adventurers ever thought of giving to uncivilized nations, in exchange for the produce of their countries! The natives crowded the ship; and Pomare, who was on board, beheld with expressions of contempt the poverty of the vessel, and the distress of her crew. In the minds of the Missionaries their circumstances awakened compassion, and they readily offered to furnish the captain with such supplies as the island afforded, and to assist him in procuring water.

The Nautilus had touched at the Sandwich Islands, and had brought away some of the natives: while the vessel remained, five of these absconded; one was brought back, but escaped again. The vessel remained five days at Tahiti, procured such supplies as the crew were most in need of, and ultimately sailed, leaving the five Sandwich Islanders on shore.

Exactly a fortnight after her departure, this vessel again entered Matavai Bay, much to the surprise of the Missionaries, who were informed by the captain and supercargo, that, in consequence of a severe gale off Huahine, she was unfitted for her voyage to Massuefero, and that they intended to proceed to Port Jackson, when they had increased their supplies. In the course of the night, two seamen absconded with the ship's boat; and the next morning the captain and supercargo addressed a letter to the Missionaries, acquainting them with the desertion of the men; and their determination, in consequence of their deficiency of hands, to recover them, cost what it would; soliciting, at the same time, aid in effecting their apprehension. The Missionaries recovered the boat, on the following day; and, anxious to afford the captain and supercargo of the Nautilus every assistance in their power, agreed to use their influence with the king, and two of the principal chiefs, to induce them to send the seamen on board. Four of the Missionaries went on this errand to the district of Pare, where the king and chiefs were residing. After walking between two and three hours, they reached the residence of Otu, the young king. The Sandwich Islanders were among his attendants, and they had reason to suspect that he had favoured the concealment of the seamen.

Desirous of disclosing their business to the chiefs when together, they remained some time, expecting the arrival of Pomare, for whom they had sent. The king was sullen and taciturn; and, after waiting nearly half an hour for Pomare, the Missionaries departed, to wait on him personally, at his own dwelling.

As they passed along, the natives tendered their usual salutations, and about thirty accompanied them. They had, however, scarcely proceeded a mile on their way, when, on approaching the margin of a river, they were each suddenly seized by a number of natives, who stripped them, dragged two of them through the river, attempted to drown them, and, after other ill-treatment, threatened them with murder. Some of the natives gave the Missionaries a few strips of cloth; and, at their request, conducted them to Pomare and Idia, whose tent was at some distance. These individuals beheld them with great concern; and, expressing no ordinary sympathy in their distress, immediately furnished them with native apparel and refreshment; and, when they had rested about an hour, accompanied them on their return to Matavai.—When they reached Otu's dwelling, Pomare called the king, his son, into the outer court, and questioned him as to the treatment the Missionaries had received. He said but little; yet there was reason to suppose, that if the assault had not been made by his direction, he was privy to it. Bent on the conquest of the whole island, and desirous, in conjunction with those attached to his interests, of depriving his father and younger brother of all authority in Tahiti, muskets and powder were articles in greatest demand, and the aid of Europeans was most earnestly desired. The Missionaries, by furnishing supplies to the vessel, had prevented his obtaining the former; and in order to be revenged on them for this act of friendship to those on board, he had allowed some of his men to follow and to plunder them. Their having applied for the return of the Sandwich Islanders, who had before absconded from the vessel, led him to suspect their business on the present occasion. The seamen, who had deserted from the Nautilus, were under the protection of the king, and appeared among his attendants. The Missionaries did not disclose the object of their visit; but Pomare insisted on the deserters being delivered up, assuring them they should be carried on board the next day. The seamen expressed their determination to remain; and one of them said, "If they take me on board again, they shall take me on board dead." The conduct of Pomare, the king's father, with that of his queen, Idia, was highly commendable: several of the articles of dress, which had been taken from the Missionaries, were restored, and the people in general appeared to compassionate them; though two of them heard the natives, who were stripping them, remark that, as they had four of them in their possession, they would go and take the fourteen remaining at Matavai. In the evening the Missionaries arrived at their dwelling, having been furnished by Pomare with a double canoe, for their conveyance home.

The impression this unpleasant occurrence produced upon the society at Matavai, was such, that eleven Missionaries, including four who were married, judged a removal from the island to be necessary; and as the captain and supercargo of the Nautilus offered a passage to any who were desirous of returning to Port Jackson, they prepared for their departure. Two days after the plunder of the Missionaries, Pomare sent the chief priest of the island with a fowl as an atonement, and a young plantain as a peace-offering, and on the following day hastened to their dwelling.

The report of the departure of the Missionaries soon spread through the island, and appeared to be regretted by many of the people. Pomare, who had ever been most friendly, manifested unusual sorrow, and used extraordinary efforts to persuade them to stay. He went through every room in their house, and every birth on board, and addressed each individual by name, with earnest entreaties to remain, and assurances of protection. *Noti, eiaha e haere,* Mr. Nott, don't go, was his language to that individual, and such was also used to others. His evident satisfaction was proportionate, when he perceived that Mr. and Mrs. Eyre, and five of the single Missionaries, resolved to continue in Tahiti.

On the 29th of March, those Missionaries who intended to leave, bade their companions farewell; and, during the night of the 30th, sailed from Matavai, and proceeded to New South Wales. It is worthy of remark, that this event, so destructive to the strength of the Mission, crippling the efforts of its members, and spreading a cloud over their future prospects, resulted not from opposition to the efforts of the Missionaries, nor from any dispute between them. and the priests or people, on subjects connected with the idolatry of the latter, but from their benevolent endeavours to serve those, whom purposes of commerce had brought to their shores, and whom adverse weather had reduced to circumstances of distress—a class of individuals whom the Missionaries, in those seas, have ever been ready to succour, but who, with some gratifying exceptions, have not always honourably requited that kindness to which, in some instances, they have owed their own preservation.

The decision of those who left Tahiti, may, to some, perhaps, appear premature, but it is not easy to form a correct estimate of the dangers to which they were exposed. They were well aware of many; but there were others, actually existing, of which they were then unconscious. Otu, called Pomare since his father's death, has often, during the latter years of his life, told Mr. Nott, that after the departure of the Duff, frequently, when he has been carried on men's shoulders round the residence of the Missionaries, Peter the Swede, who has been with him, has said, when the Missionaries were kneeling down in prayer, at their morning or evening family worship, "See, they are all down on their knees, quite defenceless; how easily your people might rush upon them, and kill them all, and then their property would be yours." And it is a melancholy fact, that the influence of unprincipled and profligate foreigners, has been more fatal to the Missionaries, more demoralizing to the natives, more inimical to the introduction of Christianity, and more opposed to its establishment, than all the prejudices of the people in favour of idolatry, and all the attachment of the priests to the interests of their gods.

However much those who remained might have been affected by the departure of so many of their companions, they felt no disposition to abandon the field, or relax their endeavours for the benefit of the people. Pomare had not only sent an atonement and a peace-offering, but, even before the Missionaries sailed, had made war upon the district, and had killed two of the men who had been engaged in assaulting them. This was, indeed, a matter of regret to the Missionaries; but it was also an evidence of his displeasure at the treatment they had received. On his assurances of protection, those who remained reposed the most entire confidence; which, during his subsequent life, his conduct uniformly warranted. Committing their persons to the merciful and watchful providence of God, and, under him, to the friendly chiefs who had manifested so much concern for their safety; they had sent all the fire-arms, ammunition, and other weapons, possessed by the Society, on board the Nautilus, excepting two muskets, which they presented to Pomare and Idia. To the former they gave up their public stores, and all the property they possessed, together with the smith's shop, and the tools. They also offered Pomare their private property, but he refused to take it; informing them, that so long as they remained, every thing in the store-room should be at their command; but that, in the event of their leaving the island, he should consider whatever remained as his own. On a subsequent occasion, when he feared, that on account of a destructive war then prevailing, they might leave, he directed them to take their property with them; hereby evincing the most disinterested friendship, and a desire to alleviate, rather than profit by, their distresses. Their situation was critical, but in a letter which they forwarded on this occasion to the Society, they express firm confidence in God, unabated attachment to their work, and contentment with such means of support as the country afforded.

Not long after the departure of the Nautilus, it was reported, that in order to avenge the death of the two men he had killed, the people of Pare had declared war against Pomare. He applied to the Missionaries for assistance, and, entering the room in which they were assembled, inquired how many of them knew how to make war. Mr. Nott replied "We know nothing of war." Pomare withdrew, and they afterwards agreed not to resort to the use of arms, either for offence or defence. Their determination was made known to their friends; and, as no dissatisfaction appeared, they were led to hope that they should be permitted peaceably to prosecute their labours, without any further solicitation on the subject. A native who had assisted in the smith's shop was enabled, after the departure of the Missionaries, who had used the forge, to make fish-hooks, adzes, and a number of useful iron articles; but the skill he had acquired, instead of being employed to promote the industry, civilization, and comfort of his countrymen, was soon applied to purposes of barbarity and murder; and the Missionaries beheld with regret that he was often employed in manufacturing not only useful tools, but weapons for battle.

Pomare subsequently made war upon the inhabitants of Pare, where the Europeans had been plundered: the people were defeated, fourteen of them killed, and forty or fifty of their houses burnt.

Five months after the departure of the Missionaries in the Nautilus, two large vessels were seen standing towards Matavai bay. As soon as they hoisted English colours, the natives were thrown into the greatest consternation, and, packing up whatever they could carry away, abandoned their houses, and were seen in every direction flying towards the mountains. Being asked their reasons for such a proceeding, they answered, that seeing two large English ships, they apprehended they were come to revenge the assault upon the Missionaries. After many assurances to the contrary, their fears seemed to be removed. When the Captains came on shore in the evening, they were welcomed by the Missionaries, and introduced to the chiefs, whose familiarity and cheerfulness soon evinced that every feeling of suspicion had subsided. These vessels were the Cornwall and the Sally of London, South Sea whalers. As the ships were in repair, and the crews in health, they remained only three days in the harbour, and sailed from the island on the 27th of August; having made a number of presents to the chiefs, they did not leave any of their crews on shore, which was a matter of great satisfaction to the Missionaries, who had beheld with regret the baneful influence of unprincipled seamen, on the minds and habits of the people.

From one of these ships, Oripaia, a chief of Papara, and rival of Pomare, had received a large quantity of gunpowder as a present. The powder being coarser in the grain than what the natives had been accustomed to receive, they imagined either that it was not powder, or that it was a very inferior kind. In order to satisfy themselves, Oripaia proposed to one of his attendants to try it. A pistol was loaded, and fired over the whole heap of powder they had received, and around which the chief and his attendants were sitting. A spark fell from the pistol, and the whole of the powder instantly exploded. As soon as the natives had recovered from the shock, perceiving the powder adhering to their limbs, they attempted to rub it off, but found the skin peel off with it; they then plunged into an adjacent river. Six of the natives were severely injured, and Oripaia with one of his attendants died. As soon as Pomare was acquainted with the accident, he begged Mr. Broomhall to visit the house in which the accident had occurred, and endeavour to relieve the sufferers. The chief appeared in a most affecting state, dreadfully scorched with the powder; Mr. Broomhall employed such applications as he supposed likely to alleviate his sufferings; these, however, increased, and both the chief and his wife attributed his pains, not to the effects of the explosion, but to the remedies applied, or rather to the poison imagined to be infused into the application by the god of the foreigners. This not only aroused the jealousy of the chief, and the rage of Otu, but had nearly cost Mr. Broomhall and his companions their lives, and made the Missionaries extremely cautious in administering medicine to any of the chiefs. Native remedies were now applied, to relieve the sufferings of Oripaia, but they were unavailing, and, after languishing for some time in the greatest agony, he expired. The body of the deceased chief was embalmed by a process peculiar to the inhabitants of the South Sea Islands. It was placed on a kind of platform; and a number of superstitious ceremonies were observed. During the performance of these rites, Pomare's orator, and some of the inhabitants of Matavai, used insulting expressions in reference to the corpse; which so incensed Otu, that, aided by the chief priest, he immediately made war upon the district of Matavai. Late in the evening, the Missionaries and people had some intimation of his intention: before daylight the next morning, the attack was commenced at one end of the district; the inhabitants fled before the assailants; and by sunrise, the warriors of Otu had scoured the district from one end to the other, driving before them every inhabitant, excepting a few in the immediate vicinity of the Missionary dwellings. Several warriors, with clubs and spears, surrounded the Missionary house, but its inmates remained unmolested; and in the course of the day, Haamanemane arrived, and assured the Mission family no evil was designed against them. In the evening they were also visited in an amicable manner by Otu and his queen.

In connexion with this attack upon the district of Matavai, which belonged to Pomare, Otu and Haamanemane declared that Pomare was deprived of all authority in the larger peninsula. The districts on the west and south side declared for Otu,. and those on the western were threatened with invasion in the event of refusal.

In the division of the territory thus seized, the chief priest received the eastern part of Matavai; but he did not long enjoy it, he was murdered, as already stated, very shortly afterwards. This event gave a new aspect to political affairs in the island, and appeared to unite in one interest Otu and Pomare his father. The inhabitants of Matavai left their places of retreat, and, having presented their peace-offering, re-occupied their lands. The Missionaries resumed their attempts to instruct the natives, but found the acquisition of the language so difficult, and the insensibility of the people so great, that they were exceedingly discouraged. Some of the natives, however, were led to inquire how it was that Cook, Vancouver, Bligh, and other early visitors, had never told them any of those things which they heard from the teachers now residing with them

Towards the close of the year 1799, the Missionaries were called to the melancholy duty of conveying to the silent grave, under very distressing circumstances, Mr. Lewis, one of their number, and the first Missionary who had terminated his life on the shores of Tahiti. He landed from the ship Duff in 1797, continued to labour with his companions, respected and useful, until about three months after the departure of the Nautilus with the families to Port Jackson, when he left the Mission house, and took up his residence with a taio, or friend, in the eastern part of the district. Three weeks afterwards, he intimated to his companions his intention of uniting in marriage with a native of the island, solemnly purposing to abide faithful towards her until death. Considering her an idolatress, the Missionaries deemed this an inconsistent and unlawful act, but Mr. Lewis, persevering in his determination, they dissolved the connexion that had subsisted between him and themselves, as members of the church of Christ, and discontinued all Christian and social intercourse with him. He was still constant in attendance on public worship. industrious in the culture of his garden, and in working for the king and principal chiefs, who were evidently much attached to him. On the 23d of November, the Missionaries heard he had died on the preceding evening. They hastened to his house, and found the corpse lying on a bed; the forehead and face considerably disfigured with wounds, apparently inflicted with a stone and a sharp instrument. The female with whom he had lived as his wife, informed them that he went out of the house on the preceding evening, and that hearing a noise shortly afterwards, she hastened to the spot whence it proceeded, and saw him on the pavement in front of the house, beating his head against the stones. On looking at that part of the pavement where he had fallen, one or two of the stones were stained with blood. Some of the natives said that he had acted as if insane, others that the evil spirit had entered into him; but, from several expressions that were used, there was reason to apprehend he had been murdered.

Assisted by two or three natives, Mr. Bicknell and Mr. Nott dug his grave in a spot near their dwelling on the north side of Matavai bay, which had been selected as a place of interment. On the evening of the 29th of November, 1799, Mr. Nott, Mr. Jefferson, Mr. Eyre, and Mr. Bicknell, bore his remains to the grave, where Mr. Harris read the xcth Psalm, and offered up an appropriate prayer to Almighty God. The circumstances of his death were truly affecting, and the feelings of the Missionaries such as it would be in vain to attempt to describe. They have since learned that he was murdered, and some of them have also regretted that after his separation, that kindness and friendly intercourse were not continued, which might perhaps, without compromise of character, have been consistently maintained. Pomare, considering himself the protector of the Missionaries, though he did not appear to think he had been murdered, yet proposed, if it appeared to the survivors that such had been the fact, to destroy the inhabitants of the district; and so much did many of the latter fear such an event, that several fled to the mountains. The Missionaries, considering that in such retaliation the innocent would suffer with the guilty, interposed, and prevailed upon the king to spare the district, but to punish the guilty whenever they might be discovered.

Scarcely were the remains of Mr. Lewis consigned to the silent grave, when an event occurred, which again reduced the number of this already weakened band. The Betsy of London, a letter of marque, arrived with a Spanish brig her prize, with which she was proceeding from South America to Port Jackson. The commander of the Betsy having intimated his intention of returning in five or six months, Mr. Harris proposed to his companions to visit New South Wales; and on the 1st of January 1800, he sailed from Matavai bay, intending to return when the ship should revisit the islands. By this conveyance, the remaining Missionaries wrote an account of their circumstances and their prospects to the directors in London, stating, that although they had not acquired a sufficient knowledge of the language to enable them publicly to preach the gospel, they had observed, whenever they had conversed with the natives, that though they could perceive the difference between Christianity and paganism, their attachment to the abominations of the latter was too strong to be removed by any other influence than that of the Spirit of God.

Anxious to avoid unnecessary expenditure, they had on a former occasion written, to prevent the Society's incurring any further expense on their account, as their remaining on the island was uncertain; but now, as there was a prospect of peaceable continuance, and the liberal supply they had taken out in the Duff, being, by plunder, presents, &c. nearly expended, they found it necessary to apply for a few articles for their own use, and others for presents to the chiefs, whom they described as daily visiting their dwellings, and treating them with kindness.

Five days after the departure of the Betsy, the Missionaries had the satisfaction to welcome again to their Society, Mr. and Mrs. Henry; who returned from Port Jackson in the Eliza, a South Sea whaler. Mr. Henry was the only one of the number who had left, that resumed his labours in Tahiti. By his arrival, the Missionaries received the pleasing intelligence of the Duff's second destination to Tahiti, and were led to expect with her arrival a reinforcement of labourers, and the various supplies of which they stood so much in need. Having repaired the vessel and recruited his stores, the captain sailed from Tahiti on the 14th of January, leaving on the island three of his seamen, whose influence among the inhabitants in general was soon found to be most unfavourable.

Hitherto, the public worship of God had been performed in one of the apartments of the Mission-house, but as it appeared expedient to erect a place for this specific object, to which also the natives might have access for the purpose of religious instruction, a spot was selected near the grave of Mr. Lewis; and on the 5th of March 1797, with the assistance of a number of Pomare's men, they commenced the erection of their chapel. The chiefs procured most of the materials, and when it was nearly finished, Pomare sent a *fish* as an *offering* to Jesus Christ, requesting that it might be *hung up* in their new chapel. This was the first building ever erected on the South Sea Islands, for the worship of the living God; and although the Missionaries were cheered with the hope of often beholding it filled with attentive hearers or Christian worshippers, they were obliged to pull it down early in the year 1802, to prevent its affording shelter to their enemies, or being set on fire by the rebels, by which their own dwelling might have been destroyed.

The pleasing anticipations which the Missionaries had been led to indulge in connexion with the second visit of the Duff, were destroyed by the arrival of the Albion, in Matavai bay on the 27th of December in the same year. Her commander, Captain Bunker, brought them no letters from England, but conveyed the melancholy tidings of the capture of the Duff by a French privateer. He also delivered from Mr. Harris, who was settled in Norfolk Island, a letter acquainting them with the murder of three of the Missionaries in the Friendly Islands, the departure of one, the flight of the rest to Port Jackson, and the total destruction of the Tonga Mission. Their own circumstances were by no means prosperous; they had heard but once from England; they were expecting every day the arrival of the Duff with cheering tidings and additional aid; but the intelligence now received, not only disappointed their hopes, but depressed their spirits, and darkened their prospects. In the letter sent at this time to the directors, they express their anxiety to hear from England, their conviction of the facilities that would be afforded towards the establishing the gospel in Tahiti and the neighbouring islands, if they were joined by a body of Missionaries and an experienced director, and recommended that a surgeon and several mechanics should be included in the number of those who might be sent.

The Albion had scarcely sailed, when large fleets of canoes, filled with fighting men, arrived, and the island was agitated with the apprehension of hostilities between the king and chiefs. The removal of Oro, the national idol, from Pare to Atehuru, was the cause of the threatened conflict: ammunition was prepared; a large assembly of chiefs and warriors met at Pare; and it was daily expected that the long concealed elements of war would there explode, and plunge the nation in anarchy and bloodshed. At this critical period, his majesty's ship, Porpoise, arrived in Matavai bay. The letter and presents Pomare received by this conveyance from the governor of New South Wales, and the attentions paid to him by the commander of the vessel, tended, in no small degree, to confirm Otu in his government, and to intimidate his enemies.

The governors of the colony of New South Wales have uniformly manifested the most friendly concern for the safety of the Missionaries, and the success of the several Missions in the South Seas. On the present occasion, Governor King, in a letter to Pomare, remarked, that he could "not too strongly recommend to his kind protection, the society of Missionaries whom he had taken under his care;" and that, "such protection could not fail to excite the gratitude of the Missionaries, and the friendship of King George." Governor Macquarie, his successor, manifested the same kindness towards the Missionaries, and an equal regard for the welfare and security of the natives. In order to protect the inhabitants of New Zealand and the South Sea Islands from the oppression, violence, and murder, of unprincipled and lawless Europeans, he issued, in December, 1813, an order, alike creditable to the enlightened policy of his administration, and the benevolence of his heart. A copy was brought to the Society Islands, and is here inserted.

Government and General Orders, dated Dec. 1, 1813.

"No ship or vessel shall clear out from any of the ports within this territory, (New South Wales,) for New Zealand, or any other island in the South Pacific, unless the Master, if of British or Indian, or the Master and Owners, if of Plantation Registry, shall enter into bonds with the Naval Officer, under £1000 penalty, that themselves and crew shall properly demean themselves towards the natives; and not commit acts of trespass on their gardens, lands, habitations, burial grounds, tombs, or properties, and not make war, or at all interfere in their quarrels, or excite any animosities among them, but leave them to the free enjoyment of their rites and ceremonies; and not take from the islands any male native, without his own and his chief's and parents' consent; and shall not take from thence any female native, without the like consent—or, in case of shipping any male natives, as mariners, divers, &c. then, at their own request at any time, to discharge them, first paying them all wages, &c. And, the natives of all the said islands being under His Majesty's protection, all acts of rapine, plunder, piracy, murders, or other outrages against their persons or property, will, upon conviction, be severely punished."

In reference to another Order resembling this, and issued Nov. 19, 1814, it is declared, that—

"Any neglect or disobedience of these Orders, will subject the offenders to be proceeded against with the utmost rigour of the law, on their return thither, (viz. New South Wales;) and, those who shall return to England, without first resorting to this place, will be reported to His Majesty's Secretary of State for the Colonies, and such documents transmitted, as will warrant their being equally proceeded against and punished."

Although the justice and humanity of the governor of New South Wales were so distinctly manifested in the foregoing Orders, these regulations were found insufficient to prevent outrage upon the natives, from the masters and crews of vessels visiting the islands: an act was therefore passed in the British parliament, in the month of June, 1817, entitled, "An Act of the 57th of the King, for the more effectual punishment of Murders and Manslaughters committed in places not within His Majesty's dominions." As it is a document important to the peace and security of the inhabitants of Polynesia, I deem no apology necessary, for inserting it nearly entire. In the preamble of the bill, it is stated,

"That grievous murders and manslaughters had been committed in the South Pacific Ocean, as well on the high seas, as on land, in the islands of New Zealand and Otaheite, and in other islands, countries, and places, not within His Majesty's dominions, by the masters and crews of British ships, and other persons, who have, for the most part, deserted from, or left their ships, and have continued to live and reside amongst the inhabitants of these islands; whereby great violence has been done, and a general scandal and prejudice raised against the name and character of British and other European traders: And, whereas, such crimes and offences do escape unpunished, by reason of the difficulty of bringing to trial the persons guilty thereof: For remedy whereof, be it enacted by the King's most excellent Majesty, by and with the advice and consent of the Lords Spiritual and Temporal, and the Commons, in this present parliament assembled, and by the authority of the same, that from and after the passing of this Act, all murders and manslaughters committed, or that shall be committed, in the said islands of New Zealand and Otaheite, or within any other islands, countries, or places, not within His Majesty's dominions, nor subject to any European state or power, nor within the territory of the United States of America, by the master or crew of any British ship, or vessel, or any of them, or by any person sailing in, or belonging thereto; or that shall have sailed in, or belonged to, and have quitted any British ship, or vessel, to live in any of the said islands, countries, or places, or either of them, or that shall be there living, shall and may be tried, and adjudged, and punished, in any of His Majesty's islands, plantations, colonies, dominions, forts, or factories, under or by virtue of the King's commission, or commissions, which shall have been or may hereafter be issued, under and by virtue, and in pursuance, of an Act passed in the forty-sixth year of His present Majesty, entitled, an Act for the more speedy trial of offences committed in distant countries, or upon the sea."

By the Porpoise, they also received the agreeable intelligence that a ship, with a reinforcement of Missionaries, and necessary supplies from England, was on her way to the islands. In the afternoon of the 10th of July, 1801, the Royal Admiral, commanded by Captain W. Wilson, anchored in the bay, having a number of Missionaries on board, together with supplies and letters from their friends and the directors, from whom they had heard only once, during the four years they had dwelt on the island. Mr. Shelly, one of the Missionaries who had been stationed in the Friendly Islands, but had escaped to New South Wales, returned to Tahiti in this ship, and was cordially welcomed by the Missionaries, along with those who had arrived from England.

On the 13th of July, 1801, Captain Wilson, and the eight Missionaries from England, landed near Point Venus, and were introduced to Otu, Pomare, and other principal chiefs, by whom they were welcomed to Tahiti. Pomare said he was pleased with their arrival, and expressed his willingness that others should join them. The gratification he expressed on their landing, however, did not arise from any desire after religious instruction, for in this interview he spoke of their engaging in war with him, and probably rejoiced in their arrival only as a means of increasing the strength of his influence, and the stability of his government. After remaining about three weeks at Tahiti, and assisting the society in their regulations by his counsel, and in the preparation of their houses by the carpenters of the ship, Captain Wilson sailed from Matavai on the 31st of July. With him, Mr. Broomhall left Tahiti for China or India. He had been above five years on the island, having arrived in the Duff, in 1797. He was an intelligent, active young man, 24 years of age, had been highly serviceable to the Mission, and was respected by the natives until about twelve months prior to the arrival of the Royal Admiral, when he intimated his doubts as to the reality of Divine influence on the mind, and the immortality of the soul. His companions endeavoured to remove his scepticism; but failing in their efforts, he was separated from their communion, having on several occasions publicly declared his sentiments to be deistical. He then lived some time with a native female, as his wife, but was soon left by her; and, on the arrival of Captain Wilson, requested permission to leave the island in his ship. His departure from the island, under such circumstances, although desirable on account of the influence of his principles and conduct on the minds of the inhabitants, could not but be peculiarly distressing to those he left behind. They followed him with their compassionate regard and their prayers, and, after a number of years, learned that he had been engaged in a vessel trading in the Indian seas; that he had at length made himself known to the Baptist Missionaries at Serampore, from whom they heard that he had renounced his erroneous sentiments, and professed his belief in the truth of the Christian revelation.

The circumstances which follow, relative to the penitence of this unhappy man, are taken from the "Circular Letters" published by the Baptist Missionary Society. In one of these, dated Calcutta, May, 8, 1809, the writer says,

"We have lately seen the gracious hand of God stretched out in a most remarkable manner, in the recovery of a backsliding *Missionary*, after nine years of wandering from God. This person had been chosen with others for an arduous undertaking; had been set apart to the great work, and had engaged in it to a considerable extent; having acquired a tolerable knowledge of the language in which he was to preach to the heathen. At this period, he fell into open iniquity; and embraced a gloomy state of infidelity, the frequent consequence of backsliding from God."

Having left the Mission and gone to sea, several alarming incidents, particularly the breaking of his thigh at Madras, and a severe illness in Calcutta, tended to awaken him to a sense of his danger. But, although he held a correspondence with several serious persons, he studiously concealed his previous character and his name. At length, after writing a long letter, in which he describes the anguish of his mind with dreadful minuteness, he obtained a private interview with Dr. Marshman and Mr. Ward, of which the following is the result.

"At the time appointed, he called on brother Marshman, at brother Carey's rooms, and, after a little conversation on the state of his soul, he added, Your now behold an apostate Missionary. I am ———, who left his brethren nine years ago. Is it possible you can behold me without despising me?—The effect which this discovery of Divine mercy displayed to a backslider, had on brother Marshman's mind, can better be conceived than described. It for the moment took away the anguish occasioned by a note that instant received from Serampore, saying that brother Carey was at the point of death! Brother Marshman entreated this returning prodigal to be assured of the utmost love on our part; encouraged him in his determination to return to his Missionary brethren, and promised to intercede on his behalf, both with his brethren, and those who sent him out."

Soon after the above interview, Mr. Broomhall embarked on another voyage to some port in India, purposing, on his return, to dispose of his vessel, and devote the remainder of his days to the advancement of that cause which he had abandoned; but from that voyage he never returned: neither Mr. Broomhall nor his vessel was ever afterwards heard of,—it being supposed the vessel foundered, and all on board perished.

CHAP. V.

First preaching in the native language—National council in Atehuru—Seizure of the idol Oro—Rebellion of the Oropa—Introduction of useful foreign fruits and vegetables—Providential arrival of two vessels—Battle of Pare—King's camp attacked, Oro retaken—Mission-house garrisoned with seamen, &c.—Desolation of the war—Death of the king's brother—Ravages of foreign diseases—Death of Pomare—Sketch of his character—Otu assumes the name of his late father—Origin of the regal name—Efforts to instruct the children—Death of the queen—Compilation of the first spelling-book—First school for teaching reading and writing—Arrival of the Hawkesbury—Death of Mr. Jefferson—Mr. Nott's visit to the Leeward Islands—Rebellion in Matavai—Defeat of the king—Departure of the majority of the Missionaries—Abandonment of the Mission.

ANXIOUS to increase the resources of the islands, those who had arrived in the Royal Admiral had brought with them a variety of useful seeds, with plants of the vine, the fig, and the peach-tree, from Port Jackson, which were planted in the Mission garden. Many of the seeds grew, and the vegetables produced added a pleasing variety to the indigenous productions of the country. The vine, the peach, and the fig, appeared to thrive very well; but in the war which broke out shortly after, the fences were broken down, the plants torn up, or trodden under foot, and the garden entirely destroyed. Pineapples and water melons, of which the natives seemed remarkably fond, were preserved amidst the general devastation. The pineapple grew luxuriantly in several parts of Tahiti; and though the natives were told it was palatable food, they were so mistaken in the nature of the fruit, that they baked numbers of them, in their native ovens, before they attempted to eat any undressed.

The Missionaries who had arrived in the Duff, had now acquired so much of the language as to be able to preach to the natives in their own tongue, and to engage in the catechetical instruction of the children. In these exercises they did not confine themselves to the inhabitants of their own vicinity, but visited the adjacent districts; and, in the month of March, 1802, Mr. Nott, accompanied by Mr. Elder, made the first Missionary tour of Tahiti, for the purpose of instructing the inhabitants. They were, in general, hospitably entertained, and had many opportunities of speaking to the people, who frequently listened with attention, and often made inquiries, either while the preacher was speaking, or after the service had ended. They seemed interested in the account of the creation, and deeply affected with the exhibition of Jesus Christ, as the true atonement for sin; instead of pearls, or pigs, or other offerings, which they had been accustomed to consider as the best means of propitiating their deities. Some said they desired to pray to the true God, but were afraid the gods of Tahiti would destroy them if they did: others remarked, that the Duff came last among the ships, and that, if the gospel had been conveyed by the first ship, the gods of feathers, as they denominated their idols, would long ago have been destroyed: and one of the principal chiefs, at whose residence they spent the night, observed to the natives around, that he believed they had the true foundation, or source of knowledge.

On their return home, they passed through the district of Atehuru, and found the king, Pomare, and all the chiefs and warriors of the land, assembled at the great Marae, where a number of ceremonies were performing in honour of Oro, the great national idol. As they passed the Marae, they saw a number of hogs on the altar, and several human sacrifices placed in the trees around; and when they reached the spot where the chiefs were assembled, they found Pomare offering five or six large pigs to Oro, on board a sacred canoe, in which the ark, or residence of the idol, was placed. Notwithstanding his being thus engaged, they told him Jehovah alone was God, that pigs were not acceptable to him as offerings, that Jesus Christ was the true atonement for sin, and that God was offended with them for killing men. The chief at first seemed unwilling, but at last said he would attend to their religion.

On the following day, when the king, chiefs, and people, were assembled within the temple, Otu and his father, pretending to have received intimation that Oro wished to be conveyed to Tautira, in Taiarabu, Pomare addressed the chiefs of Atehuru, requesting them to give him up; but the orators of the Atehuruan chiefs resisted. Otu then demanded him, but the chiefs still refused compliance. Pomare then recommended his son, the king, to allow the Atehuruan chiefs to retain the idol until a certain ceremony had been performed. This the king declined, and again insisted that Oro should be given up. This was still refused; and, having asked for some time without effect, he rose up in anger, and ordered his party to withdraw. A number of his attendants rushed upon the canoes, others seized the god by force, tore him away from the people of Atehuru, and bore him towards the sea. This was not only the signal for war, but the commencement of hostilities. The Atehuruans fled to the valley, and the king and Pomare set sail with their fleet to the place of rendezvous; and, lest Oro should feel indignant at the treatment he had received, a human sacrifice was ordered; and, as no captive was at hand, one of Pomare's own servants was murdered, and offered, as soon as the king reached the shore. The next morning, the fleet sailed with the idol for Tautira, and the Missionaries returned to their companions, with the tidings of these threatening events. When the fleet reached Papara, Pomare sent them word that it was probable the Atehuruans would attack them, and advised them to be upon their guard. Ten days after, they heard that the inhabitants of Atehuru had invaded the district of Faa, murdered those who had not escaped by flight, burnt down the houses, and continued their murderous and desolating course into the district of Pare, which joins Matavai on the south. Here they drove out the inhabitants, burned their habitations, and then returned to their own territory; not, however, without threatening to enter the district of Matavai, assault the Missionaries, and plunder their property.

This rebellion, called in the annals of Tahiti, *Te tamai ia Rua*, The war of Rua, (Rua being the name of the principal leader of the rebellion,) was the most powerful and alarming that had yet taken place; and the circumstances by which God providentially preserved the Missionaries from its rage, and from inevitable ruin, were remarkable. About six weeks before Mr. Nott commenced his tour of Tahiti, the Norfolk, an armed brig from Port Jackson, arrived at Matavai, and brought

Mr. and Mrs. Shelly to join the Mission. About a week after the arrival of the Norfolk, the Venus, another colonial vessel, came into the bay, and left on shore Captain Bishop and six seamen, to purchase pigs and salt pork for Port Jackson, while Captain Bass pursued his voyage to the Sandwich Islands, on the same errand. About the 30th of March the Norfolk was wrecked in Matavai bay, having been driven on shore by a heavy gale of wind. The hull was destroyed, but all the stores were preserved. Seventeen Englishmen were thus cast ashore, and added to the number of those already residing there. These, together with Captain Bishop and his men, exposed to one common enemy, united with the Missionaries for mutual defence; and to them, under God, the Missionaries owed their preservation. Two or three hundred warriors came from Eimeo to Pomare's aid. They encamped in the northern part of Pare, where they were joined by a number of the inhabitants of those districts, favourable to his cause; but they were attacked and driven in confusion before the rebels towards Matavai, which had now become the frontier district.

On the day of the engagement, Captain Bishop, with a strong party, occupied the pass on the top of One-tree Hill, arrested the progress of the victors, and favoured the retreat of the vanquished, whose courage appeared to have forsaken them, under the conviction that the god Oro had fought with their enemies, and rendered them invincible. The rebels did not attempt to enter the district, but sent a messenger with proposals of alliance, offering the English the government of Matavai, and the two districts to the southward, which they had already ravaged. If this was not agreed to, they demanded permission to march through the district to attack their enemies beyond Matavai, and, in the event of refusal, declared their intention of forcing a passage with the club and the spear. The refugees from the conquered districts had already sheltered themselves under the protection of the Missionaries and their companions, and they would have fallen a sacrifice to the cruelty of their enemies, had these been allowed to pass through the district. The English, therefore, acceded to the first proposition. The Atehuruans ratified the treaty, returned to their own land, and thus afforded the foreigners at Matavai, and those under their protection, a short respite from the dread of immediate attack. Had the Missionaries been the only Englishmen residing on the island at the time, it is most probable the victors would not have been checked by them in their career of conquest. They would have prosecuted their march of destruction; and, as the Missionaries remark, they must have retreated, or fallen a sacrifice to their fury.

Flushed with success, and animated with the belief that the god fought with them, the rebels, having offered in sacrifice the bodies of the slain, and united in their confederacy the districts of Papara, and the whole of the south-west side of the larger peninsula, crossed the isthmus, marched at once to Tautira, and attacked the king and Pomare; who, ever since their arrival with the idol they had seized in Atehuru, had been engaged in offering human sacrifices, and, by other acts of worship, propitiating the favour of Oro. The rebels conducted their expedition with so much secrecy and despatch, that the king was taken by surprise. Notwithstanding this, the assailants were, in their first onset, repulsed; but, renewing their attack in the night, although Pomare's party had forty muskets, and those in the hands of the rebels were not more than fourteen, they threw the king's forces into confusion, killed a chief of influence, a near relative of Pomare's, and, driving his warriors to their canoes, retook the object of their murderous contention, the image of Oro, and remained masters of the whole of Tairabu, as well as of the south and western side of the large peninsula.

Pomare, with his vanquished forces, pursued their voyage to Matavai, where he and his son were received with respect by Captain Bishop and his companions. His affairs appeared desperate, and he entertained no thoughts of security, but by flight to Eimeo. When, however, he beheld the manner in which the English had prepared to defend themselves, if attacked; and was assured by Captain Bishop, and his companions, that if he was conquered, they were not; and that they would support him in the present critical state of the nation, and assist in the restoration of his government, his prospects appeared to brighten, and he again indulged the hope that his affairs might be retrieved.

The rebels were now masters of the greater part of the island; and, as the Missionaries had every reason to believe they would attempt the conquest of the remainder, and knew that their establishment was the only point where they were likely to meet with the slightest resistance, they neglected no means of defence. The Mission-house was converted into a garrison. The enclosures of the garden were destroyed, the bread-fruit and cocoa-nut trees cut down, to prevent their affording shelter to the enemy, and the means of annoyance from their muskets or their slings. Their chapel was also pulled down, lest the enemy should occupy it or burn it, and from it set fire to their own dwelling. A strong paling, or stockade, was planted round the house; boards, covered with nails, were sunk in the paths leading to it; and thither the Missionaries, Captain Bishop, Captain House, commander of the vessel that had been wrecked, and the seamen under their orders, now retired, as they daily received the most alarming accounts of the intention of the rebels to make their next attack upon them. The veranda in front of their dwelling was protected by chests, bedding, and other articles, so as to afford a secure defence from musket-balls; and the sides of the house, which were only boarded, were fortified with similar materials. Four brass cannon, which had been saved from the wreck of the Norfolk, were fixed in two of the upper rooms, and the inmates of the dwelling were placed under arms, as far as the number of muskets would admit. The Missionaries, as well as the seamen, stood sentinels in turn, night and day, in order to prevent surprise. Their situation at this time must have been most distressing. Independently of the desolation that surrounded them, and the confusion and disquietude that must necessarily have attended their being all confined in one house, together with the two captains and their seamen, they were daily expecting an attack. Sometimes they heard that the rebels were entering Matavai from the east, at other times from the west, and sometimes they received intelligence that they had divided their forces, and intended to commence the attack from two opposite points at the same time.

Pomare erected some works on One-tree Hill, to arrest their progress, should they attempt the district in that direction; and, hearing they were still ravaging the peninsula of Tairabu, sent a strong party to *tabu-te ohua*, strike their encampment at home. His party reached Atehuru, without molestation, late at night; and, after a short concealment, falling upon the unconscious and defenceless victims, under the cover of the darkness of midnight, in two hours destroyed nearly two hundred men, women, and children. The men who remained at home, in times of war, were generally either aged or sick, and incapable of bearing arms. This unprovoked act of cruelty, on the part of Pomare, heightened to such a degree the rage of the rebels, that they vowed the entire destruction of the reigning family.

While the affairs of the island remained in this unsettled state, the Nautilus arrived, and Pomare prevailing on the captain to furnish him with a boat manned by British seamen armed, went to Atehuru to present some costly offering to Oro, whose favour he still considered to be the only means of restoring his authority. Although that idol was now in the hands of his enemies, yet, as his errand was of a sacred character, the Atehuruans, notwithstanding they would not admit him to the temple, allowed him to present his offerings, which he deposited on a part of the beach near the temple, and peaceably retired.

When Pomare returned, he solicited from the captains, men and arms to go against the insurgents; and on the 3d of July, Captain Bishop and the mate of the Nautilus, with twenty-three Europeans, well supplied with ammunition, arms, and a four-pound cannon, accompanied Pomare's forces to the attack. All the Missionaries remained at Matavai, excepting one, who accompanied Captain Bishop as surgeon. On reaching Atehuru, they found the rebels had taken refuge in their *Pare* or natural fortress, about four miles and a half from the beach. This retreat was rendered by nature almost impregnable to the native warriors, and the only avenues leading to it being defended by the barriers its occupants had thrown up, it appeared difficult, if not impossible, to take it by storm, even with the foreign aid by which the king was supported. After spending the day in almost harmless firing at the enemy, the English and the natives were on the point of embarking to return, when the rebels having been decoyed from their encampment by the daring and challenges of an active and courageous young man, who had assumed the name of *To-morrow morning*, chased him and his companions down to the sea-side. Here they were checked by Pomare's musketeers, and retreated a few moments, when they halted, and faced their pursuers; but on the arrival of the English, they were seized with a panic, and fled. Seventeen of the rebel warriors, including Rua, one of their leaders, were taken, and killed on the spot by Pomare; whose followers, according to their savage rules of war, treated their bodies with the most wanton brutality.

Pomare and his English allies marched the next morning to the strong-hold of the natives, and were much disappointed at finding it filled with men determined to defend it to the last. A female was sent, as a herald, with a flag of truce to the warriors in the fortress, informing them of the number slain, and proposing to them the king's terms of peace. Taatahee, the remaining chief of the rebels, who was related to Pomare, directed her to tell him that when they had done to him, as they had done to Rua the slain chief, then, and not till then, there would be peace. As it appeared improbable that the place could be attacked with advantage to the assailants, and equally improbable that its occupants would accept any terms of capitulation that the king would offer, Captain Bishop returned to Matavai, and on the day following Pomare sailed about twelve miles towards Pare. Here he fixed his encampment; and, although peace was not concluded, hostilities appear to have been for some time suspended.

Soon after the return of Captain Bishop, the Nautilus sailed; and the Venus having returned to Tahìti, on the 19th of the following month, Captain Bishop with his men left the island.

Dreadful and alarming as these superstitious and bloody contests had been, and though still exposed to the horrors of savage war, the Missionaries, protected in their work by the care of God, felt that they were

> "—————devote to God and truth,
> And sworn to man's eternal weal, beyond
> Repentance sworn, or thought of turning back.

and determined, in dependence on Divine protection and support, to maintain their station; diligently to labour and patiently to wait for the reward of their toil. They beheld, with deepest distress, their gardens destroyed, their trees cut down, the fences they had reared with so much care demolished, the country around a desolate wilderness, and the inhabitants reduced to a state of destitution and wretchedness; yet they could not contemplate the remarkable interposition of Providence, in affording them the means of perfect security amidst the surrounding destruction, without unmingled emotions of admiration and gratitude.

The cessation of hostilities afforded the Missionaries a re ite from anxious watching, and allowed them to pursue their former avocations. Their gardens were again enclosed, and such seeds as they had preserved were committed to the ground. The study of the language, which, under the guidance and assistance of Mr. Nott, had been regularly pursued one or two evenings every week, was resumed. In the instruction of the children, the greatest difficulties had been experienced from the restless unrestrained dispositions and habits of the scholars, who, unaccustomed to any steady application or to the least control, seldom attended to their lessons long enough to derive any advantage from the efforts of their teachers. As opportunity offered, the Missionaries also preached to the people, and catechized the children. The natives, however, continued their depredations on the little remaining property of the Mission; and, in order to deter others, one of them, who had been detected, was publicly flogged by the king's order.

Towards the close of the year 1802, Mr. Jefferson and Mr. Scott made the tour of Tahiti, for the purpose of preaching to the people. In most of the places they were hospitably entertained, though, on one occasion, the chief refused them lodging, because a former Missionary had not rewarded him for his attentions. In some instances, the natives appeared to listen with attention and interest to their message, but they frequently found great difficulty in inducing them to attend and often observed with pain that their instructions were received with indifference or with ridicule. At one place, though the people on their first arrival welcomed them cordially, yet when they understood the object of their visit, a very marked, and by no means pleasing change, appeared in their behaviour.

For many years, the first Missionaries were annoyed

in almost all their attempts to preach to the people. Sometimes, when they had gone to every house in a village, and the people promising to attend had left their houses, they often found, on reaching the appointed place, that only two or three had arrived there; at other times they either talked all the while about their dress, complexion, or features, and endeavoured to irritate them by false insinuations as to the objects of their visit; or to excite the mirth of their companions by ludicrous gestures, or low witticisms on the statements that were made. Brainard remarks, that while he was preaching, the Indians sometimes played with his dog: but the first teachers in Tahiti were often disturbed by a number of natives bringing their dogs, and setting them to fight on the outside of the circle they were addressing; or they would bring their fighting cocks, and set them at each other, so as completely to divert the audience, who would at once turn with avidity from the Missionary, to the birds or the dogs. On some occasions, while they have been preaching, a number of *Areois*, or strolling players, passing by, have commenced their pantomime or their dance, and drawn away every one of the hearers. At such times, those who had stood round the Missionary only to insult him by their insinuations, ridicule him by their vulgar wit, or afflict his mind by their death-like apathy and indifference to the important truths he had declared, have instantly formed a ring around the areois, and have gazed on their exhibitions of folly and of vice with interest and pleasure.

In addition to these sources of disturbance, they were sometimes charged with being the authors of all the disasters and suffering of the people, in consequence of praying to their God, whom the natives called a bad God when compared with Oro. Under these circumstances, it required no small degree of forbearance and self-possession, as well as patient toil, to persevere in preaching the gospel among a people whose spirit and conduct afforded so little encouragement to hope it would ever be by them received.

Hitherto their labours had been confined to Tahiti; but in December 1802, Mr. Bicknell, accompanied by Mr. Wilson, made a voyage to Eimeo, and, travelling round it, preached "the unsearchable riches of Christ" to its inhabitants, many of whom appeared to listen with earnestness, and desired to be more fully instructed.

The same year, in the month of November, *Teu*, an aged and respected chief, the father of Pomare, and the grandfather of the king, died at his habitation not far from the Mission-house. He was remarkably venerable in his appearance, being tall and well made, his countenance open and mild, his forehead high, his hair blanched with age, and his beard, as white as silver, hanging down upon his breast.* He had led a quiet and peaceful life ever since the arrival of the Mission, was probably the oldest man in the island, and, what is rather unusual, died apparently from the exhaustion of nature, or old age. He was esteemed by the natives, and supposed to be a favourite with the gods. But whenever the Missionaries had endeavoured to pour into his benighted mind the rays of divine light and truth, revealed in the sacred volume, it was a circumstance deeply regretted by them, that he had generally manifested indifference or insensibility.

* In the plate of the Cession of Matavai, he appears standing on the right hand of the king, and immediately behind Pomare.

The family at Matavai were exposed to trials not only from the evils of war, the opposition of the heathen to their instructions, but also from the false reports which were circulated against them. An instance of this occurred early in the following year, 1803, when the Unicorn, a London ship, arrived on her return from the north-west coast of America. Otu the king suddenly left Matavai, and repaired to his dwelling in Pare, incensed against some of the Missionaries, who, he had been informed, had been endeavouring to excite prejudices in the mind of the captain against him, that he might not receive any presents, and had prevented him from giving the natives the price they had asked for their pigs. This report was most unfounded, and it was hoped the effects were soon removed.

About this time the Margaret, in which Captain Byers and Mr. Turnbull had visited the islands for purposes of commerce, was wrecked on a reef about 200 miles distant; Mr. Turnbull had remained in Tahiti; Captain Byers, his officers, and crew, consisting of sixteen individuals, with the mate's wife and child, safely reached that island in a long kind of chest, or boat, which they had built with the fragments of the wreck.

Towards the close of the last year, Otu's brother Teariinavahoroa, the young prince of Tairabu, removed from the smaller peninsula in consequence of the increase of his disorder, which appeared to be consumption. Pomare, his mother, Idia, his brother and sister, and the chiefs, paid him every attention; human sacrifices were offered; and both Pomare and Otu frequently invoked their gods in his favour, and presented the most costly offerings. For a number of days no fires were allowed to be lighted, in order that these might be effectual: but all were unavailing; the young chief, who had scarcely arrived at the age of manhood, died in the district of Pare on the 19th of June 1803. The Missionaries frequently visited him after his arrival in Pare, and, as far as their scanty means would allow, administered cordials suited to his languid state. They were, however, most anxious to direct his mind to the great Physician of souls, and to lead him to apply for those remedies that would heal his spiritual maladies, and prepare him for his approaching dissolution. On this subject, they noticed with distress not only the unwillingness of his friends that any thing should be said, but also the insensibility of the young chieftain himself. It was supposed by the people, that his illness and death were occasioned by the incantations of Metia, a priest of Oro, a famous wrestler and sorcerer, whose influence, ceremonies, and prayers, had induced the evil spirits to enter into the young prince, and destroy him. Counter ceremonies were performed; prayers, called *faatere*, were offered, to drive the evil spirits from him, and these, it was imagined, would all be unavailing, should the Europeans direct his mind to any other source, or offer on his behalf prayers to any other god, and hence in part may have proceeded the aversion of his friends to the presence and efforts of the Missionaries.

Another large meeting of chiefs, priests, and warriors, was held during the summer of 1803 at Atehuru, and rumours of war were again spread through the land. Here Otu once more demanded the body or image of the great god Oro, which the chiefs agreed ultimately to give up to the custody of the king, but which they were not so ready at once to surrender.

The state of the people was at this time most affecting. Diseases, introduced by Europeans, were spreading, unmitigated, their destructive ravages, and some members of almost every family were languishing under the influence of foreign maladies, or dying in the midst of their days. The survivors, jealous of the Missionaries, viewed them as the murderers of their countrymen, under the supposition that these multiplied evils were brought upon them by the influence of the foreigners with their God. They did not scruple to tell them that He was killing the people; but that by and by, when Oro gained the ascendency, they should feel the effects of his vengeance. In addition to the diseases resulting from their immorality, there were others of a contagious and often fatal character, to which the natives were formerly strangers. These had been conveyed to the islands either by the visits of ships, or the desertion of seamen afflicted with them; they produced the most distressing sickness and mortality among the people; and, although nothing could be more absurdly imagined, yet, according to their ideas of the causes of disease and death, that they originated in the displeasure of some offended deity, or were inflicted in answer to the prayers of some malignant enemy, they were, from the representations of some, and the conjectures of others, led to suppose that these diseases were sent by the God of the Missionaries, in answer to their prayers, and because they would not reject Oro, and join in their worship.

At this time an event transpired, which threatened at first a revival of all the confusion and desolation of war. This was the demise of Pomare, the father of Otu the king. His death was sudden; he had taken his dinner, and was proceeding with two of his attendants in a single canoe towards the Dart, a vessel on the point of sailing from the bay. While advancing towards the ship, he felt a pain in his back, which occasioned him involuntarily to start in his seat; and, placing his hand on the part affected, he fell forward in the canoe, and instantly expired. The suddenness and circumstances of his death, taken in connexion with the troubles in which he had recently been engaged with the greater part of the people of the island, on account of his violent seizure of the idol at Atehuru, strengthened in no small degree the idolatrous veneration with which the natives regarded their god; and the anger of Oro was by them supposed to be the direct cause of Pomare's death.

In person, Pomare, like most of the chiefs of the South Sea Islands, was tall and stout; in stature he was six feet four inches high, his limbs active and well proportioned, his whole form and gait imposing. He was often seen at Matavai, walking along with firm steady steps, and using with ease as a walking-stick a club of polished iron-wood, that would have been almost sufficient for an ordinary native to have carried. His countenance was open and prepossessing, his conversation affable, though his manner was grave and dignified. He was originally only a chief of the district of Pare, but his natural enterprise and ambition, together with the attention shewn him by the commanders of British vessels, their presents of fire-arms and ammunition, and the aid of European seamen, especially the mutineers of the Bounty, had enabled him to assume and maintain the supreme authority in Tahiti. Though not possessed of the greatest personal courage, he was a good politician, and a man of unusual activity and perseverance. He laboured diligently to multiply the resources of the island,

and improve the condition of the people, and his adherents were always well furnished with all that the island afforded. The uncultivated sides of the mountains, and the low flat sandy parts of the shore, seldom tilled by the natives, were reclaimed by his industry; and many extensive groves of cocoa-nut trees in Tahiti and Eimeo, which the inhabitants say were planted by Pomare, remain as monuments of his industry, and yield no small emolument to their present proprietors. In all these labours he endeavoured to infuse his own spirit into the bosom of his followers, and to animate them by his example, usually labouring with his people, and planting with his own hands many of the trees.

To the Mission families he was uniformly kind. Shortly before his death, he recommended them to the protection of his son; though the more he understood the chief object of their Mission, the greater aversion he seemed to manifest to it. To the favour of the gods he considered himself indebted for all the aggrandisement of his person and family; and if the Missionaries would have allowed the claims of Oro or Tane to have received an equal degree of attention to that which they required for Jehovah, or Jesus Christ, Pomare would readily have admitted them; but when required to renounce his dependence upon the idols of his ancestors, and to acknowledge Jehovah alone as the true God, he at once rejected their message. He was justly considered as the principal support of the idolatry of his country. In patronizing the idols, and adhering to all the requirements of the priests, &c. he appears to have been influenced by the constant apprehension of the anger of his gods. Teu, his father, was a Tahitian prince; his mother was a native of Raiatea; he was born in the district of Pare; and at the time of his death, which took place on the 3d of September, 1803, was between fifty and sixty years of age.

In the circumstances attending the formation of his character, and in the commencement, progress, and result of his public career, there was a striking resemblance between Pomare, the first king of that name in Tahiti, and his contemporary, Tamehameha, the first king of the Sandwich Islands. Both rose from a comparatively humble station in society, to the supreme authority; both owed their elevation principally to their own energies, and the aid they derived from their intercourse with foreigners; both appeared the main pillars of the idolatry of their respective countries; and both left to their heirs the undisputed government of the islands they had conquered. Each appeared to have possessed natural endowments of a high order, and both were probably influenced by ambition. Pomare was distinguished by laborious and patient perseverance; Tamehameha, by bold and daring enterprise. The characters of their immediate descendants were in some respects similar to each other, though both were very different persons from their respective predecessors.

Otu the king was at Atehuru at the time of his father's death. He sent several messengers to Pare, commanding the body to be brought to him; but to this the raatiras, or resident chiefs, objected. When the Missionaries paid a visit of condolence, Idia requested them to tell her son it was her wish that the body should remain at Pare; and to this the king consented.

The death of Pomare did not alter the political state of Tahiti; its only influence on the people was such as tended to confirm them in their superstition; for, on the occasion of a religious ceremony, wherein his spirit was invoked, and which took place shortly after his decease, it was declared that he was seen by Idia, and one of the priests. To the latter it was said he appeared, above the waters of the sea, having the upper part of his person bound with many folds of finely braided cinet. From this circumstance his favourite wife assumed the name of *Tane rurua*, from Tane, a husband, and rurua, bound round, or bound repeatedly.

Towards the middle of the year 1804, the king went over to Eimeo, taking with him the great idol Oro, to propitiate whom, so many of the inhabitants had been sacrificed. About the same time, Mr. Caw, a shipwright from England, joined the Mission. Otu now assumed the name of Pomare, which has ever since been the regal name in Tahiti. Its assumption by his father was, as many names are among the Tahitians, perfectly accidental. He was travelling, with a number of his followers, in a mountainous part of Tahiti, where it was necessary to spend the night in a temporary encampment. The chiefs' tent was pitched in an exposed situation; a heavy dew fell among the mountains; he took cold, and the next morning was affected with a cough; this led some of his companions to designate the preceding night by the appellation of *po-mare*, night of cough, from *po*, night, and *mare*, cough. The chief was pleased with the sound of the words thus associated, adopted them as his name, and was ever afterwards called Po-ma-re. With the name he also associated the title of majesty, styling himself, and receiving the appellation of, "His Majesty Pomare."

Peace continued during the remainder of the year, and

the Missionaries were enabled to persevere in their labours, although they were cheerless, and apparently useless. Great attention had, during the last year, been paid to the instruction of the children in the short catechism, in which the first principles of Christianity were familiarly exhibited to the minds of the young people. Mr. Davies, in particular, had devoted much of his time to this work; and although it had hitherto been found impracticable to teach the children letters, a number had committed the catechism to memory. The gospel was preached, not only in the immediate neighbourhood of Matavai, but in every district in Tahiti and Eimeo; yet the people seemed more than ever disposed to neglect or ridicule the message. Sometimes they said, We will hear our own gods; at other times they scoffingly asked the Missionaries, if the people of Matavai had attended to their word; if the king, or any of his family, had cast away Oro; declaring, that when the king and chiefs heard the word of Jehovah, then they would also.

Early in January, 1805, the Missionaries prepared a larger catechism; and, on the 6th of March, they adopted their Tahitian alphabet. In forming this, the Roman characters were preferred; sounds in the Tahitian language attached to them, and a native name affixed to each, for the purpose of facilitating the introduction of letters among the people. It was, however, a long time before any, among the native inhabitants of Tahiti, could be induced to learn the letters of the alphabet. The Missionaries continued their labours in preaching to the people, and teaching the catechism to the children. One or two vessels arrived, but brought no letters or supplies; and, towards the close of the year, they experienced a heavy loss, in the destruction of a large and flourishing plantation.

Three of the Missionaries had cleared, enclosed, and cultivated it; and had rendered it, as far as the productions of the island were available, subservient to their interests. They had stocked it with cocoa-nuts, oranges, limes, and citrons, of which, not fewer than six hundred plants, with other productions, were growing remarkably well. In one hour, however, the whole of the fence was burnt to the ground, and the plantation destroyed, or the few plants that remained were so much injured as to be nearly useless. Great as was the loss experienced on this occasion, they had reason to fear it was caused by some of their neighbours, who had designedly set fire to the long dry grass immediately to windward of the plantation. This was probably done from motives of jealousy, lest, by cultivating the land, and reaping the fruits of it, the foreigners should suppose it had become theirs, and the natives cease to be its proprietors. On this acount, much as they suffered by its destruction, they deemed it inexpedient to complain to the king.

In the month of January, 1806, Pomare returned from Eimeo, bringing with him the idol Oro, which was kept in his sacred canoe; while the human sacrifices, offered on his arrival, were suspended on the trees around. The Missionaries paid a visit to the king, soon after his return; and, as he had become remarkably fond of using his pen, he intimated his wish that they should build him a small plastered house, near their own, in which he could attend to his writing without the interruptions he experienced in his own dwelling.

Early in the year 1806, the Mission was again

weakened by the departure of Mr. Shelly, with his family. He relinquished Missionary pursuits, and sailed for Port Jackson on the ninth of March.

In the month of July, following, the queen of Tahiti died, in the district of Pare, after an illness of nearly eight weeks. About the time her indisposition commenced, she had become the mother of a still-born child; the sickness that followed, and the fatal termination to which it led, were supposed to be the results of a cruel and unnatural practice, that cannot be described—a species of infanticide often resorted to by females of high rank in the island, although not unfrequently issuing, as was imagined on the present occasion, in the death of the perpetrator. Pomare had offered his prayers to the gods of his family, and many ceremonies had been performed, but to no purpose. The queen was in person about the middle stature; mild and affable in her behaviour; addicted to all the vices of her country; and was cut off in the prime of life, being about twenty-four years of age at the time of her death. The king and his mother appeared affected with their loss; and the grief of his relatives was severe, as the death of so many members of Pomare's family threatened, at no very remote period, its total extinction. Pomare was left a widower and childless, all the children of the late queen having been destroyed.

Although reports of war were heard during the year, there was no actual hostility; and, under discouragements every day increasing, the Missionaries were enabled to prosecute their labours. Having found it difficult to engage the attention of the children, while attempting to teach them in the presence of the adults, who ridiculed the idea of their learning letters, they opened a school in a part of their own dwelling. In October, Mr. Davies proposed to begin with the boys attached to their own houses, and met them three nights in the week for the purpose of instructing them in the catechism, and teaching them to read those few specimens of writing they had been able to prepare. At the same time, Messrs. Nott and Davies were requested to draw up a brief summary of the leading events, and a short account of the principal persons mentioned in the Old Testament, in the form of a scripture history, for the use of these scholars. In the course of the following year, a spelling book, which Mr. Davies had composed and used, was sent to England. Here it was printed, and afterwards transmitted to the islands, for the use of the schools.

No long period had elapsed since the first establishment of the Mission, without a vessel's touching at Tahiti. By many of these the Missionaries had been able to write to the directors and their friends in England, and from several they had secured a small supply of such articles as they most needed. But since the arrival of the Royal Admiral, in July, 1801, although the directors had repeatedly sent out articles to Port Jackson for Tahiti, yet the Missionaries had received neither supplies nor letters from England. Many vessels had sailed from Port Jackson, where the supplies were lying, and had afterwards touched at the island; but the captains, having no intention of doing so when they sailed, had refused to take the goods on board. Tea and sugar, and many other comforts, they had long been destitute of; and their apparel was scarcely such as to enable them to appear respectably in the company of any of their countrymen who might visit the island. Several of them were some years with only one pair of shoes; and often, in many of their journeys undertaken for the purpose of preaching and instructing the natives, they had travelled barefoot. In addition to these privations, the gloom and discouragement that depressed their spirits, on account of the total want of success attending their labours, must have been increased, in no ordinary degree, by the uncertainty and anxiety of remaining, at that remote distance from home, five years without even once hearing by letter from their native country, or their friends. From this distressing state of feeling, they were in a great measure relieved by the arrival of the Hawkesbury, a colonial vessel, which anchored in Matavai bay on the 26th of November, 1806.

Since the year 1804, the Society in England had authorized Mr. Marsden to expend annually, for the support of the Missionaries, two hundred pounds, and had also sent out supplies. Unable to meet, in Port Jackson, with any vessel proceeding to Tahiti, Mr. Marsden had at length engaged the Hawkesbury, a small sloop of about twenty tons burden, to take out the letters and articles that had been so long delayed. The communications from England conveyed to the Missionaries the welcome assurance that they were not forgotten by their friends at home; but most of the articles, especially the clothing, from the length of time it had been lying at Port Jackson, and the wretched state of the vessel in which it was sent, were so injured as to be almost useless; the packages were wet with the sea-water, and their contents consequently spoiled.

The repeated trials with which the Missionaries were exercised, the privations they endured, and the painful and protracted discouragements by which, at this period, they were depressed, were of no ordinary character. Few among modern Missionaries have been called to endure such afflictions; and it is matter of devout acknowledgment, that, notwithstanding the darkness of their prospects and the destitution of their circumstances, they were still enabled to persevere, and leave the event with Him, at whose command they had entered on their work.

Peace continuing in the island during the close of 1806, and the beginning of 1807, allowed the teachers to pursue uninterruptedly their endeavours to plant Christianity among the inhabitants, although at that time with little prospect of success.

The ravages of diseases originating in licentiousness, or nurtured by the vicious habits of the people, and those first brought among them by European vessels, appeared to be tending fast to the total desolation of Tahiti. The survivors of such as were carried off by these means, feeling the incipient effects of disease themselves, and beholding their relatives languishing under maladies of foreign origin, inflicted, as they supposed, by the God of the foreigners, were led to view the Missionaries as in some degree the cause of their suffering; and frequently, not only rejected their message, but charged them with being the authors of their misery, by praying against them to their God. When the Missionaries spoke to them on the subject of religion, the deformed and diseased were sometimes brought out and ranged before them, as evidences of the efficacy of their prayers, and the destructive power of their God. The feelings of the people on this subject, were frequently so strong, and their language so violent, that the Missionaries have been obliged to hasten from places where they had intended to have addressed the people. Instead of listening with attention, the natives seemed only irritated by being, as they said, mocked with promises of advantage from a God by whom so much suffering had been inflicted. Under these circumstances, their distresses were somewhat relieved by the arrival of Mr. Warner; who, after due preparation, had been sent from England in the capacity of surgeon to the Mission, which he joined on the 12th of May, 1807. The strength, however, which his arrival added to their establishment, was somewhat counterbalanced by the removal of Mr. Youl, one of those who had arrived in the Royal Admiral, and who departed in the vessel that conveyed Mr. Warner to Tahiti.

In the month of June, the flame of war was rekindled in Taiarabu, and the district of Atehuru, where the king's party suddenly attacked the inhabitants; and, after killing upwards of one hundred, including their principal chiefs, covered the country with all the murder and desolation that usually attended the march of the infuriated bands through the territories of those who were too weak to oppose their progress. Having driven to the mountains such as had escaped the slaughter in the assault, plundered their houses, and afterwards reduced them to ashes, the king took the bodies of the slain on board his fleet; and, sailing to Tautira, offered them in sacrifice to Oro.

Towards the close of the year, the Mission sustained a heavy loss in the death of Mr. Jefferson. He was one of those Missionaries that arrived in the ship Duff; he had borne "the heat and burden of the day," and finished his course on the 25th of Sept., 1807. He was a man of intelligence and ability, possessing extraordinary devotedness and patient zeal. He had laboured unremittingly for ten anxious years; filling, with credit to himself and advantage to the Mission, the most important station among his brethren, by whom he was highly and justly respected. He maintained an arduous post among the pioneers of the little army of Christian Missionaries; who, "unarmed with bow and sword," had ventured to attack idolatry in its strongest holds among these distant islands; and,

> "High on the pagan hills, where Satan sat
> Encamped, and o'er the subject kingdoms threw
> Perpetual night, to plant Immanuel's cross,
> The ensign of the gospel, blazing round
> Immortal truth."

And, though he fell upon the field before he heard or uttered the shout of victory, his end was peaceful, and his hopes were firm. On a visit to Matavai, in the early part of 1821, conducted by Mr. Nott, I made a pilgrimage to his grave. I stood beside the rustic hillock on which the tall grass waved in the breeze, and gazed upon the plain stone that marks the spot where his head reposes, with feelings of veneration for his character. I felt, also, in connexion with the change that has since taken place, that he had indeed desired to see the things that I beheld, but he had died without witnessing, on earth, the gladdening sight; and that, in reference to his unremitted exertions, I and my junior companions had entered into his labours, and were reaping the harvest for which he had toiled.

Shortly after Mr. Jefferson's death, Mr. Nott, accompanied by Mr. Hayward, visited the islands of Huahine, Raiatea, and Borabora; travelled round each, preaching and teaching the people; and thus, for the first time, published among their inhabitants the great truths of Christianity. Many of the natives listened with attention and apparent interest. The illness of the king terminated, for a time, the war which he had commenced against the people of Atehuru, and allowed the Missionaries uninterruptedly to pursue their labours in Tahiti.

Early in 1808, Mr. Elder left this island for Port Jackson. Peace at that period every where prevailed, but it was of short duration. The dissatisfaction of the farmers, inferior chiefs, and lower orders of the people, with Pomare's conduct, was daily increasing, and his recent massacre of the Atehuruans had greatly strengthened their determination to destroy his authority, and revive the ancient aristocratical form of government. In the month of October, the Missionaries received a note from the king, informing them of the probability of war, recommending them to be upon their guard, and not to be deceived or taken by surprise. In consequence of this intimation, and the increasing signs of approaching hostilities, they established a strict watch every night, and seldom went far from their dwelling. The preparations for battle were continued on both sides; every morning it was expected that hostilities would commence before the close of the day, and every night it was apprehended that an attack would be made before morning. In this state of distressing anxiety, without any means of flying from the gathering storm, all the families continued till the 25th of October, when a vessel from Port Jackson providentially anchored in the bay, and, by ensuring a safe retreat in the event of sudden assault, afforded no small alleviation to their minds.

On the Sabbath-day, the 6th of November, the district

of Matavai was thrown into great confusion, and numbers of men appeared in arms. The king, who was on board the ship at the time, hastened on shore, and was only restrained from commencing an immediate attack by the counsel of his uncle, who urged the necessity of invoking the favour of the gods before commencing hostilities. This afforded the people of Matavai time to retire, and encamp in the adjoining district with the people of Apaiano. Proposals of peace were sent by the king, but the rebels, being reinforced from the districts to the eastward, refused to meet Pomare, or negociate with him; and war appeared inevitable.

The king, expecting that his camp, which was at Matavai, would be immediately attacked, recommended that the wives and children of the Missionaries should take shelter in the vessel. They embarked on the 7th amid much confusion, but with the sincerest gratitude to God for the refuge so seasonably provided. The night passed without any attack; several leading chiefs, whom the rebels expected, had not arrived, and the Europeans were thus permitted to pack up a few articles for their use on board. The next morning a letter was addressed to the captain, requesting him to delay his departure forty-eight hours, that they might deliberate on the steps necessary to be taken. On the following day the Missionaries Nott and Scott went alone to the rebel camp at Apaiano, and invited the leaders to an interview with Pomare. The chiefs treated them with every mark of friendship, regretted that their establishment should suffer from the quarrel between them and the king, and requested them not to leave the island. The leaders of the rebels refused, however, to meet Pomare except in battle, and every hope of accommodation now vanished.

This disastrous war is called, in the Tahitian traditions, the *Tamai rahi ia Arahuraia,* The great war of Arahuraia. It was headed by Taute, who had long been the king's prime-minister, and who was one of the most powerful chiefs and successful warriors on the islands. His name inspired terror through the ranks of his enemies; and, when the king heard that he had joined the rebels, he was so affected, that he burst into tears. Pomare advised the married Missionaries to leave the island. They were unanimous in opinion, that there was no prospect of safety or usefulness, even should the rebel chiefs prove their friends; and this, together with the consideration of the little success that had attended the labours of so many years, occasioned their determination to remove. Four of the unmarried Missionaries offered to remain with the king, that they might be upon the spot, should any favourable change take place; the others, with most of the Europeans on the island, sailed from Tahiti on the 10th of November, 1808, and arrived the following day at the island of Huahine. Here they were hospitably received by the chiefs and people.

The affairs of Tahiti continued in the same state until the 22d of December; when the king, influenced by Metia the prophet of Oro, attacked the rebels; who were not only superior in numbers, but favoured in the conflict by the occupation of an advantageous position. Notwithstanding the prophet's prediction of victory, Pomare was defeated, and fled with precipitation to Pare; leaving a number of muskets in the hands of his enemies, and several principal warriors among the slain. Convinced, that though the chiefs of the victorious army might be friendly to them, yet that they could not restrain their followers, who, in time of war, threw off all subordination; and expecting that the victors, after this success, would instantly attack their dwelling, and that their lives were no longer secure, the Missionaries remaining at Tahiti fled to Eimeo, where they were shortly after joined by the king. Some months afterwards, three others were compelled to follow their companions to Huahine. During their residence here, some among them had made the tour of the island, and endeavoured, with but little prospect of success, to instruct the inhabitants.

The melancholy prospect of affairs, their expulsion from Tahiti, the total destruction of the settlement, and the little probability of a restoration of peace, induced them to determine on removing by the first opportunity to Port Jackson. This occurred in the course of the year; and on the 26th of October 1809, they all sailed from the islands, excepting Mr. Hayward, who remained in Huahine, and Mr. Nott, who still resided in Eimeo with the king.

After the victory of the 22d of December 1808, the rebels plundered the district of Matavai and Pare, and, devoting to destruction every house and plantation, reduced the whole country to a state of the wildest desolation and ruin. The Mission houses were ransacked and burnt, and whatever the insurgents were unable to carry away was destroyed. Every implement of iron was converted into a weapon of war. The most valuable books were either committed to the flames, or distributed among the warriors for the purpose of making cartridge papers, and the printing types were melted into musket balls.

During such seasons, it was not merely apprehension, but actual danger, to which all the Europeans were exposed. On one occasion, Mr. Nott, returning from a visit to the king, was resting in a native house, when a party of the rebels approached the spot; his native companion, one of Pomare's warriors, observing them, touched him on the shoulder, and urged him to fly to the canoe lying on the beach: he and his fellow-traveller had scarcely pushed off from the shore, when the men came up, and, finding they had escaped, invited them to land, or requested the native to allow the foreigner to walk. Mr. Nott's companion assured him, however, that if he landed, his life would certainly be taken, merely because he was a friend to the king. The natives followed the canoe for some miles, but Mr. Nott was mercifully preserved, and reached Matavai in safety, indebted, under God, to the vigilance and promptitude of his Tahitian friend for his life. Before this time, a musket ball (aimed at a native who had taken shelter in his house) was fired through the window of the room in which he was sitting; and during another war, the spear of one of the king's enemies was already poised, and would in all probability have inflicted a fatal wound in his body, when the interference of one of Mr. Nott's friends at the moment,saved him from the deadly thrust.

It is not easy to form an accurate idea of the distress of the last Missionaries who reluctantly left Tahiti, when they beheld their gardens demolished, their houses plundered and burnt, their pupils engaged in all the barbarity of a savage war; and the people, among whom they had hoped to introduce order, and peace, and happiness, doomed to the complicated miseries attending anarchy, idolatry, and all the varied horrors of cruelty and of vice. The enterprise in which they had embarked, had at its commencement united in bonds of disinterested philanthropy, parties before but seldom associated; and had, by a vigorous and combined movement, in force and magnitude surpassing any thing that had been hitherto attempted by British Christians, introduced a new era in the Missionary efforts of modern times.—It had excited among all classes the liveliest interest, called forth the most splendid efforts of sacred eloquence, and the noblest deeds of Christian benevolence; but, painful and deeply humiliating as it was, it now appeared to those devoted servants of God, who had, amidst protracted and severe privations, maintained their ground till life was no longer secure—after having engaged the prayers of the people of God, and waited in vain for the results of patient and self-denying toil, during twelve eventful years—that the scene of their labour must be abandoned.

Their enemies became bold in denouncing the enterprise as the wild project of extravagance and folly, and stamping upon its projectors and conductors the impress of the blindest fanatacism. Even those who, though they had not condemned the scheme as Utopian and visionary, had withheld their sanction and their aid, now pointed to the deserted field as a demonstration of the soundness of their judgment, and an explanation of their conduct. There were others also, who, whatever might be their opinion of the measure itself, and however they might approve or disapprove of the choice of those with whom it originated, in the selection of the most distant, isolated, and, as it regarded the moral character of its inhabitants, the most unpromising parts of the world, for the first field of their labours, considered its projectors as influenced in a great degree by self-confidence, and a desire of aggrandisement or applause. It has sometimes been unwarrantably insinuated, that the founders of the Missionary Society expected to convert the heathen to Christianity by their own energy; and the allegation has been occasionally repeated since those days,—perhaps in some instances, to increase the impression produced by the accounts of the recent changes which have taken place in those islands, contrasting the former and latter results of Missionary labours, and representing them as demonstrations of the impotency of man, and the power of the Most High. The lively feeling that attended the establishment of the Missionary Society, the liberality of the principles recognized as its basis, and the combination of different parties in its support, were at that time adapted to excite in minds of a cautious and deliberative habit, and fearful of innovation, the apprehension that it had originated in a desire, on the part of its projectors, to signalize themselves, and secure a name and influence in the Christian world, to which they were not otherwise entitled. Individuals, whose minds were deeply imbued with the subject, who had identified themselves with its progress and its results, and had embarked not only their influence, but much of their property, in the undertaking, might, and probably did, under the ardour of their feelings, indulge on some occasions in a splendour of imagery, and a richness of description, that exceeded the sober realities of fact: but they never imagined that they could subvert any system of idolatry by their own agency; or, that their efforts would be in any degree effectual for the conversion of the people, but as they were attended by the influence of the Holy Spirit. There might be, and perhaps was, a more confident hope of the speedy accomplishment of the object than now prevails; but the appeals and addresses, delivered at that period, manifest a deep conviction of human insufficiency, and breathe a spirit of entire dependence upon the blessing of God.

But although Tahiti was, by the departure of the Missionaries, surrendered, for a season, as a prey to the spoiler, and subjected to the rule of ignorance, barbarism, and idolatry, it was not abandoned by Him, in obedience to whose command to "go and teach all nations," the Mission had been undertaken. He had still "thoughts of mercy" towards its inhabitants, and was, by this distressing event, teaching those who had undertaken the work—and instructing his church, in regard to all their future efforts to extend his gospel—that singleness of aim, purity of motive, and patient diligence in labour, were of themselves insufficient for the work; that it was by His Spirit that the heathen were to be converted; and that without His blessing, Paul might plant, and Apollos might water, in vain.

The rebels were no sooner masters of the island, than they determined to murder the captain and officers, and seize the first vessel that should arrive. The Missionaries, aware of this, wrote a letter, which they gave to a native, to hand to the master of the first ship that might touch there. The Venus schooner, however, arrived, and was seized by the people, before the native could deliver his letter: the master and seamen were not murdered, but kept prisoners, to be offered in sacrifice to Oro. The Hibernia, Captain Campbell, also arrived shortly afterwards; but Captain Campbell, receiving the letter, was warned of his danger, and not only secured his own vessel, but succeeded in rescuing the schooner and her crew.

Although most of the Missionaries returned to the islands, and resumed their labours in Eimeo in 1811 and 1812, yet their efforts in Tahiti were not resumed till the close of 1817, so that on my arrival I found no one here. Hence, I have been induced to give the foregoing brief historical sketch of the leading facts connected with the establishment and termination of the first Mission to Tahiti, in connexion with my first visit to Matavai.

CHAP. VI.

Anchorage in Matavai—Visit from Pomare—Landing his horse—Interview with the queen and princess—Astonishment of the natives on viewing the horse and his rider—Description of Eimeo—Opunohu, or Taloo harbour—Landing at Eimeo—Welcome from the natives—First night on shore—Present from the chiefs—Visit to the schools—First Sabbath in the islands—Appearance and behaviour of native congregations—Voyage to Afareaitu—Native meal—Description of Afareaitu—Removal thither—Means of conveyance—Description of the various kinds of canoes used in the Society Islands—Origin of the name—Account of Tetuaroa, the watering-place of Tahiti—Methods of navigating their canoes—Danger from sharks—Affecting wreck—Accident in a single canoe—Length of the voyages occasionally made.

THE sea had been calm, the morning fair, the sky without a cloud, and the lightness of the breeze had afforded us leisure for gazing upon the varied, picturesque, and beautiful scenery of this most enchanting island. We had beheld successively, as we had slowly sailed along its shore, all the diversity of hill and valley, broken or stupendous mountains, and rocky precipices, clothed with every variety of verdure, from the moss of the jutting promontories on the shore, to the deep and rich foliage of the bread-fruit tree, the Oriental luxuriance of the tropical pandanus, or the waving plumes of the lofty and graceful cocoa-nut grove. The scene was enlivened by the waterfall on the mountain's side, the cataract that chafed along its rocky bed in the recesses of the ravine, or the stream that slowly wound its way through the fertile and cultivated valleys, and the whole was surrounded by the white-crested waters of the Pacific, rolling their waves of foam in splendid majesty upon the coral reefs, or dashing in spray against its broken shore.

The cataracts and waterfalls, though occasionally seen, are by no means so numerous on any part of the Tahitian coast, as in the north-eastern shores of Hawaii. The mountains of Tahiti are less grand and stupendous than those of the northern group—but there is a greater richness of verdure and variety of landscape; the mountains are much broken in the interior, and deep and frequent ravines intersect their declivity from the centre to the shore. As we advanced towards the anchorage, I had time to observe, not only the diversified scenery, but the general structure and form, of the island. Tahiti, excepting a border of low alluvial land, by which it is nearly surrounded, is altogether mountainous, and highest in the centre. The mountains frequently diverge in short ranges from the interior towards the shore, though some rise like pyramids with pointed summits, and others present a conical, or sugar-loaf form, while the outline of several is regular, and almost circular. Orohena, the central and loftiest mountain in Tahiti, is six or seven thousand feet above the sea. Its summit is generally enveloped in clouds; but when the sky is clear, its appearance is broken and picturesque.

The level land at the mouth of Matavai valley is broad, but along the eastern and southern sides the mountains approach much nearer to the sea. A dark-coloured sandy beach extends all round the bay, except at its southern extremity, near One-tree Hill, where the shore is rocky and bold. Groves of bread-fruit and cocoa-nut trees appear in every direction, and amid the luxuriance of vegetation, every where presented, the low and rustic habitations of the natives gave a pleasing variety to the delightful scene.

Most of the islanders who had boarded us in the morning continued in the ship, others arrived as we approached the bay, and long before we anchored, our decks were crowded with natives. Our prepossessions in their favour continued to increase, and we viewed them with no ordinary interest, as those among whom we were to spend the remainder of our days. Many of them wore some article of European dress, and all were attired in native cloth, though several had only a maro, or broad girdle, round the waist. There was a degree of openness in their countenances, and vivacity in their manners, which was not unpleasing.

We had not been long at anchor, before Pomare sent us a large albicore, and a variety of provisions, and shortly after came on board. I was struck with his tall and almost gigantic appearance; he was upwards of six feet high, and seemed about forty years of age. His forehead was rather prominent and high, his eyebrows narrow, well defended, and nearly straight; his hair, which was combed back from his forehead, and the sides of his face, was of a glossy black colour, slightly curled behind; his eyes were small, sometimes appearing remarkably keen, at others rather heavy; his nose was straight, and the nostrils by no means large, his lips were thick, and his chin projecting. He was arrayed in a handsome tiputa of native manufacture. His body was stout, but not disproportioned to his height; and his limbs, though well formed, were not firm and muscular. He welcomed me to Tahiti; but, at the same time, appeared disappointed when he learned that only one Missionary had arrived, having been led to expect several. His acquaintance with English was very partial, and mine with Tahitian much more so; our conversation was, consequently, neither very free nor animated. He inquired after King George, Governor Macquarrie, and Mr. Marsden, the time of our departure from New Holland, the nature of our voyage, &c. These inquiries I answered, and handed him a number of small presents which I had brought from England, adding a curious penknife of my own, which he had appeared desirous to possess. He had a small English Bible, and, at his request, I read to him one or two chapters. He appeared to understand, in some degree, the English language, although unable to speak it. After spending some time in the cabin, the king went to see the cattle we had brought from New South Wales, and particularly a horse, which the owners of the ship had sent him as a present.

Pomare was greatly delighted with the horse; and, in the course of the afternoon, the poor animal, after having been hung in slings, and unable to lie down during the greater part of the voyage, was hoisted out of the hold, to be taken ashore in a large pair of canoes which the king had ordered alongside for that purpose. During this transition, while the horse was suspended midway between the gangway and the yard-arm, some of the bandages gave way; when the animal, after hanging some time by the neck and fore-legs, to the great terror both of Pomare and the captain, slipped through the slings, and, clearing the ship's side, fell into the sea. He instantly rose to the surface; and, snorting, as if glad, even under these circumstances, to gain his freedom, swam towards the shore; but the natives no sooner saw him at liberty, than they plunged into the water, and followed like a shoal of sharks or porpoises after him. Some seizing his mane, others his tail, endeavoured to hold him, till the terrified creature appeared in great danger of a watery grave. The captain lowered down the boat; the king shouting, directed the natives to leave the horse to himself; but his voice was lost amid the din and clamour of the crowds that accompanied the exhausted and frightened animal to the land. At length he reached the beach in safety; and, as he rose out of the water, the natives on the shore fled with precipitation, climbing the trees, or crouching behind the rocks and the bushes for security. When, however, they saw one of the seamen, who had landed with the captain from the ship, take hold of the halter that was on his neck, they returned, to gratify their curiosity. Most of them had heard of horses, and some of them had, perhaps, seen those belonging to Mai, (Omai,) landed on the island by Captain Cook, forty years before; but it was undoubtedly the first animal of the kind the greater part of them had ever seen.

The king had not been long on board, when the queen arrived, and was ushered into the cabin. Her person was about the middle stature; her complexion fairer than any other native I have ever seen; her form elegant, and her whole appearance remarkably prepossessing. Her voice, however, was by no means soft, and her manners were less engaging than those of several of her companions. She was attired in a light loose and flowing dress of beautifully white native cloth, tastefully fastened on the left shoulder, and reaching to the ankle; her hair was rather lighter than that of the natives in general; and on her head she wore a light and elegant native bonnet, of green and yellow cocoa-nut leaves; each ear was perforated, and in the perforation two or three flowers of the fragrant Cape jessamin were inserted. She was accompanied by her sister, Pomare Vahine. Aimata, the young princess, only daughter of Pomare and the queen, who appeared about six years of age, was brought by her nurse, and followed by her attendants into the cabin. We delivered the few presents we had brought for them, regretting that we could not enter into conversation. They spent about two hours on board; and then, followed by their numerous retinue, returned to the shore.

Soon after sunrise the next morning, our vessel was surrounded with canoes, and provisions in abundance were offered for barter. Pomare also sent us a present.

About nine o'clock, I saw crowds of natives repairing towards the place where the horse had been tied up, in charge of one of Pomare's favourite chiefs; and shortly afterwards he was led out, while the multitude gazed at him with great astonishment. Soon after breakfast, our captain landed with the saddle and bridle, and other presents, which Mr. Bernie, of Sydney, had sent out with the horse. They were delivered to Pomare, who requested that the saddle and bridle might be put on the horse, and that the captain would ride him. His wishes were complied with, and the multitude appeared highly delighted when they saw the animal walking and running along the beach, with the captain on his back. They called him *buaa-horo-fenua* and *buaa-afai-taata;* land-running pig, and man-carrying pig. About midday the captain returned to the ship; and we shortly afterwards weighed anchor, and sailed for the island of Eimeo.

Moorea, the name most frequently given by the natives to this island, was discovered by Captain Wallis, and by him called Duke of York Island. It is situated about twelve or fourteen miles west from Tahiti, and is twenty-five miles in circumference. In the varied forms its mountains exhibit, the verdure with which they are clothed, and the general romantic and beautiful character of its scenery, this island far exceeds any other, in either the Georgian or Society groups. A reef of coral, like a ring, surrounds the island; in some places one or two miles distant from the shore, in others united to the beach. Several small and verdant islands adorn the reef: one lies opposite the district of Afareaitu on the eastern side; two others, a few miles south of Papetoai; the latter are covered with the elegantly growing Casuarina, or Aito trees, and were a favourite retreat of Pomare the Second. Eimeo is not only distinguished by its varied and beautiful natural scenery, but also by the excellence of its harbours, which are better than those in any of the other islands.

On the north side is Taloo harbour, in lat. 17° 30′ north, long. 150° west: one of the most secure and delightful anchoring places to be met with in the Pacific; Opunohu is the proper name of this harbour; near the mouth of which, on the right-hand side, there is a small rock, called by the natives *Tareu*, towards which, it is possible, Captain Cook was pointing, or looking, when he inquired of the natives the name of the harbour his ship was then entering. Tareu might be easily understood as if spelled Taloo, and the name of the rock thus mistaken for that of the harbour. Separated from Opunohu by a high mountain, is another capacious bay, called, after its discoverer, Cook's harbour; it is equally convenient for anchorage with the former, but is rather more difficult of access.

On the north-eastern side of Eimeo, between the mountain and the sea, is an extensive and beautiful lake, called Tamai, on the border of which stands a sequestered village, bearing the same name. The lake is stocked with fish, and is a place of resort for flocks of wild ducks, which are sometimes taken in great numbers. The rivers of Eimeo are but small, and are principally mountain streams, which originate in the high lands, roll down the rocky bottoms of the deep ravines, and wind their way through the valleys to the sea. The mountains are broken, and considerably elevated, but by no means so high as those of Tahiti, which are probably 7000 feet above the level of the sea.

We enjoyed a most delightful sail along the northern part of Eimeo, the next morning, and soon after twelve o'clock anchored in the spacious and charming bay of Opunohu, or, as it is usually called, by foreigners, The harbour of Taloo.

Long before we anchored, Messrs. Bicknell, Wilson, Henry, and Davies, came on board, followed by the other members of the Mission, who greeted our arrival with satisfaction. We accompanied them to the shore, and landed on the western side of the bay, in the afternoon of the 13th of February, 1817, happy, under circumstances of health and comfort, to enter upon our field of future labour, and grateful for the merciful providence by which we had been conducted in safety to the end of our long and eventful voyage.

On reaching the habitations of the Missionaries, we were cordially welcomed to their society, and were rejoiced to behold them cheered by the intelligence we had brought, and the prospect of receiving a still greater accession to their numbers. The evening passed pleasantly and rapidly away; many of the pious inhabitants and chiefs, in the neighbourhood, came to greet our arrival, with evident emotions of delight; among them was one, whose salutation I shall never forget: "*Ia ora na oe i te Atua, Ia ora oei te haere raa mai io nei, no te Aroha o te Atua oe i tae mai ai,*" "Blessing on you from God, peace to you in coming here, on account of the love of God are you come." These were his words. His person was tall and commanding, his hair black and curling, his eyes benignant, and his whole countenance beamed with a joy that declared his tongue only obeyed the dictates of his heart. His name was AUNA, a native of Raiatea, formerly an *areoi* and a warrior, who had arrived, with numbers of his countrymen, to the support of Pomare, after his expulsion from Tahiti, but whose heart had been changed by the power of the gospel of Christ. He was afterwards associated with us at Huahine, subsequently became my fellow-labourer in the Sandwich Islands, and was, when I last heard from the islands, about to be ordained pastor of a Christian church in Sir Charles Sanders's Island.

At a late hour we retired to rest, but not to sleep. We needed and sought repose, but the incidents of the day had produced a degree of excitement that did not speedily subside; in addition to which, the constant and loud roaring of the surf kept us awake till nearly daybreak. The house in which we lodged was near the shore; and the long heavy billows of the sea rolling in successive surges over the coral reefs that surround the island, kept up, through the night, a hollow and heavy sound, resembling that produced by the rumbling of carriages in a vast city, heard at a distance in the stillness of evening. The wall, or outside of the dwelling, was composed only of large sticks, or poles, placed perpendicularly from the floor to the roof, two or three inches apart, so that we could see the ocean on one side, and the dark outline of the inland mountains on the other; while looking up *through* the roof, we could easily discern the stars twinkling in a blue and cloudless sky. We did not, however, feel the air too cool; and our lodging was quite as good as that in which the Missionaries to the Sandwich Islands passed their first night in Honoruru; and much better than Mr. Marsden, and his companion, procured in New Zealand. The first night he passed on shore, he slept on the earthern floor, by the side of a warrior, the murderer of the crew of the Boyd, and a cannibal; and the spot on which he lay was encircled by native spears fixed in the ground.

In the morning we arose somewhat refreshed; and, in the course of the day, landed our goods from the vessel. A house had been prepared, by the king, for the expected Missionaries; but, as it was damp, and our residence at Papetoai was not likely to be permanent, we took up our abode in a dwelling already occupied in part by Mr. Crook and his family.

I was astonished at the accounts I now received, of the change that had taken place among the people. The profession of Christianity was general, many had learned to read, and were teaching others; all were regular in their exercises of devotion; and, in many of the small gardens attached to the native houses, it was pleasing to see the little *fare bure huna,* house for hidden prayer. The greater part of the Missionaries, who had fled to Port Jackson, when expelled from Tahiti in 1808, having been invited by Pomare, returned in 1812. In 1816 they were joined by Mr. Crook, who had been stationed by Captain Wilson in the Marquesas: they had visited Tahiti, for the purpose of preaching to the inhabitants, but they had not been able to re-establish the Mission in their original station, and were, consequently, all residing at Eimeo when we arrived.

The chiefs of the district, and island, soon visited us, received a few articles as presents, and appeared highly gratified with what they saw, especially with some engravings of natural history. They sent us a present of food; or, as they call it, "*faaamua,*" a feeding; consisting of two or three large pigs, which were dragged along by force, squalling terribly all the way, and tied to a stick near the door; a number of bunches of plantains, bananas, cocoa-nuts, and bread-fruit, were also brought, and piled up in three heaps on the sand, near the pigs. I was then called out, and a native repeated the names of the chiefs who had sent us the food; and, pointing to the heaps of fruit and the pigs, said one was for me, and another for Mrs. Ellis, and the third for our infant daughter. He then directed the native servants of the house to take care of it, and departed.

Soon after my arrival, I visited the school, and was greatly delighted to behold numbers of adults, as well as children, under the direction of Messrs. Davies and Tessier, learning their alphabet and their spelling, or reading with distinctness their lessons, which were principally extracts from Scripture.

The building, in which they were taught, stood near the sea-beach, under the shade of a clump of cocoa-nut trees. Though of no very durable kind, it appeared well adapted to the purpose to which it was appropriated. It was upwards of sixty feet long, and rather narrow. The thatch was composed of the leaves of the pandanus, neatly fastened on rafters of purau or hibiscus, and the walls, or sides and ends, were formed with straight branches resembling the rafters, and planted in the ground about two inches asunder. There was a door at each end; windows were altogether unnecessary in such a building, as the space between the poles, forming the outside, admitted light and air in abundance; and wind, with rain, sometimes in larger quantities than was quite agreeable. The floor, which was of sand, was covered with long dry grass. A rustic sort of table, or desk, between three and four feet high, stood on one side, equally distant from each end, and the whole of the building was filled with low forms, on which the natives were sitting; while, on one side I saw one or two forms longer and broader than the rest, with small ledges on the sides, filled with sand, for the purpose of teaching writing, after the manner of the national schools in England. A number of pillars in the centre supported the ridge pole, or rather the different ridge poles, which unitedly sustained the roof of the building. The different joints in these, and the narrow horizontal boards supporting the bottoms of the rafters, presented a kind of chronological index to the history of the place. It was first erected by the liberality of a gentleman in London. He presented to Tapioi, the Marquesan youth who accompanied Mr. Bicknell to England, the articles with which the natives were hired to build this first school and chapel in Eimeo. It was then much more compact, and the width better proportioned than it now appeared. It had always been employed, not only as a school, but also as a chapel. When the number of scholars and worshippers of the true God increased, so as to render accommodation difficult, one of the ends had been taken down, a new piece of timber joined to the ridge pole, the building lengthened about twelve or fifteen feet, and the end then closed up. When the place became again too small, a similar enlargement had been made; and, as the new piece which supported the roof, was laid upon the former ridge pole, it distinctly marked the increase of Christian worshippers at the place within the last four or five years.

The first Sabbath I spent in the islands, was a day of deep and delightful interest. The Missionaries were accustomed to meet for prayer at sun-rise, on the morning of the Sabbath. This service I attended, and was also gratified to find, that not fewer than four or five hundred of the natives, imitating their teachers in this respect, met for the purpose of praise and supplication to the true God, during the interval of public worship, which was held early in the morning, and four in the afternoon.

About a quarter before nine in the morning, I accompanied Mr. Crook to the public worship of the natives, held in the same house in which I had visited the school a day or two before. It was, indeed, a rude and perishable building, totally destitute of every thing imposing in effect, or exquisite in workmanship; yet I beheld it with emotions of pleasure, as the first roof under which the natives of Tahiti had assembled, in any number, to receive the elements of useful knowledge, to listen with sincerity and satisfaction to the word of God, and to render publicly unto Him the homage of their grateful praise; for,

> "Though gilded domes, and splendid fanes,
> And costly robes, and choral strains,
> And altars richly dress'd;
> And sculptur'd saints, and sparkling gems,
> And mitred priests, and diadems,
> Inspire with awe the breast:
>
> "'Tis not the pageantry of show
> That can impart devotion's glow,
> Nor sanctify a prayer.
> The soul enlarged, devout, sincere,
> With equal piety draws near
> The holy house of God,
> That rudely rears its rustic head,
> Scarce higher than the Indians' shed;
> By Indians only trod."

The place was thronged with people, and numbers were standing or sitting round the doors and the outside of the building. When we arrived, they readily made way for us to enter; when a scene, destitute indeed of magnificence and splendour as to the structure itself, or the richness in personal adornment of its inmates, but certainly the most delightful and affecting I had ever beheld, appeared before me. Between five and six hundred native Christians were there assembled, to worship the true God. Their persons were cleanly, their apparel neat, their countenances either thoughtful, or beaming with serenity and gladness. The heads of the men were uncovered, their hair cut and combed, and their beards shaven. Their dress was generally a pareu round the waist; and a native tiputa, over their shoulders, which covered the upper part of the body, excepting the arms. The appearance of the females was equally interesting; the greater part of them wore a neat and tasteful bonnet, made with the rich yellow-tinted cocoa-nut leaf. Their countenances were open and lively; many of them had inserted a small bunch of the fragrant and delicately white gardinia, or Cape jessamine flowers, in their hair; in addition to which, several of their chief women wore two or three fine native pearls fastened together with finely braided human hair, and hanging pendent from one of their ears, while the other was adorned with a native flower. Their dress was remarkably modest and becoming, being generally what they term *ahu bu*, which consists of large quantities of beautifully white native cloth, wound round the body, then passed under one arm, and fastened on the other shoulder, leaving uncovered only the neck and face, and part of one arm.

The assembly maintained the most perfect silence, until Mr. Davies, who officiated on the occasion, and was seated behind the table, which answered the double purpose of a desk for the schoolmaster, and a pulpit for the minister, rose up, and gave out a hymn in the native language. The whole congregation now rose, and many of them joined in the singing. A prayer was then offered, during which the congregation remained standing; another hymn was sung; the people then sat down, and listened attentively to a discourse, delivered by the Missionary standing on the ground behind the desk. When this was ended, a short prayer was offered, the benediction pronounced, and the service closed. The assembly dispersed with the utmost propriety and order; many of them, as they passed by, cordially shook me by the hand, and expressed their joy at seeing me among them. My joy, and excitement of feeling, was not less than theirs. There was something so pleasing and novel in their appearance, so peculiar in their voices when singing, and in their native language, both during the prayers and sermon, and something so solemn and earnest in their attention, with such an air of sincerity in devotion during the whole service, that it deeply affected my heart. I was desirous of speaking to them in return, and expressing the grateful satisfaction with which I had beheld their worship; but the scene before me had taken such a powerful hold of my feelings, that I returned home in silence, filled with astonishment at the change that had taken place, and deeply impressed with the evidence it afforded of the efficacy of the gospel, and the power of the Almighty. At eleven o'clock I attended public worship in the English language.

At four in the afternoon the natives again assembled, and I attended at their worship. Though I could not understand their language, I was pleased with the large attendance, and the serious and earnest manner in which the people listened to an animated discourse delivered by Mr. Nott. In the evening several of the Missionaries met for social worship, and with this sacred exercise we closed our first Sabbath in the Society Islands, under a deep impression of the advantages of Christianity, and the pleasing effects which we had that day witnessed, of Divine influence over the hearts of the most profligate idolaters.

In the afternoon of the succeeding Sabbaths, I visited a number of Christian chiefs at their own houses. We usually found them either reading together, conversing on the contents of their books, or some other religious subject. At Hitoti's dwelling which, I visited on the second Sabbath after my arrival, the household were about to kneel down for prayer when we entered; we joined them, and several of the petitions which the chief offered up to God, appeared, when interpreted by my companion, remarkably appropriate and expressive.

In the course of my first week on shore, I made several excursions in different parts of the district. The soil, in all the level part of the valley, was a rich vegetable mould, with a small portion of alluvial, washed down from the surrounding hills, which are generally covered with a stiff kind of loam or brownish-red ochre. Several large plantations were well stocked with the different productions of the island; but a large portion of the valleys adjacent to the settlement, were altogether uncultivated, and covered with grass or brush-wood, growing with all the rank luxuriance that a humid atmosphere, a tropical sun, and a fertile soil, would combine to produce.

I also accompanied one of the Missionaries on a voyage to the opposite side of the island, about twenty miles distant from the settlement at Papetoai. Two natives paddled our light single canoe along the smooth water within the reefs till we reached Moru, where we landed, to take some refreshment at the house of a friendly chief. This was the first native meal I had sat down to, and it was served up in true Tahitian style. When the food was ready, we were requested to seat ourselves on the dry grass that covered the floor of the house. A number of the broad leaves of the purau, hibiscus tileaceus, having the stalks plucked off close to the leaf, were then spread on the ground, in two or three successive layers, with the downy or underside upwards, and two or three were handed by a servant to each individual, instead of a plate. By the side of these vegetable plates, a small cocoa-nut shell of salt water was placed for each person. Quantities of fine large bread-fruit, roasted on hot stones, were now peeled and brought in, and a number of fish that had been wrapped in plantain leaves, and broiled on the embers, were placed beside them. A bread-fruit and a fish was handed to each individual, and, having implored a blessing, we began to eat, dipping every mouthful of bread-fruit or fish into the small vessel of salt water,—without which, to the natives, it would have been unsavoury and tasteless. I opened the leaves, and found the fish nicely broiled; and, imitating the practice of those around me, dipped several of the first pieces I took into the dish placed by my side: but there was a bitterness in the sea water which rendered it rather unpalatable, I therefore dispensed with the further use of it, and finished my meal with the bread-fruit and fish.

About two o'clock in the afternoon, we resumed our journey; travelling sometimes along the sea-beach, and at other times availing ourselves of the canoe until near sunset, when we reached Afareaitu—and created by our arrival no small stir among the people.

The next morning we examined the district, and were delighted with its fertility, extent, and resources. Afareaitu is on the eastern side of Eimeo, opposite the district of Atehuru in Tahiti, and is certainly one of the finest districts in the island. It comprises two valleys, or rather one large valley partially divided by a narrow hilly ridge extending from the mountains in the interior, towards the shore. The soil of the bottom of the valley is rich and fertile, well stocked with cocoa-nuts and bread-fruit trees. The surrounding hills are clothed with shrubs or grass, and the lofty and romantic mountains forming the central boundary, are adorned with trees or bushes even to their summits. Several broad cascades flowed in silvery streams down the sides of the mountain, and, broken occasionally by a jutting rock, presented their sparkling waters in beautiful contrast with the rich and dark foliage of the stately trees, and the flowering shrubs that bordered their course. A number of streams originating in these water-falls pursued their course through the valley, and one, receiving in its way the tributary waters of a number of sequestered streamlets, swelled at times into what in these islands might be called a river, and flowed along the most fertile portions of the district into the sea.

A small bay was formed by an elliptical indentation of the coast, an opening in the reef opposite the bay admitted small vessels to enter, and a picturesque little coral island, adorned with two or three clumps of hibiscus and cocoa-nut trees, added greatly to the beauty of its appearance. There was no swamp or marshy land between the shore and the mountains; the ground was high, and the whole district not only remarkably beautiful, but apparently dry and healthy. The abundance of natural productions, the apparent salubrity of the air, the convenience of the stream of water, the facility of the harbour, combined to recommend it as an eligible spot for at least the temporary residence of a part of the Missionaries. We therefore waited on the principal chiefs, one of whom had accompanied us from Papetoai, and inquired if it would be agreeable to them for us to come and reside there. They expressed themselves pleased with the prospect of such an event, and promised every assistance in the erection of our houses, &c. Having accomplished the object of our visit, we left Afareaitu, and returned to Papetoai the same evening.

The circumstances of the inhabitants of the windward and leeward islands, most of whom had renounced idolatry, and their earnest desire to receive religious instruction, rendered it exceedingly desirable, that the Missionaries should no longer remain altogether at Papetoai, but establish themselves in the different islands; but the vessel which they had commenced building in 1813, being still unfinished, and the anticipation of a considerable accession to their numbers, induced them to defer forming any new station, until such reinforcement should arrive.

The natives in the several islands were in want, not only of teachers, but also of books. I had taken out a printing-press and types, and having, at the request of the Directors, learned the art of printing in England, it was proposed, that as a temporary measure, to supply the existing demand for books, the press should be set up at Afareaitu. By this arrangement two stations would be formed in Eimeo, and the whole of the inhabitants be brought more fully under religious instruction. In order to carry these plans into effect, we left Papetoai on the 25th of March, with Mr. Davies, Mr. and Mrs. Crook and family. Mrs. Ellis, and myself, with an infant and her nurse, set out in a native canoe, having most of our goods and luggage on board. Mr. Crook and family preceded us in a fine large double canoe, called "*Tiaitoerau*," literally "wait for the west wind," from *tiai* to wait, and *toerau* west wind. It was between thirty and forty feet in length, very strong, and, as a piece of native workmanship, well built. The keel, or bottom, was formed with a number of pieces of tough Tamanu wood, *inophyllum callophyllum,* twelve or sixteen inches broad, and two inches thick, hollowed on the inside, and rounded without, so as to form a convex angle along the bottom of the canoe; these were fastened together by lacings of tough elastic cinet, made with the fibres of the cocoa-nut husk. On the front end of the keel, a solid piece, cut out of the trunk of a tree, so contrived as to constitute the forepart of the canoe; was fixed with the same lashing; and on the upper part of it, a thick board or plank projected horizontally, and formed a line parallel with the surface of the water. This front piece, usually five or six feet long, and twelve or eighteen inches wide, was called the *ihu vaa,* nose of the canoe, and, without any joining, comprised the stem, bows, and bowsprit of the vessel.

The sides of the canoe were composed of two lines of short plank or board, an inch and a half or two inches thick. The lowest line was convex on the outside, and nine or twelve inches broad; the upper one straight. The stern was considerably elevated, the keel was inclined upwards, and the lower part of the stern resembled the bottom of a pointed shield, while the upper part of the noo, or stern, was nine or ten feet above the level of the sides. The whole was fastened together with cinet, not continued along the seams, but by two, or at most, three holes made in each board, within an inch of each other, and corresponding holes made in the opposite piece, and the cinet passed through from one to the other. A space of nine inches or a foot was left, and then a similar set of holes made. The joints or seams were not grooved together, but the edge of one simply laid on that of the other, and fitted with remarkable exactness by the adze of the workman, guided only by his eye: they never used line or rule. The edges of their planks were usually covered with a kind of pitch or gum from the bread-fruit tree, and a thin layer of cocoa-nut husk spread between them. The husk of the cocoa-nut swelling when in contact with the water, fills any apertures that may exist, and, considering the manner in which they are put together, the canoes are often remarkably dry. The two canoes forming Tiaitoerau, which was a double one, were fastened together by strong curved pieces of wood, placed horizontally across the upper edges of the canoes, to which they were fixed by strong lashings of thick cinet.

Skreened Canoe.

The space between the two bowsprits, or broad planks projecting from the front of our canoe, was covered with boards, and furnished a platform of considerable extent; over this a kind of temporary awning of platted cocoa-nut leaves was spread, and under it the passengers sat during the voyage. The upper part of each of the canoes was not above twelve or fifteen inches wide; little projections were formed on the inner part of the sides, on which small moveable thwarts or seats were fixed, whereon the men sat who paddled it along, while the luggage was either placed in the bottom, piled up against the stern, or laid on the elevated stage between the two canoes. The heat of the sun was extreme, and we found that our rustic awning afforded a grateful shade.

The rowers appeared to labour hard. Their paddles, being made of the tough wood of the hibiscus, were not heavy; yet, having no pins in the sides of the canoe, against which the handles of the paddles could bear, but leaning the whole body over the canoe, first on one side, and then on the other, and working the paddle with one hand near the blade, and the other at the upper end of the handle, and shovelling as it were the water, appeared a great waste of strength. They often, however, paddle for a time with remarkable swiftness, keeping time with the greatest regularity. The steersman stands or sits in the stern, with a large paddle; the rowers sit in each canoe two or three feet apart, the leader sits next, the steersman gives the signal to start, by striking his paddle violently against the side of the canoe, every paddle is then put in and taken out of the water with every stroke at the same moment; and after they have thus continued on one side for five or six minutes, the leader strikes his paddle, and the rowers instantly and simultaneously turn to the other side, and thus alternately working on each side of the canoe, they go along at a considerable rate. There is generally a good deal of striking the paddle when a chief leaves or approaches the shore, and the effect pretty much resembles that of the smacking of the whip, or sounding of the horn, at the starting or arrival of a coach.

The isolated situation of the islanders, and their dependence on the sea for a large proportion of the means of subsistence, necessarily impart a maritime character to their habits, and render the building, fitting, and managing of the vessels one of the most general and important of their avocations. It also procures no small respect and emolument for the *Tahua tarai vaa,* builder of canoes. *Vaa waa,* or *vaka,* is the name of a canoe, in most of the islands of the Pacific; though by foreigners they are uniformly called canoes, a name first given to this sort of boat by the natives of the Caribbean Islands,* and adopted by Europeans ever since, to designate the rude boats used by the uncivilized natives in every part of the world.

The canoes of the Society Islanders are various, both in size and shape, and are double or single. The canoes belonging to the principal chiefs, and the vaa mataaina, public district canoes, were in general large—fifty, sixty, or nearly seventy feet long, and each about two feet wide, and three or four feet deep; the sterns remarkably high, sometimes fifteen or eighteen feet above the water, and frequently ornamented with rudely carved hollow cylinders, square pieces, or grotesque figures, called *tiis.* The rank or dignity of a chief was supposed, in some degree, to be indicated by the size of his canoe, the carving and ornaments with which it was embellished, and the number of his rowers.—Next in size to these was the pahi, or war canoe. I never saw but one of these: the stern was low, and covered, so as to afford a shelter from the stones of the assailants; the bottom was round, the upper part of the sides

* After his first interview with the natives of the newly discovered islands, in the Caribbean sea, we are informed by Robertson, that Columbus returned to his ship, accompanied by many of the islanders in their boats, which they called *canoes*; and though rudely formed out of the trunk of a single tree, they rowed them with surprising dexterity.

was narrower, and perpendicular; a rude imitation of the human head, or some other grotesque figure, was carved on the stern of each canoe. The stern, often elevated and curved like the neck of a swan, terminated in the carved figure of a bird's head, and the whole was more solid and compact than the other vessels. There was a kind of platform in the front, or generally near the centre, on which the fighting men were stationed: these canoes were sometimes sixty feet long, between three and four feet deep, and with their platforms in front, or in the centre, were capable of holding fifty fighting men.*

War Canoe.

The vaatii, or sacred canoe, was always strong and large, more highly ornamented with carving and feathers than any of the others. Small houses were erected in each, and the image of the god, sometimes in the shape of a large bird, at other times resembling a hollow cylinder ornamented with various coloured feathers,

* In Cook's voyages, a description is given of some, one hundred and eight feet long; but I never saw any so large.

was kept in these houses. Here the prayers were frequently preferred, and the sacrifices offered.

Their war canoes were generally strong, well built, and highly ornamented. They formerly possessed large and magnificent fleets of these, and other large canoes; and, at their general public meetings, or festivals, no small portion of the entertainment was derived from the regattas, or naval reviews, in which the whole fleet, ornamented with carved images, and decorated with flags and streamers, of various native coloured cloth, went through the different tactics with great precision. On these occasions the crews by which they were navigated, anxious to gain the plaudits of the king and chiefs, emulated each other in the exhibition of their seamanship. The vaatii, or sacred canoes, formed part of every fleet, and were generally the most imposing in appearance, and attractive in their decorations.

The peculiar and almost classical shape of the large Tahitian canoes, the elevated prow and stern, the rude figures, carving, and other ornaments, the loose folding drapery of the natives on board, and the maritime aspect of their general places of abode, are all adapted to produce a singular effect on the mind of the beholder. I have often thought, when I have seen a fleet of thirty or forty approaching the shore, that they exhibited no faint representation of the ships in which the Argonauts sailed, or the vessels that conveyed the heroes of Homer to the siege of Troy.

Every canoe, of any size, had a distinct name, always arbitrary, but frequently descriptive of some real or imaginary excellence in the canoe, or in memory of some event connected with it. Neither the names of any of their gods, or chiefs, were ever given to their vessels; such an act, instead of being considered an honour, would have been deemed the greatest insult that could have been offered. The names of canoes, in some instances, appear to have been perpetuated, as the king's state canoe was always called Anuanua, or the rainbow. The most general and useful kind of canoe is the tipairua, or common double canoe, usually from twenty to thirty feet long, strong and capacious, with a projection from the stem, and a low shield-shaped stern. These are very valuable, and usually form the mode of conveyance for every chief of respectability or influence, in the island. They are also used to transport provisions, or other goods, from one place to another.

They have also a remarkably neat double canoe, called Maihi, or twins, each of which is made out of a single tree, and are both exactly alike. The stem and stern are usually sharp; although, occasionally, there is a small board projecting from each stem. These are light, safe, and swift, easily managed, and seldom used but by the chiefs. A canoe of this kind was a favourite conveyance with the late king Pomare.

The single canoes are built in the same manner, and with the same materials, as the double ones. Their usual name is *tipaihoe*, single landing, and they are more various in their kind than the others. The small *buhoe*, the literal name of which is single shell, is generally a trunk of a tree, seldom more than twenty feet in length, rounded on the outside, and hollow within; sometimes sharp at both ends, though generally only at the stem. It is used by fishermen among the reefs, and also along the shore, and in shallow water, seldom carrying more than two persons. The single maihi is only a neater kind of buhoe.

The *vaa motu*, island-canoe, is generally a large, strong, single vessel, built for sailing, and principally used in distant voyages, from one island to another. In addition to the ordinary edge, or gunwale, of the canoe, planks, twelve or fifteen inches wide, are fastened along their sides, after the manner of wash-boards in a European boat. The same are also added to double canoes, when employed on long voyages. A single vaa is never used without an outrigger, varying in size with the vessel; it is usually formed with a light spar of the hibiscus, or of the erythrina, which was highly prized as an *ama*, or outrigger, on account of its being both light and strong. This is always placed on the left side, and fastened to the canoe by two horizontal poles, from five to eight feet long; the front one is straight and firm, the other curved and elastic; it is so fixed, that the bark, when empty, does not float upright, being rather inclined to the left; but, when sunk into the water, on being laden, &c. it is generally erect, while the outrigger, which is firmly and ingeniously fastened to the sides by repeated bands of strong cinet, floats on the surface. In addition to this, the island-canoes have a strong plank, twelve or fourteen feet long, fastened horizontally across the centre, in an inclined position, one end attached to the outrigger, and the other extending five or six feet over the opposite side, and perhaps elevated four or five feet above the sea. A small railing of rods is fastened along the sides of this plank, and it is designed to assist the navigators in balancing the keel, as a native takes his station on the one side or the other, to counteract the inclination which the wind or sea might give to the vessel. Sometimes they approach the shore with a native standing or sitting on the extremity of the plank, and presenting a singular appearance, which it is impossible to behold without expecting every undulation of the sea will detach him from his apparently insecure situation, and precipitate him into the water.

Single, or Island Canoe.

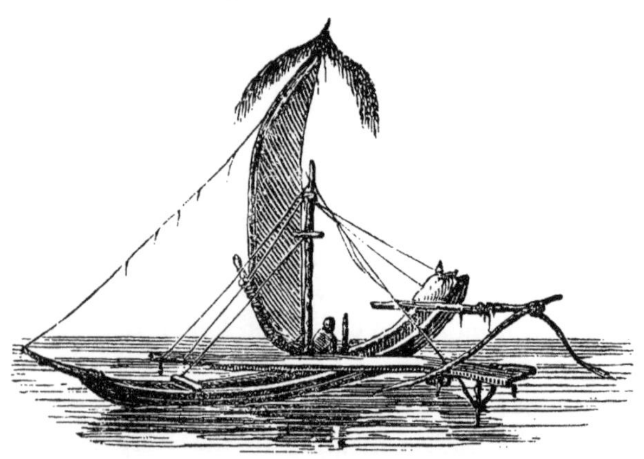

This kind of canoe is principally employed in the voyages which the natives make to *Tetuaroa*, a cluster of islands, five in number; the names of which are Rimatu, Onehoa, Moturua, Hoatère, and Reiona; these are enclosed in one reef, in which there is an opening on the north-west, but only such as to admit, and that with difficulty, their own canoes. The whole cluster is called Tetuaroa. They are low coralline islands, the highest parts being not more than three or four feet above the water, and the only soil they contain is composed of sand and fragments of coral, with which is mingled vegetable mould produced on the islands, or carried from Tahiti. The chief article of food produced in these islands is the fruit of the cocoa-nut tree; with extensive and verdant groves of which they are adorned. They seem, at a distance, as if they were growing on the surface of the water, and the roots and stems of many are washed by the spray, or by the tide, when it rises a few inches higher than usual. Upon the kernel of the cocoa-nut, and the fish taken among the reefs, the inhabitants principally subsist.

Tetuaroa, the long, or distant, sea, is part of the hereditary possessions of the reigning family, and is attached to the district of Pare. Most of the inhabitants of these little islets occupy, under the king, a part of his own land, from which they are supplied with bread-fruit and taro. They are much employed in fishing, and formerly brought over large quantities of fish, carrying to the islands in return bread-fruit, and other edible productions of Tahiti. In the wars which disturbed the conclusion of the reign of Pomare the First, and the commencement of that of his successor, many of the inhabitants were cut off; and the decrease of population, thus occasioned, has diminished the intercourse between these islands and Tahiti.

In addition to the fishery carried on here, Tetuaroa has long been a kind of watering-place for the royal family, and a frequent resort for what might be called the fashionable and gay of Tahiti.—Hither the areois, dancers, and singers, were accustomed to repair, together with those whose lives were professedly devoted to indolent pleasures. It was also frequented by the females of the higher class, for the purposes of *haapori*, increasing the corpulency of their persons, and removing, by luxurious ease under the embowering shade of the cocoa-nut groves, the dark tinge which the vertical sun of Tahiti might have burnt upon the complexions. So great was the intercourse formerly, that a hundred of these island-canoes have been seen at a time upon the beach of Tetuaroa.

In navigating their double canoes, the natives frequently use two sails, but in their single vessels only one. The masts are moveable, and are only raised when the sails are used. They are slightly fixed upon a kind of step placed across the canoe, and fastened by strong ropes or braces extending to both sides, and to the stem and stern. The sails were made with the leaves of the pandanus split into thin strips, neatly woven into a kind of matting. The shape of the sails of the island-canoes is singular, the side attached to the mast is straight, the outer part resembling the section of an oval, cut in the longest direction. The other sails are commonly used in the same manner as sprit or lugger sails are used in European boats. The ropes from the corners of the sails are not usually fastened, but held in the hands of the natives. The rigging is neither varied nor complex; the cordage is made with the twisted bark of the hibiscus, or the fibres of the cocoa-nut husk—of which a very good *coiar* rope is manufactured.

The paddles of the Tahitians are plain, having a smooth round handle, and an oblong-shaped blade. Their canoes have no rudder, but are steered by a man in the stern, with a paddle generally longer than the rest. In long voyages they have two or three steering paddles, including a very large one, which they employ in stormy weather, to prevent the vessel from drifting to leeward. The *tataa*, or scoop, with which they bale out the leakage, is generally a neat and convenient article, cut out of a solid piece of wood. Their canoes were formerly ornamented with streamers of various coloured cloths, and tufts of fringe and tassels of feathers were attached to the masts and sails, though they are now seldom used. A small kind of house or awning was erected in the centre, or attached to the stern, to skreen the passengers from the sun by day and the damp by night. The latter is still used, though the former is but seldom seen. They do not appear ever to have ornamented the body or hull of their vessels with carving or painting; but, notwithstanding this seeming deficiency, they had by no means an unfinished appearance.

In building their vessels, all the parts were first accurately fitted to each other, the whole was taken to pieces, and the outside of each plank smoothed by rubbing it with a piece of coral and sand moistened with water; it was then dried, and polished with fine dry coral. The wood was generally of a rich yellow colour, the *cinet* nearly the same, and a new well-built canoe is perhaps one of the best specimens of native skill, ingenuity, and perseverance, to be seen in the islands. Most of the natives can hollow out a buhoe, but it is only those who have been regularly trained to the work, that can build a large canoe, and in this there is a considerable division of labour,—some laying down the keel and building the hull, some making and fixing the sails, and others fastening the outriggers, or adding the ornaments. The principal chiefs usually kept canoe-builders attached to their establishments, but the inferior chiefs generally hired workmen, paying them a given number of pigs, or fathoms of cloth, for a canoe, and finding them in provision while they are employed. The trees that are cut down in the mountains, or the interior of the islands, are often hollowed out there, sometimes by burning, but generally by the adze, or cut into the shape designed and then brought down to the shore.

Idolatry was interwoven with their naval architecture, as well as every other pursuit. The priest had certain ceremonies to perform, and numerous and costly offerings were made to the gods of the chief, and of the craft or profession, when the keel was laid down, when the canoe was finished, and when it was launched. Valuable canoes were often among the national offerings presented to the gods, being ever afterwards sacred to the service of the idol.

The double canoes of the Society Islands were larger, and more imposing in appearance, than most of those used in New Zealand or the Sandwich Islands, but by no means so strong as the former, nor so neat and light as the latter. I have, however, made several voyages in them. In fine weather, and with a fair wind, they are tolerably safe and comfortable; but when the weather is rough, and the wind contrary, they are miserable sea-boats, and are tossed about completely at the mercy of the winds. Many of the natives that have set out on voyages from one island to another have been carried from the group altogether, and have either perished at sea, or drifted to some distant island.

In long voyages, single canoes are considered safer than double ones, as the latter are sometimes broken asunder, and are then unmanageable; but, even though the former should fill or upset at sea, as the wood is specifically lighter than the water, there is no fear of their sinking. When a canoe is upset or fills, the natives on board jump into the sea, and all taking hold of one end, which they press down, so as to elevate the other end considerably above the sea, a great part of the water runs out; they then suddenly loose their hold of the canoe, which falls upon the water, emptied in some degree of its contents. Swimming along by the side of it, they bale out the rest, and then climb into it again, and pursue their voyage. This has frequently been the case; and, unless the canoe is broken by upsetting or filling, they are seldom prevented from accomplishing their voyage. The only evil they fear in such circumstances is that of being attacked by sharks, which have sometimes made sad havock among those who have been wrecked at sea.

An instance of this kind occurred a few years ago, when a number of chiefs and people, all together thirty two, were passing from one island to another, in a large double canoe. They were overtaken by a severe tempest, the violence of which tore their canoes asunder, and separated them from the horizontal spars by which they were united. It was in vain for them to endeavour to place them upright, or empty out the water, for they could not keep them in an erect position, nor prevent their incessant overturning. As their only resource, they collected the scattered spars and boards, and constructed a raft, on which they hoped they might drift to land. The weight of the whole number, who were now collected on the raft, was so great as to sink it so far below the surface, that they sometimes stood above their knees in water. They made very little progress, and soon became exhausted by fatigue and hunger. In this condition they were attacked by a number of sharks. Destitute of a knife, or any other weapon of defence, they fell an easy prey to these rapacious monsters. One after another was seized and devoured, or carried away by them; and the survivors, who with dreadful anguish beheld their companions thus destroyed, saw the number of assailants apparently increasing, as each body was carried away, until only two or three remained. The raft, thus lightened of its load, rose to the surface of the water, and placed them beyond the reach of the voracious jaws of their relentless destroyers. The voyage on which they had set out, was only from one of the Society Islands to another, consequently they were not very far from land. The tide and the current now carried them to the shore, where they landed, to tell the melancholy fate of their fellow-voyagers.

But for the sharks, the South Sea Islanders would be in comparatively but little danger from casualties in their voyages among the islands; and although when armed they have sometimes been known to attack a shark in the water, yet when destitute of a knife or other weapon, they become an easy prey, and are consequently much terrified at such merciless antagonists.

Another circumstance also, that added to this dread of sharks, was, the superstitious ideas they entertained relative to some of the species. Although they would not only kill, but eat certain kinds of shark; the large blue sharks, *squalus glaucus*, were deified by them, and, rather than attempt to destroy them, they would endeavour to propitiate their favour by prayers and offerings. Temples were erected, in which priests officiated, and offerings were presented to the deified sharks, while fishermen and others, who were much at sea, sought their favour. Many ludicrous legends were formerly in circulation among the people, relative to the regard paid by the sharks at sea, to priests of their temples, whom they were always said to recognize, and never to injure. I received one from the mouth of a man, formerly a priest of an *akua mao*, shark god; but it is too absurd to be recorded. The principal motives, however, by which the people appear to have been influenced in their homage to these creatures, was the same that operated on their minds in reference to other acts of idolatry; it was the principle of fear, and a desire to avoid destruction, in the event of being exposed to their anger at sea.

The superstitious fears of the people have now entirely ceased. I was once in a boat, on a voyage to Borabora, when a ravenous shark approaching the boat, seized the blade of one of the oars, and being detached from that, darted at the keel of the boat, which he attempted to bite. While he was thus employed, the native whose oar he had seized, leaning over the side of the boat, grasped him by the tail, succeeded in lifting him out of the water, and, with the help of his companions, dragged him alive into the boat, where he began to flounder and strike his tail with great rage and violence. We were climbing up on the seats out of his way, but the natives, giving him two or three blows on the nose with a small wooden mallet, quieted him, and then cut off his head. We landed the same evening, when I believe they baked and ate him.

The single canoes, though safer at sea, were yet liable to accident, notwithstanding the outrigger, which required to be fixed with care, to prevent them from upsetting. To the natives this is a matter of slight inconvenience, but to a foreigner it is not always pleasant or safe. Mrs. Osmond, Mrs. Barf, Mrs. Ellis, and myself, with our two children, and one or two natives, were once crossing the small harbour at Fare, in Huahine; a female servant was sitting in the fore part of the canoe, with our little girl in her arms, our little boy was at his mother's breast, and a native, with a long light pole, was paddling the canoe along, when a small buhoe, with a native youth sitting in it, darted out from behind a bush that hung over the water, and before we could turn, or the youth could stop his canoe, it ran across our outrigger. This in an instant went down, our canoe was turned bottom upwards, and the whole party precipitated into the sea. The sun had set soon after we started from the opposite side, and the twilight being very short, the shades of evening had already thickened around us, and prevented the natives on the shore from perceiving our situation. The native woman held our little girl up with one hand, and swam with the other towards the shore, aiding, as well as she could, Mrs. Osmond, who had caught hold of her dark hair, which floated on the water behind her; Mrs. Barf, on rising to the surface, caught hold of the outrigger of the canoe that had occasioned our disaster, and, calling out for help, informed the people on the shore of our danger, and speedily brought them to our assistance.

Mr. Osmond no sooner reached the beach, than he plunged into the sea; Mrs. O. leaving the native by whom she had been supported, caught hold of her husband, and not only prevented his swimming, but sunk him so deep in the water, that, but for the timely arrival of the natives, both would probably have found a watery grave. Mahine-vahine, the queen, sprang in, and conveyed Mrs. Barf to the shore. I came up on the side opposite to that on which the canoe had turned over, and found Mrs. Ellis struggling in the water, with the child still at her breast. I immediately climbed upon the canoe, and raised her so far out of the water, as to allow the little boy to breathe, till a small canoe came off to our assistance, into which she was taken, when I swam to the shore, grateful for the deliverance we had experienced.

It was not far from the beach where this occurred, yet the water was deep, and several articles which we had in the canoe, were seen the next day lying at the bottom, among coral and sand, seventeen or eighteen fathoms below the surface. Accidents of this kind, however, occur but seldom; and though we have made many voyages, this is the only occasion on which we have been in danger.

The natives of the eastern isles frequently come down to the Society Islands in large double canoes, which the Tahitians dignify with the name of *pahi*, the term for a ship. They are built with much smaller pieces of wood than those employed in the structure of the Tahitian canoes, as the low coralline islands produce but very small kinds of timber, yet they are much superior both for strength, convenience, and sustaining a tempest at sea. They are always double, and one canoe has a permanent covered residence for the crew. The two masts are also stationary, and a kind of ladder, or wooden shroud, extends from the sides to the head of the mast. The sails are large, and made with fine matting. Several of the principal chiefs possess a pahi paumotu, which they use as a more safe and convenient mode of conveyance than their own canoes. One canoe, that brought over a chief from Rurutu, upwards of three hundred miles, was very large. It was somewhat in the shape of a crescent, the stem and stern high and pointed, and the sides deep; the depth from the upper edge of the middle to the keel, was not less than twelve feet. It was built with thick planks of the Barringtonia, some of which were four feet wide; they were sewn together with cocoa-nut cinet, and although they brought the chief safely, probably more than six hundred miles, they must have been very ungovernable and unsafe in a storm or heavy sea.

CHAP. VII.

Account of the remarkable change in the South Sea Islands—Discouraging impressions under which the Missionaries abandoned the islands—Invitation from Pomare to return—State of the king's mind during his exile in Eimeo—His reception of the Missionaries—Death of three of their number—Influence of domestic bereavement on the Missionary life—Pomare's profession of Christianity—Application for baptism—Demonstration of the impotency of their idols—Proposal to erect a place of worship—Extracts from his correspondence—Influence of his steady adherence to Christianity—Ridicule and persecution to which he was exposed—Visit of Missionaries to Tahiti—Oitu and Tuahine—Description of the scenery of the valleys in Tahiti—Explanations of the plate of Matavai.

PREVIOUS to our embarkation from England, we had heard that a favourable change, in regard to Christianity, had taken place, in the minds of the king of Tahiti and a few of the people. On our arrival in Port Jackson, this intelligence was confirmed, and we were also encouraged by the accounts we received of the abolition of idolatry by the whole of the inhabitants of the Georgian or Windward Islands.

Here we also saw the family idols of Pomare, which had been sent from the islands to be forwarded to England, as specimens of the objects they had been accustomed to worship. When we reached the islands, we found, not only that the reports we had heard were correct, but that the change had progressively advanced, becoming daily more extensive in its influence, and decisive in its character, and that the whole of the inhabitants were no longer idolaters, but either professors of Christianity, or desirous to receive religious instruction.

We had now spent some weeks with the Missionaries and people at Papetoai; this had afforded us the means of learning from those who had been on the spot, many of the particulars connected with this amazing and important work. We had also witnessed something of its effects in the conversation and deportment of numbers who had experienced its moral influence, and evinced its benign and elevating power. It was naturally a matter of the deepest interest to a Missionary's mind, important in all its bearings on the object nearest to his heart, and first in the aims and the purposes of his life.

The accounts given by the Missionaries, on my first arrival, and the many interesting facts which subsequently came to my knowledge, when I had acquired such an acquaintance with the language of the people, as to be able to pursue my inquiries among them, have not only excited the highest delight, but convinced me, that, in the circumstances under which the change occurred, the agency by which it was accomplished, and the permanency of its effects, it is altogether one of the most remarkable displays of Divine power that has occurred in the history of mankind, and is, perhaps, unparalleled since the days of the apostles. Detached notices of this event have been transmitted to England in the letters of the Missionaries, and in the different publications of the Missionary Society. No connected and regular account has, however, yet been furnished; and, notwithstanding all that has been recorded, it may still be affirmed in the language of the deputation sent by the Society to the South Seas, that "God has indeed done great things here."

It is much to be regretted, that the Missionaries on the spot—who were intimately acquainted with every indication of the moral and spiritual process that was going on, even in its incipient stages, and every event which marked its gradual development, until, in the language of the natives on another but similar occasion, it burst upon them like the light of the morning—did not, at the time, prepare a full and particular account of the work which, under God, they had been instrumental in effecting: but their motto always was, to "say too little rather than too much," to persevere in labour, rather than employ their time in detailing their engagements; and to exercise the greatest caution and brevity in speaking of any thing connected with themselves, or the people around them, lest subsequent events should disappoint the anticipations which present favourable appearances might originate. This prudential reserve, on some accounts, cannot be too highly commended; yet, it is possible to carry it too far; and, in the present instance, however honourable to the individuals who maintained it, it cannot be doubted that the world has been thereby deprived of a full record of events, intimately connected with the destinies of the people among whom they transpired, and with the propagation of the gospel in the most distant parts of the world, during every future age of the Christian church.

Before proceeding to narrate the leading matters connected with our residence in Afareaitu, some account of that change may, perhaps, be neither improperly nor unacceptably introduced in this place, where our Missionary life may be said to have commenced. It was on my first arrival in Eimeo, that the accounts of this work, although partial, produced the greatest effect on my own mind, and left an impression that was only deepened by subsequent details from the natives themselves; and which, through whatever scenes I may yet pass, will never be effaced. I would, however, only offer it as a substitute for the more explicit statement which my predecessors in the islands might render; and if, by attracting their attention to the subject, I should induce them to furnish such a desideratum, my attempts will not have been altogether in vain. Should this be elicited, they will confer no ordinary benefit on the cause of Missions, and afford great satisfaction to the Christian world.

In the year 1809, Mr. Nott alone remained with the king and the people in the island of Eimeo; the other Missionaries, with the exception of Mr. Hayward, removed from Huahine to Port Jackson. Although the gospel had been fully, faithfully, and constantly preached, for some years in Tahiti, occasionally in most of the other islands, and many of the people had imbibed a tolerably clear speculative knowledge of the leading doctrines taught in the sacred volume, yet there was no individual on whom they could look, as having been benefited by their instructions—no one whose mind was savingly enlightened, or whose heart had experienced any moral change. Discouraging as these circumstances were, the Missionaries would not have abandoned their station, but for the destruction with which the civil war, and the defeat of the king, seriously threatened them; and, in addition to this darkened aspect of affairs, as it regarded the success of their enterprise, the state of feeling bordering on hopeless despair, under which they departed from the islands, greatly augmented their distress.

While in Port Jackson, they received affectionate and encouraging letters from the Society, and their friends in England, and communications of a most touching, yet confident kind, from the king, who invited their return.

The way being thus opened for the resumption of their work, and depending on the blessing of God, they again embarked, in the autumn of 1811, for the islands. During their absence, Pomare had remained excluded from his hereditary dominions, and in exile on the island of Eimeo. Whether the melancholy reverses he had experienced, and the depression of spirits consequent upon the dissolution of his government, and the desolation of his family, led him to doubt the truth of that system of idol-worship to which he had been devoted, and on which he had invariably relied for success in every military, civil, and political enterprise, or whether the leisure it afforded for contemplation and inquiry, under the influence of these feelings, inclined him to reflect more seriously on the truth of those declarations he had often heard respecting the true God, and to consider his present condition as the chastening of that Being whom he had refused to acknowledge,—it is impossible to determine; but these disastrous events had evidently subdued his spirit, and softened his heart.

When the Missionaries who returned from Port Jackson landed in Eimeo, the king received them with the warmest demonstrations of joy. Mr. and Mrs. Bicknell, the first who arrived, resided some time in the same house with him. He spent much of his time in reading and writing, in conversation, and in earnest inquiry about God, and the way of acceptance with Him,—and sometimes spoke in terms astonishing even to the Missionaries themselves. One or two other natives appeared also favourably impressed in regard to the religion of the Bible. Under these auspicious appearances, although prevented by the unsettled state of Tahiti from resuming their station in Matavai, the Missionaries were enabled to commence their labours in the island of Eimeo. They also indulged a hope of establishing a Mission in Raiatea, one of the Leeward or Society Islands, when a series of domestic trials frustrated all their plans of extended usefulness, and confined them for several years to this island.

On the 28th of July, 1812, Mrs. Henry finished her earthly career. She had accompanied her husband from her native country in the ship Duff, with the first Missionaries who landed in Tahiti. In all the trials of the Mission she had sustained her part; and, with unwavering devotedness to its interests, had endeavoured to perform with efficiency and cheerfulness the duties of her station, until her life fell a sacrifice to the privations and toils of her eventful and perilous career. It was, however, a sacrifice cheerfully offered on her part. Her memory was greatly esteemed by those who had borne with her the burden of the day, and survived her in the field. In a letter to the Directors of the London Missionary Society, under the date of June 24, 1813, the Rev. S. Marsden thus wrote of Mrs. Henry—"No woman, in my opinion, could be more sincere, and more devoted to the work, than she was. Her natural disposition was amiable, her piety unaffected, and her love for the poor heathens unfeigned. I trust she is now resting from her labours in Abraham's bosom; and that some poor heathens, amongst whom she had lived, have gone before, and that some will follow after, to glory." This afflictive bereavement was followed by another equally painful, viz. the death of Mrs. Davies,—which took place on the fourth of the following September. Her disconsolate partner had scarcely received the sympathies of his companions in exile and labour, when the newly closed grave of the mother was opened again, to receive the remains of an infant daughter, who survived its parent but three short weeks. In one week more, Mrs. Hayward terminated in death her sufferings, and was buried by the side of her departed sisters. The letters which conveyed to England the animating tidings of the first dawning of a brighter day on Tahiti, conveyed also the sad recital of these inroads of death; and well might the Missionaries on that occasion "sing of mercy and of judgment."

When death enters a family, and removes a wife and a mother from the social and domestic circle, though every alleviation which society, friendship, and religion can impart are available, there is a chasm left, and a wound inflicted on the survivors, which must be felt in order to be understood: when death repeatedly enters in this way a family connexion, the distress is proportionably augmented; but it is impossible to form an adequate idea of the desolateness of the Mission family, (for such it might be called,) at this time, and the cheerless solitude of those thus bereft of the partners of their days, and the mothers of their children. They were left to sustain alone the toils, sorrows, and privations of their remote and isolated station, and to pursue in solitary pilgrimage the arduous and rugged track in which the providence of God had called them to walk, far from the sympathy of the kindred and friends of the departed. They were equally remote from all the kind attentions of tenderest friendship, the rich consolations of Christian intercourse, and the public ordinances of that religion, which is alone adapted to impart effectual consolation. Cut off also from the endearments of home, the pleasures of delightful intercourse in civilized life, the satisfaction derived from books, and the reciprocal interchange of all the offices of friendship, the only earthly solace a Missionary enjoys among an uncivilized people, except what he derives from his work, is found in the social endearments of the domestic circle. However remote from the land of his nativity may be its locality, however rustic his abode, however rude its appendages, or limited its sources of comfort, compared with what in other parts may be enjoyed,—around his rural hearth, and in the bosom of his family, there he finds the scene of his richest earthly felicity. In any situation, bereavements such as those which befell the little band at Eimeo at this time, would have been distressing: to the Missionaries they were peculiarly so. The channels of comfort were dried up, and though they had full and free access to the fountain of all blessedness and consolation, and were enabled to say—"He hath done all things well," yet their trial must have been peculiarly poignant and severe. It is remarkable, that at a period of such unparalleled domestic distress, the most encouraging appearances of the Divine favour towards the nation around them, should have been afforded; and it is probable that the very cheering prospects under which they were at this time called upon to pursue their Missionary engagements, greatly alleviated their sorrow.

They had established public worship; Mr. Davies had opened a school; an increased and pleasing attention had been manifested, by several, to the instructions communicated; and only ten days before the death of Mrs. Henry, Pomare, the king of Tahiti, publicly professed his belief in Jehovah the true God, and his determination to serve him. He also requested to be baptized, and to become one of the disciples of the Lord Jesus Christ, assuring the Missionaries that his resolution to give himself up to God, was the result of long and increasing conviction of the truth and superiority of the religion of the Bible, expressing at the same time his desire to be more fully instructed in the matters to which it referred.

Pomare had for some time past shewn his contempt for the idols of his ancestors, and his desire to be taught a more excellent way, that he might obtain the favour of the true God. The natives had watched the change in his mind with the most fearful apprehension, as to its results upon the minds of his subjects. They were powerfully affected on one occasion when a present was brought him of a *turtle*, which was always held sacred, and dressed with sacred fire within the precincts of the temple, part of it being invariably offered to the idol. The attendants were proceeding with the turtle to the Marae, when Pomare called them back, and told them to prepare an oven, to bake it in his own kitchen, and serve it up, without offering it to the idol. The people around were astonished, and could hardly believe the king was in a state of sanity, or was really in earnest. The king repeated his direction; a fire was made, the turtle baked, and served up at the next repast. The people of the king's household stood, in mute expectation of some fearful visitation of the god's anger, as soon as the king should touch a morsel of the fish; by which he had, in this instance, committed, as they imagined, an act of daring impiety. The king cut up the turtle, and began to eat it, inviting some that sat at meat with him to do the same; but no one could be induced to touch it, as they expected every moment to see him either fall down dead, or seized with strong convulsions. The king endeavoured to convince his companions that their idea of the power of the gods was altogether imaginary, and that they had been the subjects of complete delusion; but the people could not believe him: and although the meal was finished without any evil result, they carried away the dishes with many expressions of astonishment, confidently expecting some judgment would overtake him before the morrow, for they could not believe that an act of sacrilege, such as he had been guilty of, could be committed with impunity.

The conduct and conversation of Pomare in reference to the gods, on this and similar occasions, must necessarily have weakened the influence of idolatry on the minds of those by whom he was attended; and if it produced no immediate and salutary effect on them, it doubtless confirmed his own belief in the vanity of idols, and the folly of indulging either hope or fear respecting them. A number of the principal chiefs of the Leeward Islands, as well as the adherents to his cause, and the friends of his family in Tahiti, constantly resided with the king, after his expulsion from the island of his ancestors, and accompanied him on his return to resume his former government. He spared no efforts favourably to impress them in regard to Christianity; but to no purpose for a long time. When he offered himself for baptism, he stated that he had endeavoured to persuade Tamatoa, his father-in-law, and Tapoa, the king and principal chief of Raiatea, to renounce idolatry, and become the disciples of Jesus Christ; but they had assured him, whatever he might do, they would adhere to Oro. Others expressed the same determination; and Pomare came forward alone, requesting baptism, and desiring to hear and obey the word of God, as he said "he desired to be happy after death, and to be saved at the day of judgment." He did not confine his efforts to private conversation, but in public council urged upon Tamatoa and Mahine, the chiefs of Raiatea and Huahine, the adoption of the Christian religion; hereby publicly evincing his own determination to adhere to the choice he had made.

The Missionaries had every reason to believe the king was sincere in his desires to become a true follower of Christ; but as they then deemed only those who were true converts to Christianity, proper subjects for the Christian rite of baptism, and feared that his mind might not be sufficiently informed on the nature and design of that ordinance, and that he was rather an earnest inquirer after divine truth, than an actual possessor of its moral principle and spiritual influence, they proposed to him to defer his baptism until he had received more ample instruction. They were also desirous to receive additional evidence of his sincerity, and of the uprightness and the purity of his conduct, during a longer period than they had yet observed it. The king acquiesced in their proposal, and requested their instructions.

At the same time that the king thus publicly desired to profess Christianity, he proposed to erect a large and substantial building for the worship of the true God. His own affairs remained unsettled and discouraging; he was still an exile in Eimeo, and rumours of war not only prevailed in Tahiti, but invasion threatened Eimeo. This island the Missionaries considered only as a temporary residence, till they should be able to resume their labours in Tahiti, or establish a mission in the leeward islands, and therefore recommended him to defer it. But he replied, "No, let us not mind these things, let it be built."

Shortly after this important event, which may justly be considered as the dawning of that day, and the first ray of that light, which has since shed such lustre, and beamed with such splendour and power, upon these isles of the sea, two chiefs arrived from Tahiti, inviting Pomare to return and resume his government, promising an amicable adjustment of their differences. The interests of his kingdom appeared to require his concurrence with their proposal; and, on the thirteenth of August, in less than a month after the pleasing event referred to, he sailed with them from Eimeo, followed by the chiefs and people from the Leeward Islands, and most of the inhabitants of Papetoai and its vicinity. His departure, in this critical state of mind, was much to be regretted, as it deprived him of the instructions of his teachers, exposed him to many temptations, and much persecution.

Pomare, in infancy, had been rocked in the cradle of paganism, and trained under its influence through subsequent life. His father Pomare, and his mother Idia, were probably more infatuated with idolatry, and more uniformly attached to the idols, and every institution connected with their worship, than even the priests, or perhaps any other individuals in the islands. He had been early and often initiated in all the mysteries of falsehood and abomination connected with the system, and had engaged with avidity in the bloody and murderous rites of idol worship. In addition to this, he had been nurtured amid the debasing and polluting immorality, for which his country, ever since its discovery, had been distinguished; and although his ideas of the moral perfections of the true God might be but indistinct, and his views of the purity required in the gospel but partial, yet it might naturally be expected, that the convictions of guilt in such an individual, when first awakened to a sense of the nature and consequence of sin, would be deep and severe. That this was actually the case, appears from several letters which he wrote to the Missionaries soon after his arrival in Tahiti, as well as from the conversation they had with him on the subject.

In a letter, dated Tahiti, September 25, 1812, he thus expresses himself: "May the anger of Jehovah be appeased towards me, who am a wicked man, guilty of accumulated crimes,—of regardlessness and ignorance of the true God, and of an obstinate perseverance in wickedness! May Jehovah also pardon my foolishness, unbelief, and rejection of the truth! May Jehovah give me his good Spirit to sanctify my heart, that I may love what is good, and that I may be enabled to put away all my evil customs, and become one of his people, and be saved through Jesus Christ, our only Saviour! I am a wicked man, and my sins are great and accumulated. But O, that we may all be saved, through Jesus Christ." Referring to his illness about this time, he said, "My affliction is great; but if I can only obtain God's favour before I die, I shall count myself well. But, O! should I die with my sins unpardoned, it will be ill indeed with me. O! may my sins be pardoned, and my soul saved, through Jesus Christ! May Jehovah regard me before I die, and then I shall rejoice, because I have obtained the favour of Jehovah."

In another letter, written about a fortnight afterwards, he observes, "I continue to pray to God without ceasing. Regardless of other things, I am concerned only that my soul may be saved by Jesus Christ! It is my earnest desire, that I may become one of Jehovah's people; and that God may turn away his anger from me, which I deserve, for my wickedness, my ignorance of him, and my accumulated crimes!" In February, 1813, he wrote to the following effect. "The Almighty can (or will) make me good. I venture with my guilt (or evil deeds) to Jesus Christ, though I am not equalled in wickedness, not equalled in guilt, not equalled in obstinate disobedience, and rejection of the truth, hoping that this very wicked man may be saved by Jehovah, Jesus Christ."

Such was the interesting state of Pomare's mind, at the close of the year 1812, and the commencement of 1813. At the same time that this event shed such light upon the prospects of the Missionaries, other circumstances concurred, to confirm them in the conviction, that God was about to favour in a signal manner their enterprise, to follow their labours with his blessing, and with still greater success. Of one or two other natives they had every reason to hope most favourably, while one, who died about this time, left a pleasing testimony behind, of repentance, and reliance on the pardoning mercy of God.

The king's visit to Tahiti did not succeed so well as the messengers had promised, or his friends had anticipated: rumours of war prevailed in the western and southern parts of the island, and many of the chiefs sent professions of subjection; but the continuance of such acknowledgment was uncertain. Some of his ablest allies, especially Tapoa the chief of Raiatea, was removed by death, and the others prepared to return to their own islands. Early in the following year, the district of Matavai was surrendered to Pomare, but he was justly doubtful of the sincerity of the surrender. Amidst all these unfavourable circumstances, he continued bold and uncompromising in his renunciation of the idols, and every rite of idolatry; observing the sabbath, and, on every suitable occasion, exhibiting the truth and excellency of the religion of Jesus Christ. Although this honourable conduct produced a surprising effect upon the minds of many of the inhabitants of Tahiti and Eimeo, who considered the king better acquainted both with the religion of the natives, and that of the foreigners, than any other person in the islands; it procured him many enemies, and exposed him to no ordinary degree of ridicule and persecution, not only from his idolatrous rivals, but from his allies, and the members of his household and family. These attributed all his reverses to the respect he had shewn the Missionaries, and the inclination he had indulged towards their God; and declared that he need not expect his affairs to be retrieved, since he had forsaken the gods of his ancestors, and insulted those to whom his family was indebted for the elevated distinction to which it had been raised in Tahiti, and the neighbouring islands. Pomare, however, was uninfluenced by any of these representations, and, notwithstanding the embarrassed state of his affairs, and the uncertainty of the result, to which the present agitation, and the approaching national assembly of chiefs and people, might lead, and though his friends added insult and reproach to his misfortunes, he remained "steadfast and unmoveable."

The communications between Tahiti and Eimeo were now frequent, and the repeated accounts of Pomare's persevering and laudable endeavours to enlighten the minds of his subjects, were not the only cheering tidings they received. Mr. Bicknell went over in a vessel bound to the Pearl Islands, and in a few days returned, with the pleasing report that a spirit of inquiry had been awakened among some of the inhabitants of that island, that two of those they had formerly instructed, had occasionally met to pray to God. In order to ascertain the nature and extent of the anxiety which had been excited, and to confer with the individuals under its influence, Messrs. Scott and Hayward, having been deputed by their companions to visit Tahiti, sailed over from Eimeo, on the 15th of June, 1813. Although the king was residing in Matavai, they landed in the district of Pare, and proceeding to the valley of Hautaua, they learned that the report was correct, and that in the neighbourhood there were some who had renounced idolatry, and professed to believe in Jehovah, the true God.

On the following morning, according to the usual practice when travelling among the people, they retired to the bushes near their lodgings, for meditation and secret prayer. The houses of the natives, however large they might be, never contained more than one room; and were generally so crowded with people, that retirement was altogether unattainable. While seeking this, about the dawn of the day, on the morning after their arrival, Mr. Scott heard a voice at no great distance from his retreat. It was not a few detached sentences that were spoken, but a continued address; not in the lively tone of conversation, but solemn, as devotion; or pathetic, as the voice of lamentation and supplication.

A variety of feelings led him to approach the spot whence these sounds proceeded, in order to hear more distinctly. O, what hallowed music must have broke upon his listening ear, and what rapture must have thrilled his soul, when he distinctly recognized the voice of prayer, and heard a native, in the accents of his mother-tongue, with an a dour that proved his sincerity, addressing petitions and thanksgivings to the throne of mercy. It was the first time he knew that a native on Tahiti's shores had prayed to any but his idols; it was the first native voice in praise and prayer, that he had ever heard, and he listened almost entranced with the propriety and glowing language of devotion, then employed, until his feelings could be restrained no longer. Tears of joy started from his gladdened eye, and rolled in swift succession down his cheeks, while he could hardly forbear rushing to the spot, and clasping in his arms the unconscious author of his ecstacy. He stood transfixed as it were to the spot, till the native retired; when he bowed his knees, and, screened from human observation by the verdant shrubs, offered up, under the canopy of heaven, his grateful adoration to the Most High, under all the melting of soul, and the excitement of spirit, which the unprecedented, unexpected, though long-desired events of the morning had inspired. When the Missionaries met at the house in which they had lodged, the good tidings were communicated; the individual was sought out; and they were cheered with the simple yet affecting account he gave of what God had done for his own soul, and of the pleasing state of the minds of several of his countrymen.

His name was then *Oito,* though it is now Petero; he had formerly been an inmate of the Mission family at Matavai, and had received instructions there. He has since been a useful member of the community, and is still a consistent member of a Christian society; in which he has for some years sustained, with credit to himself and advantage to the church, the office of deacon. He had occasionally been with the king since his return to Tahiti, and some remarks from Pomare had awakened convictions of sin in his conscience. Anxious to obtain direction and relief, yet having no one to whom he could unburden his mind with hopes of suitable guidance, he applied to *Tuahine,* who had for a long time lived with the Missionaries; hence Oito inferred he would be able to direct his mind aright. Tuahine has since rendered the most important services to the Mission, in aiding Mr. Nott with the translations. When the Gospel by John, and the Acts of the Apostles, were finished, and Mr. Nott left Huahine, in July 1819, he removed to Raiatea, his native island, and has since been not only a useful member of society, and an ornament to the religion he professes, but an officer in the Christian church in Raiatea.

Tuahine's mind, on the subject of the Christian religion, was in a state resembling that of Oito's. Their conversation deepened their impressions; they frequently met afterwards for this purpose, and often retired to the privacy of the sequestered valleys or verdant shrubberies adjacent to their dwellings, for conversation and prayer. The singularity of their conduct, together with the report of the change in the sentiments of the king, soon attracted observation: many derided them, but several young men and boys attached themselves to Oito and Tuahine, and this little band, without any Missionary to teach them, or even before any one was acquainted with the circumstance, agreed to refrain from worshipping the idols—from the evil practices of their country—to observe the Sabbath-day,—and to worship Jehovah alone. They had established among themselves a prayer-meeting, which they held on the Sabbath, and often assembled at other times for social worship.

This intelligence was like life from the dead to the Missionaries; they thanked God, and took courage; but before commencing their journey round Tahiti, they wrote to their brethren in Eimeo an account of what they had seen and heard: declaring all that they had heard was true, that God had "also granted to the Gentiles repentance unto life," that some had cast away their idols, and were stretching out their hands in prayer to God, &c. The effect of their letter was scarcely less on the minds of the Missionaries in Eimeo, than the recital had been to themselves in Tahiti. They were deeply affected, even unto tears. I have often heard Mr. Nott speak, with evident indications of strong feeling, of the emotions with which this letter was read. And when we consider the long and cheerless years, which he and some of his associates had spent in fruitless, hopeless toil, on that unpromising field, the slightest prospect of an ultimate harvest, which these facts certainly warranted, was adapted to produce unusual and exalted joys,—emphatically a Missionary's own,—joys "that a stranger intermeddleth not with."

Messrs. Scott and Hayward made the tour of Tahiti, preaching to the people whenever they could collect a congregation, and then returned to Eimeo with Tuahine, Oito, and their companions,—who accompanied them, in order to attend the school, and receive more full instruction in those things, respecting which, though formerly so indifferent, they were now most anxious to be informed.

Tuahine was born in the island of Raiatea, but had been some time residing in the inland parts of the district of Pare. Oito was an inhabitant, if not a native, of Hautaua, and in this lovely, verdant, and sequestered valley, the first native meeting for prayer was held, and the first associated vows were paid to heaven.

I have often passed along the mouth or opening of this valley, and regret that I never had an opportunity of traversing its interior, and visiting the abode of Oito, or the sites of the rural oratories of the first Christians in Tahiti. Hautaua valley is an interesting spot, not only on account of the events connected with the early history of Christianity, which transpired within its borders, but also from the peculiarity of its scenery.

In the exterior, or border landscapes, of Tahiti and the other islands, there is a variety in the objects of natural beauty; a happy combination of land and water, of precipices and level plains, of trees, often hanging their branches clothed with thick dark foliage over the sea, and distant mountains shewn in sublime outline and richest hues; and the whole often blended in the harmony of nature, produces sensations of admiration and delight. The inland scenery is of a different character, but not less impressive. The landscapes are occasionally extensive, but more frequently circumscribed. There is, however, a startling boldness in the towering piles of basalt, often heaped in romantic confusion near the source or margin of some cool and crystal stream, that flows in silence at their base, or dashes over the rocky fragments that arrest its progress: and there is the wildness of romance about the deep and lonely glens, around which the mountains rise like the steep sides of a natural amphitheatre, till the clouds seem supported by them—this arrests the attention of the beholder, and for a time suspends his faculties in mute astonishment. There is also so much that is new in the character and growth of trees and flowers, irregular, spontaneous, and luxuriant in the vegetation, which is sustained by a prolific soil, and matured by the genial heat of a tropic clime, that it is adapted to produce an indescribable effect. Often, when, either alone, or attended by one or two companions, I have journeyed through some of the inland parts of the islands, such has been the effect of the scenery through which I have passed, and the unbroken stillness which has pervaded the whole, that imagination, unrestrained, might easily have induced the delusion, that we were walking on enchanted ground, or passing over fairy lands. It has at such seasons appeared as if we had been carried back to the primitive ages of the world, and beheld the face of the earth, as it was perhaps often exhibited, when the Creator's works were spread over it in all their endless variety, and all the vigour of exhaustless energy, and before population had extended or the genius and enterprise of man had altered, the aspect of its surface.

The valleys of Tahiti present some of the richest inland scenery that can be imagined. Those in the southern parts are remarkable for their beauty, but none more so than those of Hautaua, Matavai, and Apaiano. Those portions of them, in which the incipient effects of the advancement of civilization appear, are the most interesting; presenting the neat white plastered cottages in beautiful contrast with the picturesque appearance of the mountains, and the rich verdure of the plains.

The accompanying plate represents a scene in the valley of Matavai, near the bank of the river which flows through the district. It was taken on the spot by Capt. Elliot, who spent some time at Matavai, in the beginning of 1821. The rustic building by the side of the stream is a Missionary's cottage, and was at that time occupied by Mr. Nott. The surrounding scenery is delineated with accuracy and care; but the effect of the lofty mountain in the centre, which often appeared encircled with clouds, through which its romantic peaks sometimes penetrated, and of the rich purple hue that glowed on its sides, with other parts of the landscape, are such as to surpass the efforts of the graphic art.

Sketched by Capt. R. Elliot. R.N. Engraved by B. Winkles.

INTERIOR OF THE DISTRICT OF MATAVAI, IN TAHITI.

CHAP. VIII.

First record of the names of the professors of Christianity—Taaroarii's rejection of idolatrous ceremonies—Determination of Patii, the priest of Papetoai—Idols publicly burnt at Uaeva, in Eimeo—Increase of the scholars—Contempt and persecution on account of the profession of Christianity—Baneful influence of idolatry on social intercourse—Humiliating circumstances to which its institutes reduced the female sex—Happy change in domestic society, attending the introduction of Christianity—Persecution of the Christians—Worshippers of the true God sought as victims, for sacrifice to the pagan idols—Notice of Abrahama—Martyrdom in Tahiti.

SOON after the return of Messrs. Scott and Hayward from Tahiti, indications of the same convictions and inquiry were occasionally manifested in Eimeo; and on the 25th of July, 1813, which was the Sabbath, the first place for public worship erected in the island of Eimeo was opened. It was also the first building in the islands ever used by the natives for this sacred purpose. The exercises of the day were highly interesting both to the Missionaries and their little band of followers. At the close of the evening service Mr. Davies gave notice, according to previous arrangements, that on the following morning a public meeting would be held; when all who had sincerely renounced their false gods, who had desired also to relinquish their evil customs, to receive Jehovah for their God, and to be instructed in his word, were invited to attend. Forty natives came at the time appointed; the design of the meeting was explained by Mr. Nott. It was, to urge those who were undecided, and wished to become sincere disciples of Jesus Christ, to make their desires known—that the Missionaries might pay them special attention, and give them suitable instructions: they listened attentively, and many appeared deeply affected. They were afterwards individually interrogated as to their desires in reference to these important matters: during this inquiry thirty-one declared they had renounced the idols, their worship, and every practice connected with idolatry; wishing to abandon every thing contrary to the word of God. These thirty-one requested to have their names written down as those that desired to worship God, and to become disciples of Christ. Others said they intended to cast away their idols, but did not wish to have their names written down at that time. All who felt inclined to come were invited, but none were urged. The names of these thirty-one were written down; and among the first of them, Oito and Tuahine's were to be seen. In writing down the names of those who thus publicly professed Christianity, the Missionaries were influenced by a desire, not only to instruct them more fully, but to become personally acquainted with them, and to exercise over them a guardian care, which they could not do without knowing their names, places of abode, &c. To their number, eleven more were soon added; and with these they afterwards held frequent meetings, for the purpose of informing their minds, and encouraging them to faithfulness in their attachment to the Redeemer. Among the last number was Taaroarii, the young chief of Huahine and Sir Charles Sanders' Island, and Matapuupuu, a principal areoi, and chief priest of Huahine, who had long been one of the main pillars of idolatry in the island to which he belonged.

On the 28th of July, 1813, a number of areois visited Taaroarii's encampment at Teataebua, five miles from Papetoae, the Missionary settlement; prepared an entertainment, invited him to attend, and, before it commenced, were about to perform some heathen rites connected with the food they were to eat, and to deliver an oration, in which his rank, descent, and connexion with the gods by origin and family, and his future place among them, were to have been detailed. This, Taaroarii strictly prohibited; declaring that he intended no longer to acknowledge the gods of Tahiti, which were no gods; that no more ceremonies should be performed on his account, as he purposed to worship Jehovah. He was anxious to know more respecting God, and wished them also to hear about Him; and, therefore, sent a message to Mr. Nott, requesting him to come down and preach to the people at his place of abode.

Mr. Nott gladly complied with his request, and, accompanied by Mr. Hayward, repaired a few days afterwards to his encampment. When they arrived at Tiataibua, Puru, the king of Huahine, and the chief of Eimeo, received them very cordially: said his son Taaroarii wished to be instructed in the word of God, to learn about Jehovah and Jesus Christ, of whom he had so frequently heard Pomare speak. The chief added, that although he had no desire after these things himself, he did not wish to oppose his son, or prevent his hearing whatever Mr. Nott might have to communicate. The finger of the Almighty was strikingly exhibited in the door thus effectually opened for the preaching of the gospel.

Puru and his adherents had not been much with the Missionaries. The people of Huahine and their chief were certainly among the most superstitious and idolatrous tribes of the Pacific. Pomare, and not the Missionary, had on this occasion been employed as the agent, under God, in influencing the mind of the young chief, who was likely to become the king of Huahine and Eimeo, and in a way which at once demonstrated that it was the purpose of God that he should be made acquainted with divine truth. Hence he was induced to prohibit an acknowledgment to the gods of his ancestors, and to invite the messengers of salvation to his camp, to speak unto him and his adherents words whereby they might be saved. While the Missionaries admired the means by which God had thus shewn them that the work was His, and not theirs, and thus deprived them of attributing any thing to their own influence, they rejoiced in the opportunity now afforded of proclaiming the tidings of mercy from the most High. Mr. Nott conversed a long time with them, and preached an instructive and affecting discourse from Isa. xlix. 7. I have often heard the young man's mother-in-law, and other members of the household, speak of this discourse as having deeply impressed their minds. When Mr. Nott left them, he invited the chief and his adherents to visit the station on the Sabbath, and cultivate an intercourse with other Christian chiefs.

On the following Sabbath, Taaroarii attended; his father also became, a few months afterwards, a sincere convert. They accompanied us to Huahine in 1818. Taaroarii died rather suddenly in 1821. His father is the venerable king of Huahine; and has, ever since his return, proved not only a father to the people, but a uniform and bright ornament to the religion of the Cross.

Besides these regular periods of instruction and times of public worship, the Missionaries frequently held special meetings with those whose names they had written down, for the purpose of unfolding more fully the sublime doctrines of revelation, and uniting with them in social worship. They had the delightful satisfaction of hearing some of the new converts engage in prayer, and were surprised and gratified, in a high degree, with their fluency and fervour, as well as the appropriateness of their language, when engaged in this sacred duty. They also learned with pleasure, that they were accustomed to retire morning and evening for secret prayer.

In one of the visits which Mr. Nott made to the residence of Taaroarii, for the purpose of preaching to his people, he was followed by Patii, the priest of the temple in Papetoai, the district in which the Missionaries resided. This individual appeared to listen most attentively to what was said; and after the conclusion of the service, he and Mr. Nott proceeded together along the beach towards the settlement. As they walked, Patii fully disclosed the feelings of his mind to Mr. Nott, and assured him that on the morrow, at a certain hour, he would bring out the idols under his care, and publicly burn them. The declaration was astounding; it was too decisive and important in its nature, and promised results almost too momentous to be true. Mr. Nott replied, "I fear you are jesting with me, and stating what you think we wish, rather than what you intend. I can scarcely allow myself to believe what you say." "Don't be unbelieving," replied

Patii, "wait till to-morrow, and you shall see." The religion of Jesus Christ was the topic of conversation until they reached the settlement; when Patii took his leave, and Mr. Nott informed his colleagues of the success of his visit to the young chief of Huahine, and the determination which the priest of the district had made known to him. The impression which the intelligence of these events produced upon their minds, was that of mingled admiration, gratitude, and hope, to a degree that may be better imagined than expressed.

The arrival of the evening of the following day was awaited with an unusual agitation and excitement of feeling. Hope and fear alternately pervaded the minds of the Missionaries and their pupils, with regard to the burning of the idols, and the consequent tumult, devastation, and bloodshed that might follow. The adherents of Christianity were but few, (less than fifty,) and surrounded by jealous and cruel idolaters—who already began to wonder "whereunto this thing might grow." Patii, however, was punctual to his word. He, with his friends, had collected a quantity of fuel near the sea-beach; and, in the afternoon, the wood was split, and piled on a point of land in the western part of Papetoai, near the large national Marae, or temple, in which he had officiated. The report of his intention had spread among the people of the district, and multitudes assembled to witness this daring act of impiety, or the sudden vengeance which they expected would fall upon the sacrilegious criminal. The Missionaries and their friends also attended. The varied emotions of hope and fear, of dread and expectation, with a strange air of mysterious foreboding, agitating the bosoms of the multitude, were strongly marked in the countenances of the spectators; resembling, perhaps in no small degree, the feeling depicted in the visages of the assembled Israelites, when the prophet Elijah summoned them to prove the power of Baal, or to acknowledge the omnipotence of the Lord God of Israel. A short time before sun-set, Patii appeared, and ordered his attendants to apply fire to the pile. This being done, he hastened to the sacred depository of his gods, brought them out, not indeed as he had been on some occasions accustomed to do, that they might receive the blind homage of the waiting populace,—but to convince the deluded multitude of the impotency and the vanity of the objects of their adoration and their dread. When he approached the burning pile, he laid them down on the ground. They were small carved wooden images, rude imitations of the human figure; or shapeless logs of wood, covered with finely braided and curiously wrought cinet of cocoa-nut fibres, and ornamented with red feathers. The accompanying representations will convey some idea of the shape and appearance of the former kind.

Patii tore off the sacred cloth in which they were enveloped, to be safe from the gaze of vulgar eyes; stripped them of their ornaments, which he cast into the fire; and then one by one threw the idols themselves into the crackling flames—sometimes pronouncing the name and pedigree of the idol, and expressing his own regret at having worshipped it—at others, calling upon the spectators to behold their inability even to help themselves. Thus were the idols which Patii, who was a powerful priest in Eimeo, had worshipped, publicly destroyed. The flames became extinct, and the sun, which had never before shed his rays upon such a scene in those islands, cast his last beams, as he sunk behind the western wave, upon the expiring embers of that fire, which had already mingled with the earth upon which it had been kindled—the ashes of the once obeyed and dreaded idols of Eimeo.

Patii on this occasion was not prompted by a spirit of daring bravado, but by the conviction of truth, deeply impressed upon his heart, and a desire to undeceive his deluded countrymen; probably considering, that as his conduct and instruction had heretofore done much to extend and propagate the influence of idolatry, so his thus publicly abandoning it, and exposing himself to all the consequences of their dreaded ire, would most effectually weaken their confidence in the gods, and lead them to desire instruction concerning that Being, who, he was convinced, was the only living and true God,—who was a spirit, and was to be worshipped, not with human or other sacrifices, save those of a broken heart and a contrite spirit, or the sacrifices of thanksgiving and of praise.

Although many of the spectators undoubtedly viewed Patii with feelings analogous to those with which the Melitians viewed the apostle Paul when the viper fastened on his hand, and were, many of them, evidently disappointed when they saw no evil befall him, they did not attempt to rescue the gods, when insulted, and perhaps riven by the axe, or stripped to be cast into the flames. No tumult followed, and no one came forward to revenge the insult offered to the tutelar deities of their country. Probably, Gamaliel-like, they thought it best not to interfere at that time, as their belief in the power of the gods had hitherto remained unshaken, and they doubtless expected that, in their own way, the gods would take signal vengeance on those by whom, in the sight of the nation, they had been thus dishonoured.

The watchful providence of God, over His infant cause in these islands, was remarkably conspicuous in preserving Patii and his friends, and allowing them, after the events of the evening, safely and peacefully to return. There were many present, who were indignant at the insult, and filled with rage at the impiety of the act, as well as convinced, that if this conduct should be imitated by others, not only would their craft and their emoluments be endangered, but they would no longer be able to exercise that unquestioned influence over the people, to which they had hitherto been accustomed; nor to indulge their base propensities, and live in that luxurious ease they then enjoyed. Had any popular tumult followed this heroic act, the idolaters were so numerous and powerful, and the Christians so weak, that their destruction would have been inevitable; and even the lives of the Missionaries, who would have been considered as the cause of all the disturbances, might not have been secure. God, however, preserved them, and they returned, to render to him the thanks and the glory due unto his name.

The conduct of Patii, when it became more extensively known, produced the most decisive effects on the priests and people. Numbers in Tahiti and Eimeo were emboldened, by his example—not only in burning their idols, but demolishing their maraes or temples; their altars were also stripped and overthrown, and the wood employed in their construction converted into fuel, and used in the native kitchens.

Patii became the pupil of the Missionaries, and a constant worshipper of the true God, persevering amidst much ridicule and persecution. Whether his mind had at this time undergone a divine and decisive change, it is not necessary now to inquire; every evidence that could be required, has since been given, of the sincerity of his profession of Christianity, and the influence of its principles on his heart. His conduct, from this period, has been uniformly moral and upright, his mind humble, his disposition affectionate and mild, and his habits of life reformed and industrious. The influence of his character in Papetoai, where he is best known, has occasioned his election to an important office in the Christian church. He is a valuable and steady friend, and an assistant, in whom the Missionaries can repose confidence. Although not a chief of the highest rank, he had been raised by the king and people to the office of a magistrate, in his own district. His conduct on the above occasion gave idolatry a stab more deadly than any which it had before received, and inflicted a wound, from which, with all the energy subsequently manifested, it never could recover.

In the month of March, 1814, Mr. Nott, accompanied by Mr. Hayward, visited Huahine, Raiatea, and Tahaa, the principal of the Society Islands, conversing with the inhabitants, travelling round the islands, and preaching to the people wherever it was convenient. In every place they were welcomed and entertained with hospitality. The inhabitants frequently assembled to hear their instructions, as soon as they knew of their arrival in a district or village; whereas, on every former occasion, it had required much time and labour, by personal application, to assemble the smallest congregations. Many appeared to listen with earnestness and satisfaction to the message they delivered, called God the good spirit, and scrupled not to designate their own gods as *varua maamaa,* and *varua ino,* foolish spirits, and evil spirits.

In the autumn of the same year, Mr. Wilson went on board a vessel at Eimeo, which was driven to the leeward islands, where contrary winds detained him and his companions for three months. During this period he was much among the people, preached to attentive congregations on the Sabbath and other days, and was happy to find that those whose names had been written down at Tahiti continued steadfast. He also added to their number thirty-nine others, whose names, at their own desire, were recorded as the professed worshippers of the true God. When he left them, they expressed the deepest regret, and requested that one of the Missionaries would come and reside among them.

Before Mr. Nott visited the Society Islands, he finished the translation of the Gospel of Luke; and, in the course of the same year, the Missionaries sent a copy of their catechism to New South Wales, to be printed there. They were exceedingly anxious to obtain a supply of elementary books, as the spelling books from England were expended, and the desire for instruction had increased to such a degree, that upwards of two hundred scholars attended their school at Papetoai.

About this time, several of the chiefs of the Society Islands, and many of their adherents, who had come up in 1811 to assist Pomare in the recovery of his government and authority in Tahiti, returned to their own dominions; not, however, without most earnestly requesting the Missionaries to send them teachers and books.

Tamatoa and his brother, with other chiefs, had been residing for some time at the Missionary station in Eimeo, they had attended the school and public instruction in the place of worship; and several, among the most promising, of whom was *Paumoana*, at present a valuable native Missionary in the Harvey islands, appeared to be under the decisive influence of Christian principle.

After an absence of two years, during which he had resided in Tahiti, vainly expecting the restoration of his government, and endeavouring to recover his authority in his hereditary dominions, Pomare returned to Eimeo in the autumn of 1814, accompanied by a large train of adherents and dependants, all professors, at least, of Christianity. These regularly attended the school, and increased the congregation to such a degree, that it was necessary to enlarge the place of worship. The king had been unable to withstand the temptation with which he had been assailed at Tahiti, to use ardent spirits; and although not addicted to entire intoxication, yet it induced the Missionaries to fear that he, like Agrippa, was but almost a Christian. They could not but indulge unfavourable apprehensions on his account; yet, considering his previous habits, that intemperance had ever been the vice to which he was most addicted, and the peculiar temptations to which his residence in Tahiti had exposed him, they could not readily relinquish the hopes they had entertained respecting him.

The numerous attendance and increasing earnestness of the people, induced the Missionaries to meet them for Divine worship twice on the Lord's day, and once during the week. In addition to these public instructions, they held a meeting every Sabbath evening with those whose names had been written down as the disciples of Christ, and spent much time in more private endeavours to direct the views, and confirm the belief, of those who were desirous to be added to their number. These sacred exercises were enlivened by the natives, who united with their teachers in celebrating the praises of Jehovah, a number of the natives having been taught to sing hymns that had been composed in the native language. The Missionaries had often, with mingled feelings of horror and pity, heard their songs of licentiousness or of war, as well as the cantillations of their heathen worship, and their songs in honour of their idols; and it is hardly possible to form an adequate idea of the delightful transport with which, at first, they must have heard the high praises of the Almighty ascend from native voices.

Upaparu, a principal chief in the eastern part of Tahiti, came over to Eimeo for the express purpose of seeking Christian instruction, and attending the assemblies for public worship. He was accompanied by twelve of his people, equally anxious with himself, and his wife, Maihota, to know more respecting these important matters. On the 15th of April they reached the Missionary station. The following day was the Sabbath. They attended public worship in the forenoon; and when they saw the congregation standing up, and heard them sing the praises of Jehovah in their native tongue, they were for some time mute with astonishment, and some of them so deeply affected, as to be unable to refrain from tears. An excellent discourse was afterwards delivered by Mr. Scott, to which they listened with mingled feelings of wonder and delight.

A variety of events occurred at this time, to confirm the attachment of those who had professed themselves favourable to Christianity, and to induce those who were undecided to join them. On one occasion, a family in Eimeo were plunged into great distress, on account of the sufferings of one of its members, and the prospect of a fatal issue. A priest was sent for, who implored the assistance of his god; but, continuing his intercession for a long time, without any apparent relief to the sufferer, he deserted, and left the family in hopeless disappointment. A native, who was a worshipper of Jehovah, was among the attending friends. He kneeled down, and offered up a fervent prayer to the true God. While he was thus engaged, relief was afforded, and the weeping and forebodings of the family turned into grateful wonder, and joyous gratulations.* I simply

* In recording this incident, it is proper to state, that the Missionaries disclaim all idea of *miraculous* interposition. At the same time, the providential coincidence of the events, and the encouragement which the word of God gives to "fervent and effectual prayer," demand attentive consideration, and grateful acknowledgment.—Psalm cvii. 43.

state the fact, as it is recorded by the Missionary in the island at the time, without making any comment; which, indeed, it neither requires nor admits. On the minds of the family, and the inhabitants of the place, it produced a powerful impression. They hastened to the idol temple of the district, which they demolished, breaking down the altars, and bringing forth their gods, which they execrated as false, and publicly committed to the flames.

A similar instance occurred early in this year. One of the scholars, the wife of an areoi, who had for some time, with her husband's consent, attended the school, was suddenly taken ill. The members of the family were alarmed; and, accustomed to attribute every calamity to the anger of the gods, immediately concluded that her illness was occasioned by their displeasure, which she had probably incurred by attending the school and the Christian worship of the Missionaries. *Patii*, the priest of the district, was instantly sent for. On his arrival, a small pig and a young plantain were procured, and handed to Patii; who, in offering them to his god, thus addressed him: *O Satani! eiaha oe e riri, faaora, faaora, Teie te hapa, ua faarue ia oe, ua haavarehia e te papaa, Teie te buaa, eiaha e riri;* "O Satan! be not angry, restore, restore; this is the sin, deceived by the foreigners (she) has forsaken you. Here is a pig (as an atonement,) be not angry." In this address it is singular to notice the application of the term Satan to the god Patii invoked. It was introduced by the Missionaries, and at this time adopted by the Christians, when speaking of any of the idols of Tahiti. Although dangerously ill at the time these efforts were made, the woman recovered; and, notwithstanding all the fearful representation of consequences, made by her friends, attended the school again, so soon as her strength admitted. Her infatuation, as they conceived it to be in this respect, not only encouraged her school-fellows, but, with other circumstances which occurred about the same time, made a considerable impression on the minds of the idolaters, and occasioned some of the priests publicly to declare their firm conviction "that the religion of the foreigners would prevail, *in spite of all opposition.*

The progress of Divine truth was so rapid among the natives, that, in the close of 1814, not fewer than 300 hearers regularly attended the preaching of the gospel. Upwards of 200 had given in their names, as professors of Christianity. Three hundred scholars attended the means of instruction in Eimeo; besides which, there were a number in Sir Charles Sander's Island, Huahine, and Raiatea; so that, at this time there is reason to believe that between five and six hundred had renounced idol-worship.

These encouraging appearances, in regard to the affairs of the new converts, only appeared to arouse the anger of their idolatrous enemies, who were no longer satisfied with simply ridiculing, and treating with contempt, the objects of their hatred, but proceeded to more alarming plans of resistance against the progress of those new principles which were daily gaining ground among the people. It was by no means an uncontested triumph, nor an undisputed possession, that Christianity acquired in those islands; every inch was reluctantly surrendered; and, at several periods, persecution raged, amid the Elysian bowers of Tahiti and Eimeo, as much as ever it had done in the valleys of Piedmont, or the metropolis of the Roman empire. Many, in Tahiti especially, were plundered of their property, banished from their homes and their possessions, their houses were burnt, and they themselves hunted for sacrifices to be offered to Oro, merely because they were *Bure Atua* prayers to God. In some places, the persecutions were so inveterate as to produce remonstrances, even from several of the inferior chiefs, who were themselves idolaters.

The commencement of the year 1815 is distinguished, in the annals of Tahiti, by changes in society, affecting deeply, not only the religious, but the domestic condition of the people, especially of the females. Idolatry had exerted all its withering and deadly influence, not only over every moment of their earthly existence, but every department of life, destroying, by its debasing and unsocial dictates, every tender feeling, and all the enjoyments of domestic intercourse. The father and the mother, with their children, never, as one social happy band, surrounded the domestic hearth, or, assembling under the grateful shade of the verdant grove, partook together, as a family, of the bounties of Providence. The nameless but delightful emotions, experienced on such occasions, were unknown to them, and all that we are accustomed to distinguish by the endearing appellation of domestic happiness. The institutes of Oro and Tane inexorably required, not only that the wife should not eat those kinds of food of which the husband partook, but that she should not eat in the same place, or prepare her food at the same fire. This restriction applied not only to the wife, with regard to her husband, but to all the individuals of the female sex, from their birth to the day of their death. In sickness or pain, or whatever other circumstances, the mother, the wife, the sister, or the daughter, might be brought into, it was never relaxed. The men, especially those who occasionally attended on the services of idol worship in the temple, were considered *ra*, or sacred; while the female sex, altogether, was considered *noa*, or common: the men were allowed to eat the flesh of the pig, and of fowls, and a variety of fish, cocoa-nuts, and plantains, and whatever was presented as an offering to the gods, which the females, on pain of death, were forbidden to touch; as it was supposed, they would pollute them. The fires at which the men's food was cooked, were also sacred, and were forbidden to be used by the females. The baskets in which their provision was kept, and the house in which the men ate, were also sacred, and prohibited to the females under the same cruel penalty. Hence the inferior food, both for wives, daughters, &c. was cooked at separate fires, deposited in distinct baskets, and eaten in lonely solitude by the females, in little huts erected for the purpose.

The most offensive and frequent imprecations which the men were accustomed to use towards each other, referred also to this degraded condition of the females. *E taha miti noa oe no to medua*, Mayest thou become a bottle, to hold salt water for thy mother; or another, Mayest thou be baked as food for thy mother; were imprecations they were accustomed to denounce upon each other: or, Take out your eye-ball, and give it to your mother to eat.

To this cheerless and debasing distinction, the female sex had been for ages subject, from the direct injunctions of their false system of religion; and as its cumbrous fabric began to give way, this barbarous and arbitrary requisition was proportionably disregarded.

Not only were the sacred materials with which the altars, and the apendages of the temple, had been constructed, converted into fuel; but the food, considered sacred, was esteemed so no longer, the invidious and debasing distinctions attached to the females were removed, and both sexes, among those who professed Christianity, sat down together to their cheerful meal.

Under the influence of these encouraging prospects, although enfeebled by frequent indisposition, the Missionaries prosecuted their work; their scholars increased in the same degree that the profession of Christianity prevailed, and a supply of four hundred copies of their abridgment of the New Testament, and a thousand copies of small elementary books, which had been printed in New South Wales, arrived very opportunely about this time; spelling books they were still much in want of, as those formerly printed in England had long been expended.

Such was the pleasing state of things in the commencement of 1815. The importance and advantages of education appeared to be more extensively appreciated, and between forty and fifty, principally adults, regularly attended the Mission school. The agents of vice, idolatry, and cruelty, were not inactive. The struggle between light and darkness, truth and error, order and anarchy, benevolence and barbarism, had never appeared more intense and conspicuous, than at this time. The little band of scholars in the Mission school, and worshippers in the chapel, unwilling to enjoy their privileges alone, employed every proper and persuasive means to induce their friends and relatives to attend to these things; at least to make a trial of the school, and to hear what was said about the true God. The latter, however, frequently became indignant at the very proposal, charging the God of the foreigners with all the maladies under which they suffered, and the disturbances that agitated the country; accusing them also of bringing down the vengeance of their own gods upon the family, by deserting their altars, and worshipping with the strangers. Frequently, however, they answered their entreaties only with ridicule and scorn, tauntingly inquiring, Where is the good of which you speak so much—the salvation of which you tell us? the foreigners themselves die, their pupils die, or suffer the same pain that we do; and what good have you derived from going to their schools? Let us see—if you go this week, and bring home a good bundle of cloth, or scissors, or knives, or any thing else worth having, then we will go too; if not, we will have nothing to do with such profitless work. The state of things resembled greatly that described by the Saviour, when speaking of the results that should follow the promulgation of his gospel. In many a family, the husband was an idolater, and the wife a Christian,—or the reverse; the parents addicted to the gods of their ancestors, and the child a disciple of Jesus Christ; and many a wife was beaten by her husband, and many a child driven from the parental roof, solely on account of their attachment to the new religion. In Tahiti, the idolaters proceeded to the greatest acts of lawless violence and horrid murder.

More than once, individuals were selected to be offered in sacrifice to the gods, only because they were Christians. Mr. Davies, in his journey round Tahiti, in 1816, met with the murderer of the young man who was offered in sacrifice by the people of Taiarabu, to insure success in their last attack upon the people of Atehuru and Papara, and whose tragical death he justly considered, ought to be recorded, because it is hoped it was "the last human sacrifice offered in Tahiti," and because the victim was selected "on account of his attachment to Christianity."

Aberahama, an interesting and intelligent young man, who was a pupil in our school at Eimeo, was marked out as a victim; and, when the servants of the priests came to take him, being obliged to fly for his life, he was pursued by the murderers, shot at, wounded, and but narrowly escaped. When he received the ball, he fell, and, unable to save himself by flight, crawled among the bushes, and hid himself so completely, as to elude the vigilant search of his enemies, although it was continued for some time, and they often passed near his retreat. Under cover of the darkness of night, he crept down to the dwelling of his friends, who dressed his wound, and conveyed him to a place of safety. But, although he recovered from the shot, and lives, not only to enjoy the blessings of the gospel in this world, and to be useful in imparting its benefits to others, he will, to adopt the language of Mr. Davies, "carry the honourable scar to his grave."

An immolation, equally affecting, was related to me by Mr. Nott. A fine, intelligent young man, on becoming a disciple of Christ, and a public worshipper of Jehovah, was ridiculed by his family; this proving ineffectual, flattering promises were made of temporal advantages, if he would again unite with those who had been his former associates in idol worship; these he also declined. He then was threatened with all their weight of vengeance; and still remaining firm to his determination, he was banished from his father's house, and forced to leave his home. Not satisfied with this, that rage and malignant hatred of Christianity, which is gendered by ignorance and idolatry, and cherished by satanic infatuation, pursued him still. A heathen ceremony was at hand, for which a human victim was required, and this young man was selected by his persecutors, because he professed to be a worshipper of the true God. A more acceptable sacrifice they thought they could not offer, as the revenge they should thereby wreak upon him, they conceived would not only gratify their own insatiate malice, but be so acceptable to the gods whom he had rejected, as certainly to render them propitious. On the evening of the day preceding that on which the ceremony was to take place, the young man, as his custom was, had retired to the brow of a hill that overlooked the valley where he dwelt; and there, seated beneath the embowering shade of an elegantly growing clump of trees, was absorbed in meditation, previous to offering up his evening supplications to his God. While thus engaged, his seclusion was invaded, and his solitude disturbed, by the appearance of a band similar, in some respects, to that which broke in upon the Saviour's retirement in Gethsemane. A number of the servants of the priests and chiefs approached the young man, and told him that the king had arrived, and, wishing to see him, had sent them to invite him down. He knew of the approaching ceremony,—that a human sacrifice was then to be offered, —and he no sooner saw them advancing to his retreat, than a sudden thought, like a flash of lightning, darted through his mind, intimating that he was to be the victim. He received it as a premonition of his doom·

and, in reply to the request, told them, calmly, that he did not think the king had arrived, and that, therefore, it was unnecessary for him to go down. They then told him that the priest, or some of his friends, wished to see him, and again invited him to descend. "Why," said he, "do you thus seek to deceive me? The priest, or friends, may wish to see me, but it is under very different circumstances from what your message would imply: I know a ceremony approaches, that a human victim is then to be offered—something within tells me *I am to be that victim*, and your appearance and your message confirms my conviction. Jesus Christ is my keeper, without his permission you cannot harm me; you may be permitted to kill my body, but, *I am not afraid to die!* My soul you cannot hurt; that is safe in the hands of Jesus Christ, by whom it will be kept beyond your power." Perceiving there was but little prospect of inducing him, by falsehood, to accompany them towards the beach, and irritated, probably, by his heroical reply, they rushed upon him, wounded, and murdered him, and then, in a long basket made with the leaves of the overshadowing cocoa-nut tree, bore his body to the temple, where, with exultation, it was offered in sacrifice to their god. They had, perhaps, beheld, with fiend-like joy, his writhing agonies in death, and listened, with equal delight, to his expiring groans. The unconscious earth had been saturated with his blood; and, when they placed his body on the rude altar, or suspended it from the sacred tree, in the presence of their god, they not only supposed they offered a sacrifice, at once acceptable and efficacious, but, doubtless, viewed the immolation as one by which they had achieved for idolatry a triumph over humanity and Christian principle. Before, however, these feelings could be exercised, and the earth had drank up his blood, or his insulted corpse was deposited on their altar, his liberated and ransomed spirit had winged its way to the realms of blessedness, received the welcome greeting of his Saviour, and, invested with the robes of victory, the palm of triumph, and the crown of glory, had joined "the noble army of martyrs;" and united in ascriptions of grateful homage unto Him who had loved him, and not only made him faithful to the end, but triumphant over death. Those who heard the young man's dying words, and witnessed his calm unshaken firmness in the moment of trial, with many, among whom the report circulated, were probably led to think differently of the religion he professed, than they had done before. The blood of the martyrs has ever been the seed of the church; and, from an exhibition of principles so unequivocal in their nature, and so happy in their effects, it is not too much to presume that it proved so on the present occasion.

CHAP. IX.

Distillation of ardent spirits—Description of a native still—Materials employed in distillation—Murderous effects of intoxication—Seizure of the Queen Charlotte—Murder of the officers—Escape of Mr. Shelly —Seizure of the Daphne—Massacre of the captain and part of the crew —Upaparu removes to Eimeo—First Christians denominated BURE ATUA—Public triumph over idolatry in Eimeo—Visit of the Queen and her sister to Tahiti—Emblems of the gods committed to the flames—Account of Farefare—Projected assassination of the Bure Atua—Manner of their escape—War in Tahiti—Pomare's tour of Eimeo.

INTEMPERANCE at this time prevailed to an awful and unprecedented degree. By the Sandwich Islanders, who had arrived some years before, the natives had been taught to distil ardent spirits from the saccharine *ti* root, which they now practised to a great extent, and exhibited, in a proportionate degree, all the demoralizing and debasing influence of drunkenness.

Whole districts frequently united, to erect what might be termed a public still. It was a rude, unsightly machine, yet it answered but too well the purpose for which it was made. It generally consisted of a large fragment of rock, hollowed in a rough manner, and fixed firmly upon a solid pile of stones, leaving a space underneath for a fire-place. The but-end of a large tree was then hollowed out, and placed upon the rough stone boiler for a cap. The baked *ti* root, *Dracanæ terminalis*, macerated in water, and already in a state of fermentation, was then put into the hollow stone, and covered with the unwieldy cap. The fire was kindled underneath; a hole was made in the wooden cap of the still, into which a long, small, bamboo cane, placed in a trough of cold water, was inserted at one end, and, when the process of distillation was commenced, the spirit flowed from the other into a calabash, cocoa-nut shell, or other vessel, placed underneath to receive it.

Tahitian Still.

When the materials were prepared, the men and boys of the district assembled in a kind of temporary house, erected over the still, in order to drink the *ava*, as they called the spirit. The first that issued from the still being the strongest, they called the *ao;* it was carefully received, and given to the chief; that subsequently procured, was drunk by the people in general. In this employment they were sometimes engaged for several days together, drinking the spirit as it issued from the still, sinking into a state of indescribable wretchedness, and often practising the most ferocious barbarities.

Travellers among the natives experienced greater inconvenience from these district stills than from any other cause, for when the people were either preparing one, or engaged in drinking, it was impossible to obtain either their attention, or the common offices of hospitality. Under the unrestrained influence of their intoxicating draught, in their appearance and actions they resembled demons more than human beings.

Sometimes, in a deserted still-house might be seen the fragments of the rude boiler, and the other appendages of the still, scattered in confusion on the ground; and among them the dead and mangled bodies of those who had been murdered with axes or billets of wood in the quarrels that had terminated their dissipation.

It was not only among themselves that their unbridled passions led to such enormities. One or two European vessels were seized, and the crews inhumanly murdered. The first was the Queen Charlotte, of Port Jackson, the vessel by which we arrived in the islands.

Towards the autumn of 1813, Mr. Shelly, formerly a Missionary in Tongatabu, and subsequently in Matavai, arrived as master of the Queen Charlotte, at Eimeo, on his way to the Paumotu, or Pearl Islands. These lie to the eastward of Tahiti, and form what is denominated the Dangerous Archipelago. The vessel was but imperfectly manned, and a number of natives, of Raiatea and Tahiti, were taken on board, to dive among the lagoon islands for the pearl oyster. They proceeded to their destination, but had scarcely commenced their pearl-fishing, when the natives attacked the crew, barbarously murdered the first and second officers, who were men of fine stature and benevolent dispositions; and killing one of the seamen, took possession of the ship. Mr. Shelly's life was threatened, and only spared at the instance of two Tahitians, who, anxious to save him, requested that he might be kept, to navigate the vessel to Tahiti, whither they intended to return. One of these natives was Upaparu, a chief of rank, present secretary to the government of Tahiti, and a steady friend to foreigners. When the vessel arrived at Tahiti, Pomare succeeded in securing to Mr. Shelly its restoration, though most of the property had been plundered. Matting was procured for sails, and the vessel reached Port Jackson in safety.

Flushed with the success that had attended the savage and daring effort of the Raiateans, the Tahitians, whom Captain Fodger had employed on board his vessel, the Daphne, for the purpose of diving among the pearl islands, rose upon the ship's company, murdered the captain and some of the men, took possession of the vessel, and brought her to Tahiti. Mr. G. Bicknell, a nephew of Mr. Bicknell, was on board at the time, but his life was spared, amidst the general carnage that attended the assault. The mutinous natives returned to their own island, but were met as they were about to enter the harbour at Tahiti, by Captain Walker of the Endeavour, who succeeded in retaking the vessel, and thus deprived them of their plunder.

These acts of daring outrage and appalling crime, on the one side, and of increasing and decided attachment to the principles of order, humanity, and religion, on the other, seemed to indicate that matters in Tahiti were fast verging to an important issue, and that, before long, some violent convulsion in society must follow. The Missionaries could not view these things with insensibility, as they saw what they had to expect, should they fall into the hands of those who had been guilty of such wanton cruelty; their support was, however, derived from the conviction, that their God was governor among the nations, and that the Lord omnipotent reigned.

Towards the close of the year 1813, one of the early scholars departed to the world of spirits, under the consolation that pure religion imparts in the hour of death. He was often heard to say, while confined to his couch, when he saw his former companions going to the school, or the place of worship, "My feet cannot follow, but my heart goes with you." He did not pretend to know much, but he knew that he was a sinner, and that Jesus Christ came into the world to save sinners, and this knowledge removed from his mind the fear of death.

Early in the same year, the number of pupils, and of those who professed Christianity, considerably increased in Eimeo, and favourable intelligence continued to arrive from the adjacent island.

The report of the increase of the Christians, and their advancement in knowledge, &c. had already circulated throughout Tahiti; the minds of many were unsettled, and numbers were halting between two opinions. Upaparu, a chief of rank and influence in the eastern part of Tahiti, with his wife, and twelve or thirteen of his people, came over to Eimeo, in order to receive instruction. The inhabitants of the Leeward Islands, whose encampment he passed when on his way to Papetoai, strongly persuaded him to join their party, and carry the flag of the gods to Raiatea, entreating him to adhere to the religion of his fathers, and to beware of *Matupuupuu,* a man of influence, an areoi, and a high-priest, from Huahine, who had recently joined the Christian converts, and Utami, a well-informed and enterprising man, chief in the island of Tahaa, who, with his wife, had also attached himself to their number.

Fifty had now given in their names, as having renounced idolatry, desiring to acknowledge Jehovah alone as the true God, and to be instructed in the obedience his word required. Others attended in such numbers, that it was found necessary to enlarge the first place of worship they had ever used in the islands. The converts were punctual and regular in their observance of the outward ordinances of religion, in frequent social meetings for prayer, and seasons of retirement for private devotion. Their whole moral conduct seemed changed; the things they once delighted in, they now abhorred, and found enjoyment in what had formerly been a source of ridicule or aversion. Their habit of invariably asking a blessing, and returning thanks after meals, and their frequent attention to prayer, attracted the notice of their countrymen, and procured for them, as a term of reproach from their enemies, the designation of *Bure Atua,* literally Prayers to God; from *Bure,* to pray, and *Atua,* God; the meaning of which was, the people who prayed to God, or the praying people. Bure Atua is a designation in no respect dishonourable to those to whom it was applied, and of which they have never been ashamed, though considered as an epithet of contempt or opprobrium, and applied in a manner similar to that in which the term Saint or Methodist is used in the present day, or the designation of Nazarene or Christian was given to the first disciples. Since the profession of Christianity has become general, it has been much less used than formerly. *Haapii parau*, learners, or brethren, friends, and disciples, are the terms most frequently employed by the converts themselves.

In the close of 1814, Pomare-vahine, the daughter of the king of Raiatea, and the sister of Pomare's queen, paid a visit to Eimeo, from the Leeward Islands, and in the month of May, 1815, made a voyage to Tahiti, in company with her sister the queen, with a numerous train of companions and attendants, most of whom professed to be Christians. Their object was to make the tour of Tahiti, with the visitor from the Leeward Islands. Previously, however, to their embarkation, a signal triumph was achieved in favour of Christianity, at a public festival, in which they were the most conspicuous party.

It has ever been considered a mark of respect due to every distinguished visitor, to prepare, soon after the arrival of such an individual, a sumptuous feast, termed by the natives a *faamuraa*, or feeding. Not, however, by furnishing a rich and splendid entertainment at the habitation of the proprietors, and inviting as guests the parties in honour of whom it was prepared, but by cooking a number of whole pigs, fowls, and fish, with a proportionate accompaniment of roots and vegetables, puddings, and what may be called their made-dishes, and carrying the whole to the encampment of the visitor, with a considerable addition of the choicest fruits the season may afford.

An expensive and sumptuous entertainment of this kind was furnished by the chiefs of Eimeo for the queen's sister. A large quantity of every valuable kind of food was dressed and presented, together with several bundles of native cloth. On such occasions, it was customary for a priest or priests to attend; and before any of it was eaten, to offer the whole to the gods, by taking parts of the animals, and particular kinds of the fruit, to the temple, and depositing them upon the altar. The king and his friends were desirous on this occasion to prevent such an acknowledgment. When, therefore, the food was presented to Pomare-vahine, before any article was touched by the attendants, and while the spectators were expecting the priests to select the customary offerings to the idols, one of her principal men, who was a Christian, came forward, uncovered his head, and, looking up to heaven, offered in an audible voice their acknowledgments and thanksgivings to Jehovah, who liberally gave them food and raiment and every earthly blessing. The assembled multitude were confounded and astonished; and the food being, by this act, offered as they considered to Jehovah, no one dared to take any part of it to the idol temple.

When the party reached Tahiti, they landed in Pare, the hereditary district of Pomare's family, where they were welcomed by the friends of the king, and the guardian of *Aimata*, his only child, who with her nurse resided here.

From the few Christians in the neighbourhood, they were happy to learn that the inhabitants of large sections of Pare, and the adjacent district of Matavai, the former residence of their teachers, had renounced idolatry, and were desirous. to receive Christian intruction.

By the queen, or her sister, the king sent over a new book to Aimata, his infant daughter, which being considered as an indication of his purpose that she should be trained up in the new religion, was a source of great encouragement to the converts, and of corresponding dissatisfaction to the idolaters, who already began to meditate on the means of their destruction.

It was not in Pare and Matavai alone that the professed worshippers of God were to be found. Some there were who openly avowed their attachment to the new order of things, maintaining, in the midst of the heathen around them, daily worship in their families, and morning and evening private devotion; others, who, for fear of giving offence to their chiefs or neighbours, maintained secretly their profession, and at the hour of midnight met together, as the persecuted Christians in England have often formerly done, in the depths of the woods, or the retired glens of the valleys, for conference or social prayer.

The state of affairs in Tahiti was such, as to prevent the queen and her sister from proceeding on their intended tour of the island; but while they remained at Pare, a circumstance occurred similar to that which had transpired in Eimeo, though probably more decisive and important in its immediate results.

When a present of food and cloth was brought to the visitors by some of the chiefs of Tahiti, the priests also attended, and, observing the party disinclined to acknowledge or render the customary homage to the gods, began to expatiate on the power of the gods, and, pointing to some bunches of *ura*, or red feathers, which were always considered emblematical of their deities, employed insulting language, and threatened with vengeance the queen's companions. One of Pomare-vahine's men, the individual who had offered their acknowledgments to

God, on the presentation of food in Eimeo, hearing this, and pointing to the feathers, said, "Are those the mighty things you so extol, and with whose anger you threaten us? If so, I will soon convince you of their inability even to preserve themselves." Running at the same time to the spot where they were fixed, he seized the bunches of feathers, and cast them into a large fire close by, where they were instantly consumed. The people stood aghast, and uttered exclamations of horror at the sacrilegious deed; and it is probable that this act increased the hatred already rankling in the bosoms of the idolatrous party.

The individual who acted so heroic and conspicuous a part on these occasions was *Farefau,* a native of Borabora, but attached to the household of Pomare-vahine, with whom he had arrived from the Leeward Islands in 1814. When he reached Eimeo, he was an idolater, but soon became a pupil in the school; and, in the close of the same year, desired that his name might be recorded among the converts. He occupied a prominent station in all the struggles between paganism and true religion; and maintained an unblemished character, and an unwavering profession, through the varied scenes of that unsettled period. He engaged with diligence in teaching the inhabitants of the remote and rocky parts of Taiarabu the catechism and the art of reading; and after a lingering illness, during which he enjoyed the presence and support which true religion alone can impart, delivered, as he expressed himself on the last day of his life, from the fear of death, and having his hopes fixed or relying on the Son of God as the only Saviour, he died in peace, at our Missionary station in Afareaitu, on the

29th of July, 1817, nearly two years after the total overthrow of idolatry in 1815.

He was a man of decision and daring enterprise; and though on the occasion in Tahiti above referred to, he may have acted with a degree of zeal somewhat imprudent, it was a zeal resulting, not from ignorant rashness, but enlightened principle, and holy indignation against the boasting threatenings and lying vanities of the priests of idolatry; to whose arts of deception he had formerly been no stranger.

The influence of the Bure Atua in the nation, from the rank many of them held, and the confidence with which they maintained the superiority of their religion, together with the accessions that were daily made to their numbers from various parts of the island, not only increased the latent enmity against Christianity which the idolaters had always cherished, but awakened the first emotion of apprehension lest this new word should ultimately prevail, and the gods, their temples, and their worship, be altogether disregarded. To avoid this, they determined on the destruction, the total annihilation, of every one in Tahiti who was known to pray to Jehovah.

A project was formed by the pagan chiefs of Pare, Matavai, and Hapaiano, to assassinate, in one night, every individual of the Bure Atua. The persecuted party was already formidable in point of numbers and rank, and the idolaters, in order to ensure success in their murderous design, invited the chiefs of Atehuru and Papara to join them. The time was fixed for the perpetration of this bloody deed. At the hour of midnight they were to be attacked, their property plundered, their houses burnt, and every prisoner secured, to be slaughtered on the spot. The parties, who for a long time had been inveterate enemies to each other, readily agreeing to the proposal, were made friends on the occasion, and cordially assented to the plan of destroying the Christians. The intended victims of this treachery were unconscious of their danger, until the evening of the 7th of July; when, a few hours only before the horrid massacre was to have commenced, they received secret intelligence of the ruin that was ready to burst upon them.

Circumstances, unforeseen and uncontrollable by their enemies, had prevented the different parties from arriving punctually at their respective points of rendezvous; otherwise, even now escape would have been impracticable, and destruction inevitable, as the Porionu, inhabitants of Pare, Matavai, and Hapaiano, would have been on the one side, and in their rear, and the party from Atehuru and Papara on the other. The delay in the arrival of some of these, afforded the only hope of deliverance.

At this remarkably critical period, the whole of the party having to attend a meeting either for public worship, or for some other general purpose, assembled in the evening near the sea. No time was to be lost. Their canoes were lying on the beach. They were instantly launched; and, hurrying away what few things they could take, they embarked soon after sunset, and reached Eimeo in safety on the following morning, grateful for the happy and surprising deliverance they had experienced. The different parties, as they arrived towards midnight, learned, with no ordinary remorse and disappointment, that their prey had been alarmed, and had escaped beyond their power.

A large body of armed and lawless warriors, belonging to different and rival chieftains, thus brought together under irritated feelings, and perhaps mutually accusing each other as the cause of their disappointment, were not long without a pretext for commencing the work of death among themselves. Ancient animosities, restrained only for the purpose of crushing what they considered a common enemy, were soon revived, and led to an open declaration of war between the tribes assembled. The inhabitants of Atehuru and Papara, who had been invited by the Porionu to join them in destroying the Bure Atua, attacked the Porionu; and, in the battle that followed, obtained a complete victory over them, killing one of their principal chiefs, and obliging the vanquished to seek their safety in flight.

After this affair, the people of Taiarabu joined the victors. The whole island was again involved in war, and the conquering party scoured the coast from Atehuru to the eastern side of the isthmus, burning every house, destroying every plantation, plundering every article of property, and reducing the verdant and beautiful districts of Pare, Faaa, the romantic valleys of Hautaua, Matavai, and Hapaiano, and the whole of the north-eastern part of the island, to a state of barrenness and desolation.

Success did not bring peace or rest to the victorious party. Proud of their triumph, insolent in crime, and impatient of control, the Atehuruans and natives of Papara quarrelled with the Taiarabuans, who had joined them in destroying the Porionu. A battle followed. The natives of Taiarabu were defeated, and fled to their fortresses in the mountains of their craggy peninsula, leaving the Oropaa masters of the island.

Numbers of the vanquished fled to Eimeo, where they were received by the king, or protected by the chiefs, who had taken no part whatever in the wars that were now desolating Tahiti, and who determined to observe the strictest neutrality; or, if they acted at all, to do so only on the defensive, should invasion be attempted.

Besides the refugees, who in consequence of defeat in Tahiti had taken shelter in Eimeo, numbers who had secretly embraced Christianity, and feared ultimate destruction from the idolaters, although religion appeared to have no influence in the present commotion, came over to Eimeo, and joined the Christians. The aggregate of those whose names were written down as such, amounted at this period to nearly four hundred, and the pupils in the school were between six and seven hundred. Want of books alone prevented its being very considerably enlarged.

Notwithstanding the Bure Atua had escaped the machinations of their enemies, and the murderous counsel of the idolaters had issued in their own defeat, yet it was impossible, that, amidst the agitation which prevailed in Tahiti, the adjacent island of Eimeo should remain free from apprehension and disquiet; and although the king had sent repeated messages of a peaceable tendency to the conquerors, and had received assurances that there was no feeling of hostility towards him and his adherents, yet they knew, by past experience, that no reliance was to be placed on such professions, and were not without daily fears that a hostile fleet might disembark an invading army on their shores.

When the queen and her sister went over to Tahiti, Pomare undertook a journey round Eimeo, purposing to travel by short stages, and, by conversation with the chiefs of the different districts, to inform them of the nature of Christianity, endeavour to induce them to receive it, and recommend it to the people. He was not at first exempt from some degree of ridicule in this undertaking; for many of the chiefs and landed proprietors in Eimeo, were by no means strongly attached to his family. They were, moreover, at that time the firm supporters of idolatry, and considered his neglect of the gods of his ancestors, as the cause of his own troubles, and the disastrous war then desolating Tahiti, his hereditary kingdom. He was not, however, discouraged; and it must have been truly gratifying to have beheld him thus usefully engaged.

Whatever may have been the influence of Christian principles on his own mind, in subsequent periods of his life, Pomare certainly was employed by the Almighty, as an instrument most effectually to promote the important process, at this time changing altogether the moral, civil, and religious aspect of the nation. The success that attended his endeavours appears from a letter which he addressed to the Missionaries while encamped in the district of Maatea, on the side of the island nearly opposite to that in which the European settlement stood. In this letter he stated his delight in beholding the chiefs inclined to obey the word of God; which, he said, Jehovah himself was causing to grow, so that it prospered exceedingly. Thirty-four or thirty-six, in one district, had, to use his own expression, "laid hold of the word of God," though there were others who paid no attention to those things.

At Haumi, the adjoining district, but few were prevailed upon to forsake paganism; but among them was an intelligent man, who was a priest.

At Maatea, the district from which the king wrote, ninety-six renounced idolatry while he was there, in addition to others who had done so before. The change appeared to be general here. The chiefs, priests, and people publicly committed their idols to the flames, attended public worship, requested to have their names written down as desirous of becoming Christians, and importuned the king and his attendants to protract their visit, that they might be more fully informed in all the matters connected with the profession they had now made.

The Bure Atua had hitherto escaped the ruin intended for them by their enemies; and, though these were masters of Tahiti, in Eimeo, and secretly in Tahiti, the number of those who had joined the Christians was greatly increased. This state of things could not long remain. The haughty and turbulent spirit of the victors was such as to prevent it: and in the event of their proceeding to the object for which they had taken up arms, viz. the suppression of Christianity, it was by no means improbable that both the native Christians and their teachers, if they were not destroyed by their enemies, might be expelled from Tahiti and Eimeo.

CHAP. X.

The refugees in Eimeo invited to return to Tahiti—Voyage of the king and his adherents—Opposition to their landing—Public worship on the Sabbath disturbed by the idolatrous army—Courage of the king—Circumstances of the battle of Bunaauïa—Death of the idolatrous chieftain—Victory of the Christians—Clemency of the king and chiefs—Destruction of the image temple and altars of Oro—Total subversion of paganism—General reception of Christianity—Consequent alteration in the circumstances of the people—Pomare's prayer—Tidings of the victory conveyed to Eimeo—Its influence in the adjacent islands—Remarks on the time, circumstances, means, and agents, connected with the change.

In the commencement of the year 1815, the affairs of Tahiti and Eimeo, in reference to the supremacy of Christianity or idolatry, were evidently tending to a crisis; and although the converts had carefully avoided all interference in the late wars which had desolated the larger island, they were convinced that the time was not very remote, when their faith and principles must rise pre-eminent above the power and influence of that system of delusion and crime, of which they had so long been the slaves. To maintain the Christian faith, and enjoy a continuance of their present peace and comfort, they foresaw would be impossible. Under the influence of these impressions, the 14th of July, 1815, was set apart as a day of solemn fasting and prayer to God, whose guidance and protection was implored. A chastened and dependent frame of mind was very generally experienced at this period by the

Christians, which led them to be prepared for whatever in the course of Divine providence might transpire.

Soon after this event, the pagan chiefs of Tahiti sent messengers to the refugees in Eimeo, inviting them to return, and re-occupy the lands they had deserted. This invitation they accepted; and, as the presence of the king was necessary in several of the usages and ceremonies observed on these occasions, Pomare went over about the same time, formally to reinstate them in their hereditary possessions. A large number of Pomare's adherents, who were professors of Christianity, and inhabitants of Huahine, Raiatea, and Eimeo, with Pomare-vahine and Mahine, the chief of Eimeo and Huahine, accompanied the king and the refugees to Tahiti. When they approached the shores of this island, the idolatrous party appeared in considerable force on the beach, assumed a hostile attitude, prohibited their landing, and repeatedly fired upon the king's party. Instead of returning the fire, the king sent a flag of truce and a proposal of peace. Several messages were exchanged, and the negociations appeared to terminate in confidence and friendship. The king and his followers were allowed to land, and several of the people returned unmolested to their respective districts and plantations. Negociations for the adjustment of the differences that had existed between the king and his friends, and the idolatrous chiefs, were for a time carried on, and at length arranged, apparently to the satisfaction of the respective parties. The king, and those attached to his interest, were not however without suspicion, that it was only an apparent satisfaction; and they were not mistaken. The idolaters had indeed joined with them in bending the wreath of amity and peace, while they were at the same time secretly and actively concerting measures for their destruction.

The 12th of November, 1815, was the most eventful day that had yet occurred in the history of Tahiti. It was the Sabbath. In the forenoon, Pomare, and the people who had come over from Eimeo, probably about eight hundred, assembled for public worship at a place called Narii, near the village of Bunaauia, in the district of Atehuru. At distant points of the district, they stationed piquets; and when divine service was about to commence, and the individual who was to officiate stood up to read the first hymn, a firing of muskets was heard; and, looking out of the building in which they were assembled, a large body of armed men, preceded and attended by the flag of the gods, and the varied emblems of idolatry, were seen marching round a distant point of land, and advancing towards the place where they were assembled. It is war! It is war! was the cry which re-echoed through the place; as the approaching army were seen from the different parts of the building. Many, agreeably to the precautions of the Missionaries, had met for worship under arms; others, who had not, were preparing to return to their tents, and arm for the battle. Some degree of confusion consequently prevailed. Pomare arose, and requested them all to remain quietly in their places; stating, that they were under the special protection of Jehovah, and had met together for his worship, which was not to be forsaken or disturbed even by the approach of an enemy. *Auna*, formerly an areoi and a warrior, now a Christian teacher, who was my informant on these points, then read the hymn, and the congregation sang it. A portion of scripture was read, a prayer offered to the Almighty, and the service closed. Those who were unarmed, now repaired to their tents, and procured their weapons.

In assuming the posture of defence, the king's friends formed themselves into two or three columns, one on the sea-beach, and the other at a short distance towards the mountains. Attached to Pomare's camp, was a number of refugees, who had, during the late commotions in Tahiti, taken shelter under his protection, but had not embraced Christianity; on these the king and his adherents placed no reliance, but stationed them in the centre, or the rear, of the column. The *Bure Atua* requested to form the *viri* or frontlet, advanced guard; and the *paparia*, or cheek of their forces; while the people of Eimeo, immediately in the rear, formed what they called the *tapono*, or shoulder, of their army. In the front of the line, *Auna, Upaparu, Hitote*, and others equally distinguished for their steady adherence to the system they had adopted, took their station on this occasion, and shewed their readiness to lay down their lives rather than relinquish the Christian faith, and the privileges it conferred. Mahine, the king of Huahine, and Pomare-vahine, the heroic daughter of the king of Raiatea, with those of their people who had professed Christianity, arranged themselves in battle-array immediately behind the people of Eimeo, forming the body of the army. Mahine on this occasion wore a curious helmet, covered on the outside with plates of the beautifully spotted cowrie, or tiger shell, so abundant in the islands; and ornamented with a plume of the tropic, or man-of-war bird's feathers. The queen's sister, like a daughter of Pallas, tall, and rather masculine in her stature and features, walked and fought by Mahine's side;

clothed in a kind of armour of net-work, made with small and strongly twisted cords of *romaha*, or native flax, and armed with a musket and a spear. She was supported on one side by Farefau, her steady and courageous friend, who acted as her squire or champion; while Mahine was supported on the other by Patini, a fine, tall, manly chief, a relative of Mahine's family; and one who, with his wife and two children, has long enjoyed the parental and domestic happiness resulting from Christianity,—but whose wife, prior to their renunciation of idolatry, had murdered twelve or fourteen children.

Pomare took his station in a canoe with a number of musketeers, and annoyed the flank of his enemy nearest the sea. A swivel mounted in the stern of another canoe, which was commanded by an Englishman, called *Joe* by the natives, and who came up from Raiatea, did considerable execution during the engagement.

Before the king's friends had properly formed themselves for regular defence, the idolatrous army arrived, and the battle commenced. The impetuous attack of the idolaters, attended with all the fury, imprecations, and boasting shouts, practised by the savage when rushing to the onset, produced by its shock a temporary confusion in the advanced guard of the Christian army: some were slain, others wounded, and Upaparu, one of Pomare's leading men, saved his life only by rushing into the sea, and leaving part of his dress in the hands of the antagonist* with whom he had grappled. Not-

* This man was afterwards an inmate of my family, and, in conversation on the subject, has often declared that he did not go to battle to support idolatry, about which he was indifferent; but from the allegiance he owed to his chief, in whose cause he felt bound to fight, and who was leader of the idolatrous army.

withstanding this, the assailants met with steady and determined resistance.

Overpowered, however, by numbers, the *viri*, or front ranks, were obliged to give way. A kind of running fight commenced, and the parties were intermingled in all the confusion of barbarous warfare.

> "Here might the hideous face of war be seen,
> Stript of all pomp, adornment, and disguise."

The ground on which they now fought, excepting that near the sea-beach, was partially covered with trees and bushes; which at times separated the contending parties, and intercepted their view of each other. Under these circumstances it was, that the Christians, when not actually engaged with their enemies, often kneeled down on the grass, either singly or two or three together, and offered up an ejaculatory prayer to God—that he would cover their heads in the day of battle, and, if agreeable to his will, preserve them, but especially prepare them for the results of the day, whether victory or defeat, life or death.

The battle continued to rage with fierceness; several were killed on both sides; the idolaters still pursued their way, and victory seemed to attend their desolating march, until they came to the position occupied by Mahine, Pomare-vahine, and their companions in arms. The advanced ranks of these united bands met, and arrested the progress of the hitherto victorious idolaters. One of Mahine's men, *Raveae*,* pierced the body of *Upufara*,

* In 1818 this individual accompanied us to Huahine, where he died a short time before I left the islands.

the chief of Papara, and the commander-in-chief of the idolatrous forces. The wounded warrior fell, and shortly afterwards expired. As he sat gasping on the sand, his friends gathered round, and endeavoured to stop the bleeding of the wound, and afford every assistance his circumstances appeared to require. "Leave me," said the dying warrior; "mark yonder man, in front of Mahine's ranks; he inflicted this wound; on him revenge my death." Two or three athletic men instantly set off for that purpose. Raveae was retiring towards the main body of Mahine's men, when one of the idolaters, who had outrun his companions, sprang upon him before he was aware of his approach. Unable to throw him on the sand, he cast his arms around his neck, and endeavoured to strangle, or at least to secure his prey, until some of his companions should arrive, and despatch him. Raveae was armed with a short musket, which he had reloaded since wounding the chief; of this, it is supposed, the man who held him was unconscious. Extending his arms forward, Raveae passed the muzzle of his musket under his own arm, suddenly turned his body on one side, and, pulling the trigger of his piece at the same instant, shot his antagonist through the body, who immediately lost hold of his prey, and fell dying to the ground.

The idolatrous army continued to fight with obstinate fury, but were unable to advance, or make any impression on Mahine and Pomare-vahine's forces. These not only maintained their ground, but forced their adversaries back; and the scale of victory now appeared to hang in doubtful suspense over the contending parties. *Tino*, the idolatrous priest, and his companions, had, in the name of Oro, promised their adherents a certain and an easy triumph. This inspired them for the conflict, and made them more confident and obstinate in battle than they would otherwise have been; but the tide of conquest, which had rolled with them in the onset, and during the early part of the engagement, was already turned against them, and as the tidings of their leader's death became more extensively known, they spread a panic through the ranks he had commanded. The pagan army now gave way before their opponents, and soon fled precipitately from the field, seeking shelter in their pari's, strong-holds, or hiding places, in the mountains; leaving Pomare, Mahine, and the princess from Raiatea, in undisputed possession of the field.

Flushed with success, in the moment of victory, the king's warriors were, according to former usage, preparing to pursue the flying enemy. Pomare approached, and exclaimed, *Atira!* It is enough!—and strictly prohibited any one of his warriors from pursuing those who had fled from the field of battle; forbidding them also to repair to the villages of the vanquished, to plunder their property, or murder their helpless wives and children.

While, however, the king refused to allow his men to pursue their conquered enemies, or to take the spoils of victory, he called a chosen band, among which was Farefau, who had offered up the public thanksgiving at the festival in Eimeo and Patini, a near relative of Mahine, who had been his champion on that day, and sent them to Tautira, where the temple stood in which the great national idol, Oro, was deposited. He gave them orders to destroy the temple, altars, and idols, with every appendage of idolatry that they might find.

In the evening of the day, when the confusion of

battle had in some degree subsided, Pomare and the chiefs invited the Christians to assemble, probably in the place in which they had been during the morning disturbed—there to render thanks to God, for the protection he had, on that eventful day, so mercifully afforded. Their feelings on this occasion must have been of no common order. From the peaceful exercise of sacred worship, they had been that morning hurried into all the confusion and turmoil of murderous conflict with enemies, whose numbers, equipment, implacable hatred, and superstitious infatuation from the prediction of their prophets, had rendered them unusually formidable in appearance, and terrible in combat. Defeat and death had, as several of them have more than once declared, appeared, during several periods of the engagement, almost certain; and, in connexion with the anticipated extirpation of the Christian faith in their country, the captivity of those who might be allowed to live, the momentous realities of eternity, upon which, ere the close of the day, it appeared to themselves by no means improbable they would enter; had combined to produce a deep agitation, unknown in the ordinary course of human affairs, and seldom perhaps experienced even in the field of battle. They now celebrated the subversion of idolatry, under circumstances that, but a few hours before, had threatened their own extermination, with the overthrow of the religion they had espoused, and on account of which their destruction had been sought. The Lord of hosts had been with them, the God of Jacob was their helper, and to him they rendered the glory and the praise for the protection he had bestowed, and the victory they had obtained. In this sacred act they were joined by numbers, who heretofore had worshipped only the idols of their country, but who now desired to acknowledge Jehovah as God alone.

The noble forbearance and magnanimity of the king and chiefs, in the hour of conquest, when under all the intoxicating influence of recent victory and conscious power, were no less honourable to the principles which they professed, and the best feelings of their hearts, than conducive to the cause of Christianity. This generous temper did not terminate with the command issued on the field of contest, but it was a prominent feature in all their subsequent conduct.

When the king despatched a select band to demolish the idol temple, he said, "Go not to the little island, where the women and children have been left for security; turn not aside to the villages or plantations; neither enter into the houses, nor destroy any of the property you may see; but go straight along the high road, through all your late enemy's districts." His directions were attended to; no individual was injured, no fence broken down, no house burned, no article of property taken. The bodies of the slain were not wantonly mangled, nor left exposed to the elements, or to be devoured by the wild dogs from the mountains, and the swine that formerly would have been allowed to feed upon them; but they were all decently buried by the victors, and the body of the fallen chief, Upufara, was conveyed to his own district, to be interred among the tombs of his forefathers.

Upufara, the late chief of Papara, was an intelligent and interesting man; his death was deeply regretted by *Tati*, his near relative, and successor in the government of the district. His mind had been for a long time wavering, and he was, almost to the morning of the battle, undetermined whether he should renounce the idols, or still continue their votary. One of his intimate companions informed me, that a short time before his death, he had a dream which somewhat alarmed him. He thought he saw an immense oven, (such as that used in preparing *opio,)* intensely heated, and in the midst of the fire a large fish writhing in apparent agony, unable to get out, and yet unconsumed, living and suffering in the midst of the fire. An impression at this time fixed itself on his mind, that perhaps this suffering was designed to shew the intensity of the torments which the wicked would endure in the place of punishment. He awoke in a state of great agitation of mind, with profuse perspiration covering his body, and was so affected, that he could not sleep again that night. The same individual who resided with Upufara stated also, that only a day or two before the battle, he said to some one, with whom he was conversing, "Perhaps we are wrong: let us send a message to the king and Tati, and ask for peace; and also for books, that we may know what this new word, or this new religion, is." But the priests resisted his proposal, and assured the chiefs, that Oro would deliver the Bure Atua into their hands, and the *hau* and *mana*, government and power, would be with the gods of Tahiti. In addition to this, and any latent conviction that still might linger in his mind, of the power of Oro, and the result of his anger should he draw back; he stood pledged to the cause of the gods, and probably might feel a degree of pride influencing his adherence to their interest, lest he should be charged with cowardice in wishing to avoid the war, on which the chiefs, who were united to suppress Christianity, had determined.

The party sent by the king to the national temple at Tautira, in Taiarabu, proceeded directly to their place of destination. It was apprehended that, notwithstanding what had befallen the adherents of idolatry in battle, the inhabitants of Taiarabu, who were at that time more zealous for the idols than those of any other part of the island, who considered it an honour to be entrusted with the custody of Oro, and also regarded his presence among them as the palladium of their safety, might, perhaps, rise *en masse*, to protect his person from insults, and his temple from spoliation. No attempt of this kind, however, was made. The soldiers of Pomare, soon after reaching the district, proceeded to the temple, acquainted the inhabitants of the place and keepers of the temple with the events of the war, and the purpose of their visit. No remonstrance was made; no opposition offered—they entered the depository of Tahiti's former god; the priests and people stood round in silent expectation; even the soldiers paused a moment, and a scene was exhibited, probably strikingly analogous to that which was witnessed in the temple of Serapis in Alexandria, when the tutelar deity of that city was destroyed by the Roman soldiers. At length they brought out the idol, stripped him of his sacred coverings and highly valued ornaments, and threw his body contemptuously on the ground. It was a rude, uncarved log of aito wood, *casuarina equisatifolia,* about six feet long. The altars were then broken down, the temples demolished, and the sacred houses of the gods, together with their covering, ornaments, and all the appendages of their worship, committed to the flames. The temples, altars, and idols, all round Tahiti, were shortly after destroyed in the same way. The log of wood, called by the natives the body of Oro, into which they imagined the god at times entered, and through which his influence was exerted, Pomare's party bore away on their shoulders, and, on returning to the camp, laid in triumph at their sovereign's feet. It was subsequently fixed up as a post in the king's kitchen, and used in a most contemptuous manner, by having baskets of food suspended from it; and, finally, it was riven up for fuel. This was the end of the principal idol of the Tahitians, on whom they had long been so deluded as to suppose their destinies depended; whose favour, kings, and chiefs, and warriors had sought; whose anger all had deprecated; and who had been the occasion of more bloody and desolating wars, for the preceding thirty years, than all other causes combined. Their most zealous devotees were in general now convinced of their delusion, and the people united in declaring that the gods had deceived them, were unworthy of their confidence, and should no longer be objects of respect or trust.

Thus was idolatry abolished in Tahiti and Eimeo; the idols hurled from the thrones they had for ages occupied; and the remnant of the people liberated from the slavery and delusion in which, by the cunningly devised fables of the priests, and the "doctrines of devils," they had been for ages held as in fetters of iron. It is impossible to contemplate the mighty deliverance thus effected, without exclaiming, "What hath God wrought!" and desiring, with regard to other parts of the world, the arrival of that promised and auspicious era, when the gods "that have not made the heavens" shall be destroyed, and "the idols shall be utterly abolished."

The total overthrow of idolatry, splendid and important as it was justly considered, was but the beginning of the amazing work that has since advanced progressively in those islands. It resembled the dismantling of some dark and gloomy fortress, or the razing to its very foundation of some horrid prison of despotism and cruelty, with the materials of which, when cut and polished and adorned, a fair and noble structure was, on its very ruins, to be erected, rising in grandeur, symmetry, and beauty, to the honour of its proprietor, and the admiration of every beholder. The work was but commenced, and the abolition of idolatry was but one of the great preliminaries in those designs of mercy, and arrangements of divine providence, which were daily unfolded, with increasing interest and importance, in their influence on the destiny of the people.

The conduct of the victors, on the memorable 12th of November, had an astonishing effect on the minds of the vanquished, who had sought shelter in the mountains. Under cover of the darkness of night, they sent spies from the retreats to their habitations, and to the places of security in which they had left their aged and helpless relatives, their children, and their wives. These found every one remaining as they had left them on the morning of the battle, and were informed, by the wives and relatives of the defeated warriors, that Pomare and the chiefs had, without any exception, sent assurances of security to all who had fled. This intelligence, when conveyed to those who had taken refuge in the mountains, appeared to them incredible. After waiting, however, some days in their hiding-places, they ventured forth, and singly, or in small parties, returned to their dwellings; and when they found their plantations uninjured, their property secure, their wives and children safe, they were utterly astonished. From the king they received assurances of pardon, and were not backward in unitedly tendering submission to his authority, and imploring his forgiveness for having appeared in arms against him. Pomare was now, by the unanimous will of the people, reinstated on the throne of his father, and raised to the supreme authority in his dominions. His clemency in the late victory still continued to be matter of surprise to all parties who had been his opponents. "Where," said they, "can the king and the Bure Atua have imbibed these new principles of humanity and forbearance? We have done every thing in our power, by treachery, stratagem, and open force, to destroy him and his adherents; and yet, when the power was placed in his hand, victory on his side, we at his mercy, and his feet upon our necks, he has not only spared our lives, and the lives of our families, but has respected our houses and our property!" While making these inquiries, many of them, doubtless, recollected the conduct of his father, in sending one night, when the warriors of Atehuru had gone over to Tautira, a body of men, who at midnight fell upon their defenceless victims, the aged relations, wives, and children of the Atehuruans, and in cold blood cruelly murdered upwards of one hundred helpless individuals; and this probably made the conduct of Pomare II. appear more remarkable. At length, they concluded that it must be from the new religion, as they termed Christianity; and hence they unanimously declared their determination to embrace it, and to place themselves and their families under the direction of its precepts.

The family and district temples and altars, as well as those that were national, were demolished, the idols destroyed by the very individuals who had but recently been so zealous for their preservation, and in a very short time there was not one professed idolater remaining. Messengers were sent by those who had hitherto been pagans, to the king and chiefs, requesting that some of their men might be sent to teach them to read, and to instruct them concerning the true God, and the worship and obedience required by his word. Those who sent them expressed at the same time their determination to renounce every evil practice connected with their former idolatrous life, and their desire to become altogether a Christian people. Schools were built, and places for public worship erected; the Sabbath was observed, divine service performed; child-murder, and the gross abominations of idolatry, were discontinued.

What an astonishing and happy change must have taken place in the views, feelings, and pursuits of the inhabitants of Tahiti, in the course of a few weeks, from the battle of Narii, or Bunaauïa! A flood of light, like the rays of the morning, had broken in upon the intellectual and spiritual night, which, like a funeral pall, had long been spread over the inhabitants of the valleys and hills of Tahiti, and had rendered their abodes, though naturally verdant and lovely as the bowers of Eden, yet morally cheerless and desolate as the region of the shadow of death!

If the spirits of departed prophets, from their seats of bliss, look down upon our globe; how must Judah's royal bard have bent with rapture, to behold the accomplishment of triumphs, which, while he swept the hallowed harp of prophecy, he had foretold :—The multitude of the isles made glad* under Jehovah's reign, and the kings of the isles bringing presents† to his Son! And what new transport must have thrilled Isaiah's ardent spirit, when he now beheld a partial accomplishment of what, in distant ages, he had delighted to sing. "The wilderness rejoiced—the desert blossomed as the rose—the sword was beaten into the ploughshare—the wolf and the lamb dwelt together—and the islands sang the praises of Jehovah!"‡

With equal transport, and with greater sympathy, those happy disembodied spirits of just men made perfect, who have more recently entered on their everlasting rest, if they have a knowledge of what passes on earth, must have viewed the change! And if angels, who have none of those sympathies which the redeemed must feel, experience an addition to their joy, in every sinner that by penitence returns to God, it seems an inference not unwarranted by revelation, that the spirits of departed believers may have a knowledge of events and moral changes, which transpire in our world, especially with those relating to the progress of the Messiah's reign among mankind. Then with what augmented joy must that honoured and distinguished saint,§ in strict obedience to whose last bequest and dying charge, the South Sea Mission was attempted, with those holy and devoted men who first matured, and subsequently aided so nobly, the plan of sending the gospel to Tahiti, have viewed the pleasing change. Those patient labourers, also, who had toiled in the field, but had been called away before the first waive-sheaf was gathered in, must have felt their joy increased, as the enlarged spiritual

* Psalm xcvii. 1. † Psalm lxxii. 10. ‡ Isaiah xlii. 10.

§ The late Countess of Huntingdon.

perceptions which they possess enabled them to look not only on the outward change in circumstances and in conduct, but on that more delightful transformation of character, which every day unfolded to their view some new and lovely features. And with what loud ecstatic songs of gratitude and praise, must they have welcomed, to those realms of happiness, the first arrivals from those clustering isles, of redeemed and purified spirits, who had been made partakers of the grace of life, and heirs with them of immortality.

The knowledge of the spiritual nature of Christianity, possessed by many of the new converts, was doubtless but imperfect, their acquaintance with the will of God but partial, and probably on many points at first erroneous, but still there was a warmth of feeling, an undisguised sincerity, and an ardour of desire, (in scripture called "the first-love") that has never been exceeded. Aged chiefs and priests, and hardy warriors, with their spelling-books in their hands, might be seen sitting, hour after hour, on the benches in the schools, by the side, perhaps, of some smiling little boy or girl, by whom they were now thankful to be taught the use of letters. Others might be often seen employed in pulling down the houses of their idols, and erecting temples for the worship of the Prince of peace, working in happy companionship and harmony with those whom they had met so recently upon the field of battle.

Their Sabbaths must have presented spectacles on which angels might look down with joy. Crowds, who never had before attended any worship but that of their demon gods, might now be seen repairing to the rustic and lowly temple erected for Jehovah's praise; amidst their throng, mothers, wives, sisters, and daughters, who never were before allowed to join the other sex in any acts of worship. Few remained behind; all the inhabitants of the district or village, who were able, attended public worship. It is true, there was no Missionary to preach the gospel to them, or to lead their public service, yet it was performed with earnestness, propriety, and devotional feeling.

The more intelligent among the natives, who had been longest under instruction at Eimeo, usually presided. They sung a hymn; a portion of their scripture history, which was entirely composed of scripture extracts, was read; and prayer, in simplicity of language but sincerity of heart, was offered up to God. Those who had not printed books, wrote out portions of scripture for these occasions, and sometimes the prayers they used. These were often remarkably simple, expressive, and appropriate: I have one of Pomare's by me, in his own hand-writing, furnished by Mr. Nott. There is no date affixed to it, but from the evident frequency with which it has been used, and the portion of scripture written on the preceding pages of the same sheet of paper, I am inclined to think it was written about this period. The prayer is excellent, and the translation, which I also received from Mr. Nott, will require from the Christian reader no apology for its insertion, as a specimen of the style and sentiments employed by the natives of Tahiti in their devotional services. It is as follows:

"Jehovah, thou God of our salvation, hear our prayers, pardon thou our sins, and save our souls. Our sins are great, and more in number than the fishes* in the

* This is, perhaps, the most natural and expressive figure, or comparison, an *Islander* could make. There is no idea of multitude more familiar to his mind than that of a shoal of fishes, by which the shores he inhabits are occasionally or periodically visited.

sea, and our obstinacy has been very great, and without parallel. Turn thou us to thyself, and enable us to cast off every evil way. Lead us to Jesus Christ, and let our sins be cleansed in his blood. Grant us thy good Spirit to be our sanctifier. Save us from hypocrisy. Suffer us not to come to thine house with carelessness, and return to our own houses and commit sin. Unless thou have mercy upon us, we perish. Unless thou save us, unless we are prepared and made meet for thy habitation in heaven, we are banished to the fire, we die; but let us not be banished to that unknown world of fire. Save thou us through Jesus Christ, thy Son, the prince of life; yea, let us obtain salvation through him. Bless all the inhabitants of these islands, all the families thereof; let every one stretch out his hands unto God, and say, Lord save me, Lord save me. Let all these islands, Tahiti with all the people of Moorea, and of Huahine, and of Raiatea, and of the little islands around, partake of thy salvation. Bless Britain, and every country in the world. Let thy word grow with speed in the world, so as to exceed the progress of evil. Be merciful to us and bless us, for Jesus Christ's sake. Amen."

While these delightful changes were advancing in Tahiti, the king and his friends were not unmindful of those who had been left behind in a state of painful uncertainty at Eimeo. As soon as possible after the battle, a canoe was despatched by Mahine, king of Eimeo and Huahine, with the tidings of its result. Matapuupuu, or, as he is now called, Taua, was the bearer of the gladdening intelligence, and was a very suitable person to be sent on such an errand. He was a native of Huahine, where he had been chief priest since the death of his elder brother, who had sustained that office before him. He came up from Huahine to Pomare's assistance in 1811; early in the year 1813, he had made a profession of Christianity, and was among the first whose names were written down at Eimeo. He was not only a priest, but an areoi, and a warrior of no ordinary prowess. When his canoe approached the shore of Eimeo, the teachers and their pupils hastened to the beach, under the conflicting emotions of hope and fear. The warrior was seen standing on the prow of his light skiff, that seemed impatiently dashing through the spray, and rushing along the tops of the waves towards the shore, which its keel scarcely touched, when, with his light mat around his loins, his scarf hanging loosely over his shoulder, and his spear in his hand, he leaped upon the sandy beach. Before they had time to ask a single question, he exclaimed, "Ua pau! Ua pau! i te bure anae;" Vanquished! vanquished! by prayer alone! His words at first seemed but as words of irony or jest; but the earnestness of his manner, the details he gave, and the intelligence he brought from the king and some of the chiefs, confirmed the declaration.

The Missionaries were almost overcome with surprise, and hastened to render their acknowledgments of grateful praise to the Most High, under feelings that it would be impossible to describe. It was, indeed, a joy unspeakable, the joy of harvest. In that one year they reaped the harvest of sixteen laborious seed-times, sixteen dreary and anxious winters, and sixteen unproductive summers. They now enjoyed the unexpected but exhilarating satisfaction resulting from the pleasure of the Lord prospering in their hands, in a degree and under circumstances that few are privileged to experience.

As soon as possible, Mr. Nott was despatched by his companions to Tahiti. On reaching the shores of this island, from which five years before he had been obliged to flee for his life, he found it was all true that had been told them, that the people were in that interesting state described by the prophet, when, enraptured by the visions of Messiah's future glories, he exclaimed, "The isles shall wait for his law." In this delightful situation, as he travelled round the islands, he literally found them not merely willing to be instructed, but anxious to hear; meeting together of their own accord, and often spending the hours of night in conversation and inquiry on the important matters connected with the religion of Jesus Christ. When he returned, Mr. Bicknell went over on the same errand; and observed every where the most encouraging attention, on the part of the people, to the instructions he communicated. The school at Papetoai was greatly increased; and hundreds, who had been early scholars there, were now stationed as teachers among the adjacent islands, imparting to others the knowledge they had received.

Not fewer than three thousand persons at this time possessed a knowledge of the books in their native language, which were in daily use. Besides eight hundred copies of the Abridgment of Scripture, and many copies of part of the Gospel of St. Luke in manuscript, about two thousand seven hundred spelling-books had already been distributed among the pupils at Eimeo, or sent over to Tahiti; still they were unable to meet the daily increasing demand of the people.

The mighty workings of the Spirit of God, in producing this remarkable change, were not confined to Tahiti, Eimeo, and the adjacent islands, forming the

Georgian group, it extended also to the Leeward or Society Islands. A simultaneous movement appears to have taken place among the rulers of the people, to throw off the yoke of pagan priestcraft, to rend asunder their fetters, and remove from the eyes of the nation, in its remote extremities, the veil of delusion by which they had so long been blinded. Tamatoa, the king of Raiatea, shortly after his return from Tahiti, publicly renounced idol-worship, and declared himself a believer in Jehovah and Jesus Christ. Many of the chiefs, and a number of the people, followed his example.

The prince of darkness, the author of paganism, whose sway had been unrivalled, and whose seat and stronghold had long been here, as well as in the other islands, did not tamely surrender his dominions. The idolatrous chiefs and inhabitants took up arms, to defend the cause of the gods, and revenge the insult offered by the king. Their efforts, however, were but as the ragings of an expiring monster, whose fangs were broken and whose heart had been pierced. The idolaters were defeated, and afterwards treated with the same clemency and lenient conduct which the Christian conquerors in Tahiti had manifested, and Christianity was firmly established. The vanquished, however, though spared and liberated by the generosity of Tamatoa, shewed themselves unworthy of the kindness with which they had been treated, by still talking of war on behalf of the idols. But as their numbers were few, their influence small, and as the great body of the people were doubtless favourable to the new order of things, hopes of success were comparatively faint, and no further attempt was made.

The chiefs and greater part of the population of Tahaa, an island included in the same reef with Raiatea, imitated the example of Tamatoa and the Raiatean Christians, and destroyed their idols.

The intelligent and enterprising chiefs of Borabora, Mai, and Tefaaora, were remarkably active in weakening the influence of the gods on the minds of the people under their government, undermining and subverting every species of idol-worship that prevailed in the islands. They succeeded, at length, in inducing the inhabitants, by their example and persuasion, to seek an acquaintance with that more excellent way revealed in the word of God, for whose worship they erected a convenient and respectable building.

Mahine sent a special message to Huahine, and the same change took place in that island; which was perhaps, for its size and population, more attached to its idols than any other. Idol-worship, with all its attendant cruelty and moral degradation, was discontinued. The temples were demolished, and the gods committed to the flames. Thus, in one year, the system of false worship, which had, from the earliest antiquity of its population, prevailed in these islands, was happily abolished, it is hoped to be revived no more.

In the course of the following year, the loss sustained by the death of Mr. Scott was repaired by the arrival of Mr. Crook from New South Wales; he reached the islands in the month of May, and rendered important service in the prosecution of the common enterprise.

During the same year, the profession of Christianity became general throughout the whole of the Society Islands. Several of the chiefs and people of Borabora and Raiatea visited Maurua, the most westerly of the Leeward Islands, and succeeded in persuading the chiefs and people to demolish their temples and idols, and receive Christian instruction. The most pleasing results continued to attend the efforts of the new converts in Tahiti. Pomare sent most of his own family idols to the Missionaries, that, as he observed in a letter accompanying them, dated February 19th, 1816, "they might either commit them to the flames, or send them to England."

These idols I saw at Port Jackson, in 1816; they are now deposited in the Missionary Museum, Austin Friars. It is impossible to behold them without sympathizing in the feelings of Pomare, when he calls them—Tahiti's foolish gods.

A number of interesting and important inquiries is naturally suggested by this amazing change; and we are anxious to be made acquainted with every fact, in the application of those means which induced its commencement, and sustained its progress. In all its departments, and under every circumstance, it bears the impress, and exhibits, in the clearest manner, the sovereignty and the power, of the Almighty, in regard alike to the time of its commencement, the circumstances of its progress, and the means of its accomplishment.

In regard to the *time* of its occurrence. During no period in the history of the Mission, could "the time to favour" the nation have appeared more unlikely than the present. The king's mind appears to have been first seriously exercised in reference to the declaration which he subsequently made, after the dispersion of the Missionaries, and their departure from the islands, when only one (viz. Mr. Nott) remained with him; and when, in consequence of the state of perpetual alarm and agitation in which the people were kept by the war, none could be induced to attend preaching or instruction.

It is probable that at that period public ordinances were altogether discontinued. The first public or open indications of the change, were given at a time which, according to human probabilities, was but little favourable to such events. The Missionaries had but recently returned from their banishment, and the work of instruction had scarcely been resumed; it was the beginning, and but the beginning, of a second attempt to plant the gospel in those islands. The Missionaries considering the whole of the twelve years spent in Tahiti as so much time lost, were commencing afresh their endeavours on another island, and could hardly expect that at this time, after such a protracted delay, God would at once prosper their undertaking.

The *circumstances* of the nation, and of the Mission, were by no means favourable to such a change. It was not a time of peace, and leisure, but of protracted, obstinate, and barbarous war—the king and his adherents were in exile, alternately agitated by the entreaties of their auxiliaries to attempt to retrieve their affairs by a descent upon Tahiti, or expecting their retreat to be invaded by their audacious and rebellious conquerors. It was a period of humiliation, darkness, and distress; while the population of Tahiti itself was torn by factions, and desolated by wars, that threatened its extinction. Their teachers were not much more favourably circumstanced. Few in number, compared with what they had been when they maintained their former station in Matavai, and suffering under the heaviest domestic bereavements; prevented by personal indisposition, and other circumstances, from engaging, either very frequently or extensively, in the main work of instructing the people; their exertions, greatly to their own regret, were exceedingly circumscribed. In addition to these discouragements, the prejudices of many of the king's most warm and valuable friends were unusually strong, as they considered the continuance of his misfortunes to result, in part, from the countenance he was giving, and the inclination he manifested towards the religion of the foreigners.

In the *means* employed there was nothing extraordinary. It is recorded, in the history of the Greenland Missions, that the Moravian brethren, for five or seven years, laboured patiently and diligently in teaching their hearers what are termed the first principles of religion,—inculcating the doctrines of the being and attributes of God, and the requirements of his law,—without making the least favourable impression upon them, or being, in many instances, able to secure the attention of the people to their instructions. The first instance of decisive and salutary effect from their teaching, was, we are informed, what would, in general, be termed accidental, and occasioned by their reading to some native visitors an account of the sufferings and death of the Saviour, which they were translating into the vernacular tongue. The attention of one of the party was arrested, his heart deeply affected, and ultimately his character entirely changed. This circumstance led to a complete alteration in the instructions they gave. The incarnation, the life, especially the sufferings and death, of the Lord Jesus Christ, were, from this time, the principal subjects brought before the minds of their hearers, and the results were such as to shew the propriety of the alteration. Where they had before been unable to make the least impression, they now beheld numbers deeply affected, and on whom these truths appeared to produce an entire change of character and deportment. I do not, however, suppose we are to infer from the account that is given of this amazing work in Greenland, that, during the first five or seven years of their labours there, the being and character of God, &c. were inculcated, to the exclusion or neglect of the way of salvation through Jesus Christ. Their teaching would, in that case, have been more defective than I am willing to suppose it was. Nor do I think we are to conclude, that, after the change in their instruction, the doctrines of the Saviour's advent, sufferings, and death, were insisted on, to the exclusion of the former; this mode of exhibiting scripture truth would have been almost as defective as the other: but I suppose that, during the earliest years of their labours, the first principles of religion were more frequent and prominent in their instructions, than the doctrines peculiar to the gospel, and that, subsequently, these points received that more frequent attention, which the character, being, and law of God, had formerly obtained. No alteration, even of this kind, however, appears to have taken place in the kind of doctrines inculcated by the Missionaries among the Tahitians. From the time of my arrival in the islands, I had always a great desire to know whether any change had been made by the early preachers in their discourses, and other means employed at this period: I have not, however, been able to learn that there was any thing extraordinary; they do not appear in any respect to have varied the manner, or the matter, of their instructions. I have often asked Mr. Nott, and others who were on the spot, if there was any alteration in the mode of instruction, or the nature of their addresses, as to the prominency of any of the doctrines of the gospel, which had not been so fully exhibited before; but I have invariably learned, that they were not aware of the least difference in the kind of instruction, or the manner of representing the truths taught at this period, and those inculcated during their former residence in Tahiti.

Their aim had always been to exhibit fully, and with the greatest possible simplicity, the grand doctrines and precepts taught in the Bible, giving each that share of attention which it appeared to have obtained in the volume of revelation. God, they had always endeavoured to represent as a powerful, benevolent, and holy Being, justly requiring the grateful homage, and willing obedience, of his creatures. Man, they had represented as the Scripture described him, and their own observation represented him to be, a sinner against his Maker, and exposed to the consequences of his guilt;—the love of God, in the gift of his only begotten Son as a propitiation for sin, and the only medium of reconciliation with God, restoration to the enjoyment of his favour, and the blessing of immortality; faith in this atonement, and the sinner's justification before God, were truths frequently exhibited. The doctrine of Divine benevolence thus displayed, was altogether new to the Tahitians; nothing analogous to it had ever entered into any part of their mythology. Its impression on their minds was at this time proportionate. The necessity also of Divine influences, to make the declaration of these truths effectual to conversion, and to meeten those who believed for the heavenly state, had ever been inculcated in the catechetical and other exercises of the school, in the meetings for reading the Scriptures and conversation, and in the discourses delivered in their assemblies for public worship.

The wonderful change that now seemed to be wrought in the minds and hearts of many, did not appear to be more the immediate result of instructions given at the time, than the remote but certain effect of truth imparted, and precious seed, which, having been scattered years before, was now revived with a power, that the individuals themselves could not comprehend, nor on ordinary principles explain. This circumstance should never be lost sight of; it is a wonderful manifestation of the faithfulness of God, who has declared that his word shall not return unto him void, but shall be found even after many days; and it is remarkably adapted to cheer the hearts of all who are called to labour and wait patiently, sowing season after season in hope, without reaping the wished-for harvest.

The universal, and in many instances permanent, moral and religious change, that has been effected in the South Sea Islands, (of the commencement, and more important parts of which, a regular, though necessarily brief account, has now been given,) appears, in whatever view we can possibly contemplate either its nature or its results, nothing less than a moral miracle. A change so important in its character, so rapid in its progress, so decisive in its influence, sublime almost in proportion to the feebleness of the agency by which it was, under God, accomplished, although effected on but a small tribe or people, is perhaps not exceeded in the history of nations, or the revolutions of empires, that have so often altered the moral and civil aspect of our world. This great and important event, confirmed in its results, and strengthened in its character, by the extension of its influence, and the increasing power of the principles it implanted, during the last fourteen years, already occupies no inferior place among the modern evidences of Christianity, and the demonstrations of its legitimate tendency to ameliorate the condition, and elevate the moral and intellectual character, of the most wretched and depraved among mankind. Emotions of astonishment, admiration, and gratitude, involuntarily arise in every mind in the least degree susceptible of humanity or religion; while increasing convictions of the divine origin of revelation must fasten on the understanding, and additional encouragement strengthen the hopes, of every individual who, according to the promise of God, is anticipating the arrival of a period, when a transformation, equally decisive and lovely, shall change the moral deserts of the earth into regions of order and beauty, and the wilderness shall become as the garden of the Lord.

In order more fully to illustrate the kind of scripture truth that appears, in connexion with others, to have affected deeply the minds of the people, one single instance, among many that might be adduced, will shew, that in the mild and verdant islands of the south, as well as the frozen and barren regions of the north, in Tahiti as well as in Greenland, the attractions of the Cross move and melt the human heart. It was the custom of the Missionaries, not only to instruct the natives in the school, preach to them in the chapel, and itinerate through the villages, but to assemble them for the purpose of reading, from manuscript, such portions of the scripture as were deemed suitable to their circumstances. On one of these occasions, Mr. Nott was reading the first portions of the Gospel of St. John to a number of the natives. When he had finished the sixteenth verse of the third chapter, a native, who had listened with avidity and joy to the words, interrupted him, and said, "What words were those you read? what sounds were those I heard? let me hear those words again." Mr. Nott read again the verse, "God so loved," &c. when the native rose from his seat, and said, Is that true? can that be true? God love the world, when the world not love him; God *so* love the world as to give his Son to die, that man might not die. Can that be true? Mr. Nott again read the verse, "God so loved the world," &c. told him it was true, and that it was the message God had sent to them, and that whosoever believed in him, would not perish, but be happy after death. The overwhelming feelings of the wondering native were too powerful for expression or restraint. He burst into tears, and as these chased each other down his countenance, he retired to meditate in private on the amazing love of God, which had that day touched his soul; and there is every reason to believe he was afterwards raised to share the peace and happiness resulting from the love of God shed abroad in his heart.

Connected with the means employed in the accomplishment of this important work, a few remarks on the *agents* who, under God, were instrumental in effecting it, may not be inappropriate. In common with the Missionaries in other parts of the world, they have been described, by the enemies of religion, as ignorant and dogmatical fanatics; more intent on the inculcation of the peculiarities of their sect or party, than promoting the well-being of the people; holding out no inducement, by precept or example, to industrious habits, &c.

The present state of the islands in which they have spent so many years, compared with what it was at the time of their arrival, and during several subsequent years, is a sufficient refutation to every charge of this kind.

But there are individuals, from whose general habits of observation, and principles of judgment, it might have been supposed a more just conclusion would have been formed, who have occasionally described them as the most unsuitable agents that could have been employed. This mode of representation, although I do not regard the Missionaries or their proceedings as perfect, I consider to be far from just or correct. It is not my intention to eulogize their diversified labours, or to lavish panegyric upon their achievements. But in the estimate of their character, qualifications, and exertions, a variety of considerations ought to have a greater influence on the minds of those by whom they are thus represented, than they are sometimes allowed to exert. Missionary effort, on the extended scale, and in the distant and comparatively unexplored field in which they attempted it, was an event as new among the British churches, as the broad, catholic principles, upon which it was undertaken, were unparalleled.

The authentic information possessed by many who combined in arranging the plan, as well as by those who attempted its execution, was not only exceedingly limited, but received through a medium* that necessarily imparted a higher glow of colouring, than those channels through which more accurate accounts have since been transmitted. Many, no doubt, embarked in the enterprise, as subsequent events fully proved, with

* Voyages of Cook, Bligh, &c.

incorrect ideas of the work, or mistaken views of the qualifications necessary for its accomplishment. It is not, however, to those who abandoned the task, that I refer so much, as to those who (except when driven from it by the approaching desolations of murderous war) maintained their post, and died in the field; or who, after having sustained the privation and toil of thirty years of exile from country and from home, are still willing to end their days among the people with whose interests and destiny they have identified themselves.

Their family connexions may not indeed have been of the highest class, neither may the individuals themselves have enjoyed the advantages of a very liberal education, nor possessed any very extensive acquaintance with the world. It is only in comparatively recent times that individuals of this class have, by embarking personally on the arduous and self-denying work of propagating Christianity amongst the pagan nations, exhibited some noble examples of Christian devotedness. Many of the first Missionaries to the South Sea Islands were acquainted with the most useful of the mechanic arts, which were adapted to produce a very favourable impression upon the minds of the people. They had obtained a creditable English, if not a classical, education, a due knowledge of the scriptures, and an experimental aquaintance with the principles of Christianity; while some, with great mental vigour combined no small degree of intellectual culture. Their own improvement, and the preparation for the work, was prosecuted contemporaneously with their efforts to instruct the people; and the numerous and respectable philological and other manuscripts which these have transmitted to England, although never published, shew that they were far from being unqualified for their work.

Had the first Mission to the South Seas been composed entirely of individuals eminent for their scientific knowledge and classical attainments, they would probably have been less suitable agents than those who actually went; as, it may be presumed, their previous habits of life would not have furnished the best preparatives for the privations and difficulties to which they would have been exposed. Yet it would undoubtedly have been highly advantageous to the Mission, had some such gifted individuals been included among its members. Such were not, however, at that time so ready, as they have subsequently been, to engage in the enterprise, and the service necessarily devolved on those who were willing, under every accompanying disadvantage, to undertake it. They were not perhaps distinguished by brilliancy of genius, or loftiness of intellect; but in uncompromising sternness of principle, unaffected piety, ardour of devotedness, uncomplaining endurance of privations, not easily comprehended by those who have always remained at home, or visited only civilized portions of foreign climes, in undeviating perseverance in exertion under discouragements the most protracted and depressing, and in plain and honest detail of their endeavours and success, they have been inferior to few who have been honoured to labour in the Missionary field. I have known some of these devoted men, who, though not insensible to the endearments of kindred and home, and the comforts of civilized life, have for years been deprived of what most would deem the necessaries of life. These self-denying individuals have been so destitute of a change of apparel, that they could not, without some sacrifice of feeling, meet any of their own countrymen by whom the island might be visited; and, often rising in the morning from the rustic bed, without knowing whence the supplies of even native food for the day were to be derived, they have sent out a native servant-boy to seek for bread-fruit in the mountains, or to solicit a supply from the trees of some friendly chief in the neighbourhood, while they have repaired to the school, and pursued their daily exercises of instruction, cheered and encouraged only by the progress of their scholars.

Such are the men who have long laboured in these islands; and though others may have been associated with them, who have turned back, or proved themselves unequal to the station, where many, who stand firm at their post at home, would perhaps have fainted, or have fallen under the discouragements inseparable from it—they have been faithful. They seek not the praise that cometh from man, but the testimony of their consciences and the approval of Heaven; and irrespective of the honour God has put upon them, they are entitled, from their steady and successful course, to be "highly esteemed for their works' sake."

CHAP. XI.

Account of the music and amusements of the islanders—Description of the sacred drum—Heiva drum, &c. Occasions of their use—The bu or trumpet—Ihara—The vivo, or flute—General character of their songs—Ballads, a kind of classical authority—Entertainments and amusements—Taupiti, or festival—Wrestling and boxing—Effects of victory and defeat—Foot-races—Martial games—Sham fights—Naval reviews—Apai, bandy or cricket—Tuiraa, or foot-ball—The haruraa puu, a female game—Native dances—Heiva, &c.—The te-a, or archery—Bows and arrows—Religious ceremonies connected with the game—Cock-fighting—Aquatic sports—Swimming on the surf—Danger from sharks—Juvenile amusements.

With the ancient idolatry of the people, their music, their dances, and the whole circle of their amusements, had been so intimately blended, that the one could not survive the other. When the former was abolished, the latter were also discontinued. Their music wanted almost every quality that could render it agreeable to the ear accustomed to harmony, and was deficient in all that constitutes excellence. It was generally boisterous and wild, and, with the exception of the soft and plaintive warblings of the native flute, was distinguished by nothing so much as its discordant, deafening sounds.

The principal musical instrument used by the South Sea Islanders, was the *pahu*, or drum. This varied in size and shape, according to the purpose for which it was designed. Their drums were all cut out of a solid piece of wood. The block out of which they were made, being hollowed out from one end, remaining solid at the other, and having the top covered with a piece of shark's skin, occasioned their frequently resembling, in construction and appearance, a kettle-drum. The pua and the reva, which are remarkably close-grained and durable, were esteemed the most suitable kinds of wood for the manufacture of their drums. The *pahu ra,* sacred drum, which was *rutu,* or beaten, on every occasion of extraordinary ceremony at the idol temple, was particularly large, standing sometimes eight feet high. The sides of one, that I saw in Tane's marae at Maeva, was not more than a foot in diameter, but many were much larger. In some of the islands, these instruments were very curiously carved. One which I brought from High Island, and have deposited in the Missionary Museum, is not inelegantly decorated; others, however, I have seen, exhibiting very superior workmanship.

Tahitian Drums.

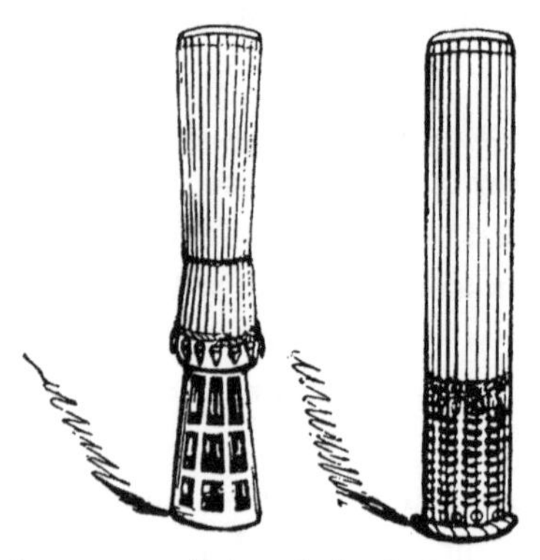

The drums used in their heivas and dances were ingeniously made. Their construction resembled that of those employed in the temple, the skin forming the head was fastened to the open work at the bottom by strings of finely-braided cinet, made with the fibres of the cocoa-nut husk. The drums beaten as accompaniments to the recital of their songs, were the same in shape, but smaller. They were all neatly made, and finely polished. The large drums were beaten with two heavy sticks, the smaller ones with the naked hand. When used, they were not suspended from the shoulders of the performers, but fixed upon the ground, and consequently produced no very musical effect. The sound of the large drum at the temple, which was sometimes beaten at midnight, was most terrific. The inhabitants of Maeva, where my house stood within a few yards of the ruins of the temple, have frequently told me, that at the midnight hour, when the victim was probably to be offered on the following day, they have often been startled from their slumbers by the dull, deep, thrilling sound of the sacred drum; and as its portentous sounds have reverberated among the rocks of the valley, every individual through the whole district has trembled with fear of the gods, or apprehension of being seized as the victim for sacrifice.

The sound of the trumpet, or shell, a species of murex used by the priests in the temple, and also by the herald, and others on board their fleets, was more horrific than that of the drum. The largest shells were usually selected for this purpose, and were sometimes above a foot in length, and seven or eight inches in diameter at the mouth. In order to facilitate the blowing of this trumpet, they made a perforation, about an inch in diameter, near the apex of the shell. Into this they inserted a bamboo cane, about three feet in length, which was secured by binding it to the shell with finely-braided cinet; the aperture was rendered air-tight by cementing the outsides of it with a resinous gum from the bread-fruit tree. These shells were blown when any procession marched to the temple, at the inauguration of the king, during the worship at the temple, or when a tabu, or restriction, was imposed in the name of the gods. We have sometimes heard them blown. The sound is extremely loud, but the most monotonous and dismal that it is possible to imagine.

The Trumpet Shell.

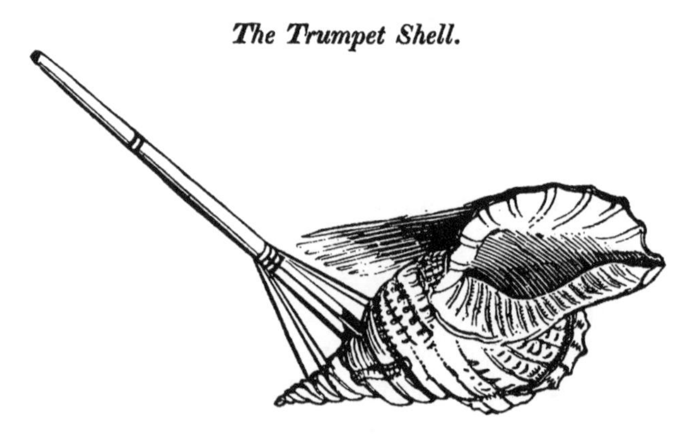

The *ihara* was another exceedingly noisy instrument. It was formed from the single joint of a large bamboo cane, cut off a short distance beyond the two ends or joints. In the centre, a long aperture was made from one joint towards the other. The ihara, when used, was placed horizontally on the ground, and beaten with sticks. It was not used in their worship, but simply as an amusement; its sounds were harsh and discordant.

The *vivo*, or flute, was the most agreeable instrument the Tahitians appear to have been acquainted with. It was usually a bamboo cane, about an inch in diameter, and twelve or eighteen inches long. The joint in the cane formed one end of the flute; the aperture through which it was blown was close to the end; it seldom had more than four holes, three in the upper side covered with the fingers, and one beneath, against which the thumb was placed. Sometimes, however, there were four holes on the upper side. It was occasionally plain, but more frequently ornamented, by being partially scorched or burnt with a hot stone, or having fine and beautifully plaited strings of human hair wound round it alternately with rings of neatly-braided cinet. It was not blown from the mouth, but the nostril. The performer usually placed the thumb of the right hand upon the right nostril, applied the aperture of the flute, which he held with the fingers of his right hand, to the other nostril, and, moving his fingers on the holes, produced his music. The sound was soft, and not unpleasant, though the notes were few; it was generally played in a plaintive strain, though frequently used as an accompaniment to their *pehes*, or songs. These were closely identified both with the music and the dances. The *ihara*, the drum and the flute, were generally accompanied by the song, as was also the native dance.

Their songs were generally historical ballads, and varied in their nature with the subjects to which they referred. They were exceedingly numerous, and adapted to every department of society, and every period of life. The children were early taught these *ubus*, and took great delight in their recital. Many of their songs referred to the legends or achievements of the gods, some to the exploits of their distinguished heroes and chieftains; while others were of a more objectionable character. They were often, when recited on public occasions, accompanied with gestures and actions corresponding to the events, or scenes described, and assumed in this respect a histrionic character. In some cases, and on public occasions, the action presented a kind of pantomime. They had one song for the fisherman, another for the canoe-builder, a song for cutting down the tree, a song for launching the canoe. But they were, with few exceptions, either idolatrous or impure; and were, consequently, abandoned when the people renounced their pagan worship. Occasionally, however, we heard parts of these songs recited, when events have occurred similar to those on which, in former times, they would have been used.

Their traditionary ballads were a kind of standard, or classical authority, to which they referred, for the purpose of determining any disputed fact in their history. The fidelity of public recitals referring to former events, was sometimes questioned by the orators or chroniclers of the party opposed to that by whom the recital had been made. The disputes which followed, were often carried on with great pertinacity and determination. As they had no records to which they could at such times refer, they could only oppose one oral tradition to another, which unavoidably involved the parties in protracted, and often obstinate debates. At such times, a reference to some distich, in any of their popular and historic songs, often set the matter in dispute at rest. On a recent occasion, two parties were disputing in reference to an event which occurred in the bay of Papara during the time Captain Bligh remained there in the Bounty, in 1788 or 1789. The fact questioned was the loss of the buoy of his anchor: after disputing it for some time without convincing his opponent, the individual who had stated the fact referred to the following lines in one of their ballads, relating that event.

> *"O mea eiá e Tareu eiá*
> *Eiá te poito a Bligh."*

> Such an one a thief, and Tareu a thief,
> Thieved (or stole) the buoy of Bligh.

The song was one well known to most, and the existence of this fact, among the others that had taken place, and the remembrance of which the ballad was designed to preserve, was conclusive, and appeared to satisfy the parties by whom it had been questioned. Most of their historical events were thus preserved. These songs were exceedingly popular for a time. The facts on which they were grounded became thus generally known; and they were, undoubtedly, one of the most effectual means they had of preserving the knowledge of the leading events of former times.

Freed, in a great degree, so far as the means of subsistence were concerned, from anxiety and labour, the islanders were greatly devoted to amusements: war, pagan worship, and pleasure, appear to have engaged their attention, and occupied the principal portion of their time. Their games were numerous and diversified, and were often affairs of national importance. They do not appear ever to have been gamblers, or to have accompanied any of their sports with betting, or staking property upon success, as the Sandwich Islanders have done from the earliest periods of their history, but seem to have followed their games simply for amusement.

The Taupiti, or Oroa, was generally a season of public festivity, when thousands, of both sexes, arrayed in their most splendid garments, assembled to witness the games. These festivals were usually connected with some religious ceremony, or cause of national rejoicing. The return of the king from a tour, or the arrival of a distinguished visitor, were among the most ordinary occasions of these games. Wrestling was the favourite, and perhaps most frequent sport; hence the taupiti, or assembly, was often called the taupiti maona, assembly for wrestling. A large quantity of food was always prepared for these occasions, and generally served out to the different parties, at the commencement of the festival, whereby the banquet was concluded before the games began. The wrestlers of one district sometimes challenged those of another, but the trial of strength and skill often took place between the inhabitants of different islands; the servants of the king of the island forming one band; and those in the train of his guest the other.

In this, as in most of their public procedings, the gods presided. Before wrestling commenced, each party repaired to the marae of the idols of which they were the devotees. Here they presented a young plantain-tree, which was frequently a substitute for a more valuable offering, and having invoked aid of the tutelar deity of the game, they repaired to the spot where the multitude had assembled. A space covered with a grassy turf, or the level sands of the sea-beach, was usually selected for these exhibitions. Here a ring was formed, perhaps thirty feet in diameter, the aufenua, people of the country, being on one side, the visitors on the other. The inner rank sat down, the others stood behind them; each party had their instruments of music with them, but all remained quiet until the games began. Six or ten, perhaps, from each side, entered the ring at once, wearing nothing but the maro or girdle, and having their limbs sometimes anointed with oil.

The fame of a celebrated wrestler was usually spread throughout the islands, and those who were considered good wrestlers, priding themselves upon their strength or skill, were desirous of engaging only with those they regarded as their equals. Hence, when a chief was expected, in whose train were any distinguished wrestlers, those among the adherents of the chief, by whom the party was to be entertained, who wished to engage, were accustomed to send a challenge previous to their arrival. If this, which was called tipaopao, had been the case when they entered the ring, they closed at once, without ceremony. But if no such arrangement had been made, the wrestlers of one party, or perhaps their champion, walked around and across the ring, having the left arm bent, with the hand on the breast; and striking the right hand violently against the left, and the left against the side, produced a loud hollow sound, which was challenging any one to the trial of his skill. The strokes on the arm were sometimes so violent, as not only to bruise the flesh, but to cause the blood to gush out.

When the challenge was accepted, the antagonists closed, and the most intense interest was manifested by the parties to which they respectively belonged. Several were sometimes engaged at once, but more frequently only two. They grasped each other by the shoulders, and exerted all their strength and art, each to throw his rival; this was all that was requisite; and although they generally grappled with each other, this was not necessary according to the rules of the game. Mape, a stout, and rather active, though not a large man, who was often in my house at Eimeo, was a famous wrestler. He was seen in the ring once, with a remarkably tall heavy man, who was his antagonist; they had grappled and separated, when Mape walked carelessly towards his rival, and on approaching him, instead of stretching out his arms as was expected, he ran the crown of his head with all his might against the temple of his antagonist, and laid him flat on the earth.

The most unbroken silence and attention was manifested during the struggle; but as soon as one was thrown, the scene was instantly changed; the vanquished was scarcely stretched on the sand, when a shout of exultation burst from the victor's friends. Their drums struck up; the women rose, and danced in triumph over the fallen wrestler, and sung in defiance to the opposite party. These were neither silent nor unmoved spectators, but immediately commenced a most deafening noise, partly in honour of their own clan or tribe, but principally to mar and neutralize the triumph of the victors. It is not easy to imagine the scenes that must often have been presented at one of their taupitis, or great wrestling matches, when not less than four or five thousand persons, dressed in their best apparel, and exhibiting every variety of costume and brilliancy of colour, were under the influence of excitement. One party were drumming, dancing, and singing, in the pride of victory, and the menace of defiance; while, to increase the din and confusion, the other party were equally vociferous in reciting the achievements of the vanquished, or predicting the shortness of his rival's triumph.

However great the clamour might be, as soon as the wrestlers who remained in the ring engaged again, the drums ceased, the song was discontinued, and the dancers sat down. All was perfectly silent, and the issue of the second struggle was awaited with as great an intensity of interest as the first. If the vanquished man had a friend or taio in the ring, he usually arose, and challenged the victor, who having gained one triumph, either left the ring, which it was considered honourable for him to do, or remained and awaited a fresh challenge. If he had retired, two fresh combatants engaged, and when one was thrown, exhibitions of feeling, corresponding with those that had attended and concluded the first struggle, were renewed, and followed every successive engagement. When the contest was over, the men repaired again to the temple, and presented their offering of acknowledgment, usually young plantain trees, to the idols of the game.

There are a number of men still living, who, under the system of idolatry, were celebrated as wrestlers through the whole of the islands. Among these, Fenuapeho, the hardy chieftain of Tahaa, is perhaps the most distinguished. He is not a large man, but broad, strong, sinewy, and remarkably firm-built. In person he appears to have beeen adapted to excel in such kinds of savage sports.

Although wrestling was practised principally by the men, it was not confined to them. Often, when they had done, the women contended, sometimes with each other, and occasionally with men, who were not perhaps reputed wrestlers. Persons in the highest rank sometimes engaged in the sport; and the sister of the queen has been seen wearing nearly the same clothing as the wrestlers wore, covered all over with sand, and wrestling with a young chief, in the midst of a ring, around which thousands of the people were assembled.

On all great public festivals, wrestling was succeeded by the Moto raa, or Boxing. This does not appear to have been so favourite an amusement with the Tahitians as wrestling; and there was generally a smaller number to engage. It was mostly practised by the lower orders and servants of the areois, and was with them, as boxing is every where, savage work; though, considering the rude and barbarous state of the people, who had little idea of influence or power, but as connected with their gods, or with mere brute strength, we are not surprised that it should have existed. The challenge was given in the same way as in wrestling; but when the combatants engaged, the combat was much sooner ended, and no time was spent in sparring or parrying the blows. These were generally straight forward, severe, and heavy; usually aimed at the head. They fought with the naked fist, and the whole skin of the forehead has been at times torn or driven off at a blow. No one interfered with the combatants while engaged; but as soon as either of them fell, or stooped, or shunned his antagonist, he was considered vanquished, the battle closed, and was instantly succeeded by the shouts and dances of triumph.

These barbarous sports, though generally followed by the common people, were not confined to them; other classes sometimes engaged; chiefs and priests were often among the most famous boxers and wrestlers. These games were not only dreadfully barbarous, but demoralizing in their influence on the people, who would set up a shriek of exultation, when the blood started, or the vanquished fell senseless on the sand. They were also often fatal. *Metia, a taura no Oro,* priest of Oro, who resided at Matavai, was celebrated for his prowess, and slew two antagonists, a father and a son, at one of these festivals, in Taiarapu. Considering the brutalizing tendency and the fatal results of boxing and wrestling, we cannot but rejoice that they have ceased with that system of barbarism and cruelty with which they were associated, and by which they were supported.

Connected with these athletic sports was another, less objectionable than either. This was the *faatitiaihemo raa,* or foot-race, in which the young men of the opposite parties engaged. Great preparation was made for this trial of strength and agility. The bodies of the runners were anointed with oil; the *maro,* or girdle, the only garment they wore, was bound tight round the loins. A wreath of flowers adorned the brows, and a light white or coloured bandage of native cloth was sometimes bound like a turban round the head. A smooth line of sandy beach was usually selected for the course. Sometimes they returned to the place from which they had started, but in general they ran the prescribed distance in a straight line. One of these races took place at Afareaitu while we resided there. It was between one of the king's servants, and a young man recently arrived from the Pearl Islands. The stranger was a tall, thin, handsome young man; and, as they walked past my house to the course, the people in general seemed to think his rival had but little prospect of equalling the swiftness of his speed, and it was thought he had already secured the *re*, or prize. The result, however, disappointed their expectations; and, as the spectators returned, I learned that, although on the first effort it was impossible to determine to whom the prize belonged, after repeated trials it was adjudged to Pomare's domestic. The *faatitiaihemo raa vaa*, or canoe-race, was occasionally practised on the smooth waters of the ocean, within the reefs, and appeared to afford a high degree of satisfaction.

Their martial games were numerous; and to these preparatory sports, the youth paid great attention. The *moto*, or boxing, and the *maona*, or wrestling, were regarded as a sort of military drilling; but the *vero patia*, throwing the spear or javelin, and the practice of throwing stones from a sling, were the principal military games. In the latter, the Tahitians excelled most of the nations of the Pacific; devoting to its practice a considerable portion of their time, and being able to cast the stone with great accuracy.

Throwing the spear, or darting the javelin, was an amusement in which they passed many of their juvenile hours. It was not a mere exercise of strength, like that exhibited in shooting with the bow and arrow, but a trial of skill. The stalk, or stem, of a plantain tree was their usual mark or target. This they fixed perpendicularly in the ground; and, retiring to a spot a number of yards distant, endeavoured to strike the mark with their missiles. These, thrown with precision and force, readily penetrated its soft and yielding substance. Although this was with some a favourite amusement, the

Tahitians do not appear to have followed it with such avidity as the Sandwich Islanders were accustomed to do, nor to have made such proficiency in the art. In order to avoid accidents while practising with the sling, the boys generally employed the fruit of the nono, *Morinda citrifolia*, instead of a stone. The mark at which they threw was a thin cane, or small white stick, fixed erect in the ground; and the force and precision with which it was repeatedly struck, were truly astonishing.

Besides these games, they often had what might be termed reviews of their land and naval forces. In these, all the appendages of battle were exhibited on land, and the fleets were equipped as in maritime war. The fighting men, in both exhibitions, wore the dress and bore the arms employed in actual combat. They also performed their different evolutions, or plans of attack and defence, advance and retreat. Sham-fights were connected with these displays of naval or military parade. In their mock engagements, they threw the spear, thrust the lance, parried the club, and at length, with deafening shouts, mingled in general and promiscuous struggle. Some of the combatants were thrown down, others captured, and the respective parties retreated to renew the contest.

Their naval reviews often exhibited a spectacle, which to them was remarkably imposing. Ninety or a hundred canoes were, on these occasions, ranged in a line along the beach, ready to be launched in a moment. Their elevated and often curiously carved sterns, their unwieldy bulk, the raised and guarded platform for the fighting men, the motley group assembled there, bearing their singularly and sometimes fantastically shaped weapons, the numerous folds of native cloth that formed their cumbrous dress, their high, broad spread turbans, the lofty sterns of their vessels, grotesque and rudely carved, together with the broad streamers floating in the breeze, combined to inspire them with the most elevated ideas of their naval prowess. The effect thus produced was greatly heightened by the appearance of the sacred canoes, bearing the images or the emblems of the gods, the flag of the gods, and the officiating or attending priests. Often, while the vessels were thus ranged along the beach, the king stood in a small one, drawn by a number of his men, who walked in the sea. In front of each canoe he paused, and addressed a short harangue to the warriors, and an *ubu*, or invocation, to the gods. After this was ended, at a signal given, the whole fleet was in a moment launched upon the bosom of the ocean, and pulled with rapidity and great dexterity to a considerable distance from the shore, where the several varieties of their naval tactics were exhibited; after which, they returned in regular order, with remarkable precision, to the shore.

Many of their games were most laborious. One at which the men played, called *apai*, or *paipai*, resembled a sport in some parts denominated "bandy." A similar game, called *palican*, was formerly a frequent amusement among the aborigines of South America, and those inhabiting the northern parts of the same continent, even as far as Canada. A ball is provided, and the players are furnished with sticks about three or four feet long, bent at one end; with these they strike the ball, each party endeavouring to send it beyond the boundary mark of their opponents. The ball is made with tough shreds of native cloth, tightly knotted together. The sticks used by the Tahitians were rude and unpolished, just as they were cut from the tree; but those used by the inhabitants of the Southern Islands are made with the *aito*, or iron-wood, the handle wrought with great care, and sometimes curiously carved, while a round protuberance is formed at the lower end, which is slightly curved, and augments the force with which they strike the ball.

The *tuiraa*, or foot-ball, is also a frequent game, though perhaps it was followed more by the women than the men: yet whole districts engaged in this amusement. In the former, they only struck the ball with a stick; in this, they employed the foot, and each party endeavoured to send it beyond the opposite boundary line, which had been marked out before they began. When either party succeeded in this, the air was rent with their shouts of success.

The *haru raa puu*, seizing of the ball, was however the favourite game of this kind. The females alone engaged in the seizing of the ball; in projecting which, neither sticks nor feet were allowed to be applied. An open place was necessary for all their sports, and the sea-beach was usually selected. The boundary mark of each party was fixed by a stone on the beach, or some other object on the shore, having a space of fifty or a hundred yards between. The ball was a large roll or bundle of the tough stalks of the plantain leaves twisted closely and firmly together. They began in the centre of the space. One party, seizing the ball, endeavoured to throw it over the boundary mark of the other. As soon as it was thrown, both parties started after it, and, in stooping to seize it, a scramble often ensued among those who first reached the ball; the numbers increased as the others came up, and they frequently fell one over the other in the greatest confusion. Amidst the shouts, and din, and disorder that followed, arms or legs were sometimes broken before the ball was secured. As the pastime was usually followed on the beach, the ball was often thrown into the sea; here it was fearlessly followed, and, with all the noise and cheering of the different parties, forty or fifty women might sometimes be seen, up to their knees or their waists in the water, splashing and plunging amid the foam and spray, after the object of their pursuit. These are only some of the games that were followed by the adults, at their great meetings or national festivals. In these, and in feasting, the hours of the day were spent.

Their dances were numerous and diversified; the *heiva* was performed by the men and women—in many the parties did not dance together. The dress of the women was remarkably curious, and not inelegant; their heads were decorated with fillets of *tamau*, or plaited human hair, and adorned with wreaths of the white sweet-scented teairi flower. The arms and neck were uncovered, the breasts ornamented with shells or coverings of curiously wrought net-work and feathers. The native cloth they wore was always white, sometimes edged with a scarlet border. Their movements were generally slow, but remarkably regular and exact; the arms, during their dances, were exercised as much as their feet. The drum and the flute were the music by which they were led; and the dance was usually accompanied by songs and ballads. There were other kinds of dances, in which smaller parties engaged; and, although sometimes held in the open air, they were more frequently performed under the cover of the spacious houses, erected in most of the districts for public entertainments. These structures were frequently large, and well-built; and consisted of a roof supported by pillars, without any shelter for the sides. A low fence, called *aumoa,* surrounded the house; and the inside was covered with mats, on which the company sat and the dancers performed. The *patau,* or prompter, sat by the drum, and regulated the several parts of the performance. After the athletic exercises of the day, the dances ensued in the evening, and were often continued till the dawn of the following morning. There were gods supposed to preside over their dances, whose sanction patronized every immorality connected with them.

The *te-a,* or archery, was also a sacred game, more so, perhaps, than any other. The bows, arrows, quiver, and cloth in which they were usually kept together, with the dresses worn by the archers, were all sacred, and under the special care of persons regularly appointed to keep them. It was usually practised as a most honourable recreation, between the residents of a place and their guests. The sport was generally followed either at the foot of a mountain, or on the sea-shore. My house, in the valley of Haamene at Huahine, stood very near an ancient *vahi te-a,* place of archery. Before commencing the game, the parties repaired to the marae, and performed several ceremonies; after which, they put on the archers' dress, and proceeded to the place appointed. They did not shoot at a mark; it was therefore only a trial of strength. In the place to which they shot the arrows, two small white flags were displayed, between which the arrows were directed.

The bows were made of the light, tough wood of the

purau; and were, when unstrung, perfectly straight, about five feet long; an inch, or an inch and a quarter, in diameter in the centre, but smaller at the ends. They were neatly polished, and sometimes ornamented with finely braided human hair, or cinet of the fibres of the cocoa-nut husk, wound round the ends of the bow in alternate rings. The string was of *romaha*, or native flax; the arrows were made of small bamboo reeds, exceedingly light and durable. They were pointed with a piece of *aito*, or iron-wood, but were not barbed. Their arrows were not feathered; but, in order to their being firmly held while the string was drawn, the lower end was covered with a resinous gum from the bread-fruit tree. The length of the arrows varied from two feet six inches, to three feet. The spot from which they were shot was considered sacred; there was one of these within my garden at Huahine. It was a stone pile, about three or four feet high, of a triangular form, one side of the angle being convex.

When the preparations were completed, the archer ascended this platform, and, kneeling on one knee, drew the string of the bow with the right hand, till the head of the arrow touched the centre of the bow, when it was discharged with great force. It was an effort of much strength in this position to draw the bowstring so far. The line often broke, and the bow fell from the archer's hand when the arrow was discharged. The distance to which it was shot, though various, was frequently three hundred yards. A number of men, from three to twelve, with small white flags in their hands, were stationed to watch the arrows in their fall. When those of one party went farther than those of the other, they waved the flags as a signal to those below. When they fell short, they held down their flags, but lifted up their foot, exclaiming, *ua pau*, beaten.

This was a sport in the highest esteem, the king and chiefs usually attending to witness the exercise. As soon as the game was finished, the bow, with the quiver of arrows, was delivered to the charge of a proper person: the archers repaired to the marae, and were obliged to exchange their dress, and bathe their persons, before they could take any refreshment, or even enter their dwellings. It is astonishing to notice how intimately their system of religion was interwoven with every pursuit of their lives. Their wars, their labours, and their amusements, were all under the control of their gods.

The arrows they employed were sometimes beautifully stained and variegated. The bows were plain, but the quivers were often truly elegant in shape and appearance. They were usually made with the single joint of a bamboo cane, three feet six or nine inches long, and about two inches in diameter. The outside was sometimes handsomely stained, and finely polished at the top and the bottom; they were adorned with finely braided cinet, and plaited human hair. The cap or cover of the quiver was usually a small, handsome, well-formed cocoa-nut, of a dark brown chocolate colour, highly polished, and attached to the quiver by a braided cinet passing up the inner side of the quiver, and fastened near the bottom.

The bow and arrow were never used by the Society Islanders, excepting in their amusements; hence, perhaps, their arrows, though pointed, were not barbed, and they did not shoot at a mark. In throwing the spear, and the stone from the sling, both of which they used in battle, they were accustomed to set up a mark, and practised, that they might throw with precision, as well as force. In the Sandwich Islands, they are used also as an amusement, especially in shooting rats, but are not included in their accoutrements for battle; while in the Friendly Islands, the bow was not only employed on occasions of festivity, but also used in war; this, however, may have arisen from their proximity to the Feejee Islands, where it is a general weapon. In the Society and Sandwich Islands, it is now altogether laid aside, in consequence of its connexion with their former idolatry. I do not think the Missionaries ever inculcated its discontinuance, but the adults do not appear to have thought of following this or any other game, since Christianity has been introduced among them.

The most ancient, but certainly not the most innocent game among the Tahitians, was the *faatitoraamoa*, literally, the causing fighting among fowls, or cock-fighting. The traditions of the people state, that fowls have existed in the islands as long as the people, that they came with the first colonists by whom the islands were peopled, or that they were made by Taaroa at the same time that men were made. The traditions and songs of the islanders, connected with their amusements, are as ancient as any in existence among them. The Tahitians do not appear to have staked any property, or laid any bets, on their favourite birds, but to have trained and fought them for the sake of the gratification they derived from beholding them destroy each other. Long before the first foreign vessel was seen off their shores, they were accustomed to train and to fight their birds. The fowls designed for fighting were fed with great care; a finely carved *fatapua*, or stand, was made as a perch for the birds. This was planted in the house, and the bird fastened to it by a piece of curious cinet, braided flat that it might not injure the leg. No other substance would have been secure against the attacks of his beak. Their food was chiefly *poe*, or bruised bread-fruit, rolled up in the hand like paste, and given in small pieces. The fowl was taught to open his mouth to receive his food and his water, which was poured from his master's hand. It was also customary to sprinkle water over these birds, to refresh them.

The natives were universally addicted to this sport. The inhabitants of one district often matched their birds against those of another, or those of one division of a district against those of another division. They do not appear to have entertained any predilection for particular colour in the fowls, but seem to have esteemed all alike. They never trimmed any of the feathers, but were proud to see them with heavy wings, full-feathered necks, and long tails. They also accustomed them to fight without artificial spurs, or other means of injury. In order that the birds might be as fresh as possible, they fought them early in the morning, soon after day-break, while the air was cool, and before they became languid from the heat. More than two were seldom engaged at once, and so soon as one bird avoided the other, he was considered as *vi*, or beaten. Victory was declared in favour of his opponent, and they were immediately parted. This amusement was sometimes continued for several days successively.

Like the inhabitants of most of the islands of the

Pacific, the Tahitians are fond of the water, and lose all dread of it before they are old enough to know the danger to which we should consider them exposed. They are among the best divers that are known, and spend much of their time in the sea, not only when engaged in acts of labour, but when following their amusements. One of their most favourite sports, is the *faahee*, or swimming in the surf, when the waves are high, and the billows break in foam and spray among the reefs. Individuals of all ranks and ages, and both sexes, follow this pastime with the greatest avidity. They usually selected the openings in the reefs, or entrances of some of the bays, for their sport; where the long heavy billows of the ocean rolled in unbroken majesty upon the reef or the shore. They used a small board, which they called *papa faahee*—swam from the beach to a considerable distance, sometimes nearly a mile, watched the swell of the wave, and when it reached them, resting their bosom on the short flat pointed board, they mounted on its summit, and, amid the foam and spray, rode on the crest of the wave to the shore: sometimes they halted among the coral rocks, over which the waves broke in splendid confusion. When they approached the shore, they slid off the board, which they grasped with the hand, and either fell behind the wave, or plunged toward the deep, and allowed it to pass over their heads. Sometimes they were thrown with violence upon the beach, or among the rocks on the edges of the reef. So much at home, however, do they feel in the water, that it is seldom any accident occurs.

I have often seen, along the border of the reef forming the boundary line to the harbour of Fare, in Huahine, from fifty to a hundred persons, of all ages, sporting like so many porpoises in the surf, that has been rolling with foam and violence towards the land, sometimes mounted on the top of the wave, and almost enveloped in spray, at other times plunging beneath the mass of water that has swept in mountains over them, cheering and animating each other; and, by the noise and shouting they made, rendering the roaring of the sea, and the dashing of the surf, comparatively imperceptible. Their surf-boards are inferior to those of the Sandwich Islanders, and I do not think swimming in the sea as an amusement, whatever it might have been formerly, is now practised so much by the natives in the south, as by those in the north Pacific. Both were exposed in this sport to one common cause of interruption; and this was, the intrusion of the shark among them. The cry of a *mao* among the former, and a *manó* among the latter, is one of the most terrific they ever hear; and I am not surprised that such should be the effect of the approach of one of these voracious monsters. The great shouting and clamour which they make, is principally designed to frighten away such as may approach. Notwithstanding this, they are often disturbed, and sometimes meet their death from these formidable enemies.

A most affecting instance of this kind occurred very recently in the Sandwich Islands, of which the following account is given by Mr. Richards, and published in the American Missionary Herald:

"At nine o'clock in the morning of June 14th, 1826, while sitting at my writing-desk, I heard a simultaneous scream from multitudes of people, *Pau i ka mano! Pau i ka mano!* (Destroyed by a shark! Destroyed by a shark!) The beach was instantly lined by hundreds of persons, and a few of the most resolute threw a large canoe into the water, and alike regardless of the shark, and the high rolling surf, sprang to the relief of their companion. It was too late. The shark had already seized his prey. The affecting sight was only a few yards from my door, and while I stood watching, a large wave almost filled the canoe, and at the same instant a part of the mangled body was seen at the bow of the canoe, and the shark swimming towards it at her stern. When the swell had rolled by, the water was too shallow for the shark to swim. The remains, therefore, were taken into the canoe, and brought ashore. The water was so much stained by the blood, that we discovered a red tinge in all the foaming billows, as they approached the beach.

"The unhappy sufferer was an active lad about fourteen years old, who left my door only about half an hour previous to the fatal accident. I saw his mother, in the extremity of her anguish, plunge into the water, and swim towards the bloody spot, entirely forgetful of the power of her former god."

"A number of people, perhaps a hundred, were at this time playing in the surf, which was higher than usual. Those who were nearest to the victim heard him shriek, perceived him to strike with his right hand, and at the same instant saw a shark seize his arm. Then followed the cry which I heard, which echoed from one end of Lahaina to the other. All who were playing in the water made the utmost speed to the shore, and those who were standing on the beach saw the surf-board of the unhappy sufferer floating on the water, without any one to guide it. When the canoe reached the spot, they saw nothing but the blood with which the water was stained for a considerable distance, and by which they traced the remains, whither they had been carried by the shark, or driven by the swell. The body was cut in two, by the shark, just above the hips; and the lower part, together with the right arm, were gone.

"Many of the people connect this death with their old system of religion; for they have still a superstitious veneration for the shark, and this veneration is increased rather than diminished by such occurrences as these.

"It is only about four months since a man was killed in the same manner at Waihee, on the eastern part of this island. It is said, however, that there are much fewer deaths by the shark than formerly. This, perhaps, may be owing to their not being so much fed by the people, and therefore they do not frequent the shores so much."

Besides the *faahee*, or surf-swimming, in which the adults principally engaged, there were a number of aquatic pastimes peculiar to the children; among these, the principal was erecting a kind of stage near the margin of a deep part of the sea or river, leaping from the highest elevation into the sea, and, chasing each other in the water, diving to an almost incredible depth, or skimming along the surface. Large companies of children, from nine or ten to fifteen or sixteen years of age, have often been seen, the greater part of the forenoon, eagerly following this apparently dangerous game, with the most perfect confidence of safety. Another amusement, which appears to afford high satisfaction to the chiidren of the islanders, is the construction of small canoes, boats, or ships, and floating them in the sea. Although they are rude in appearance, and soon destroyed, many of the boys display uncommon ingenuity in constructing this kind of toy. The hull is usually made with a piece of light wood of the hibiscus, the cordage of bark, and the sails either of the leaflets of the cocoa-nut, or the native cloth. The owners of these little vessels frequently go in small parties, and, taking their small-craft in their hands, wade up to their waist or arm-pits in the sea, and sometimes swim still further out; and then, launching their miniature fleets, consisting of ships, brigs, sloops, boats, canoes, &c. return towards the shore. They usually fix a piece of stone at the bottom of the little barks, which keeps them upright; and as the wind wafts them along the bay, their owners run along up to their knees in the sea, splashing and shouting as they watch their progress.

Such were some of the amusements of the natives in the South Sea Islands. In these, when not engaged in war, they spent much of their time. There were also others, of a less athletic kind, and of less universal prevalence. Among these, the *aperea* was one of the most prevalent; it consisted in jerking a reed, two feet and a half or three feet in length, along the ground. The men seldom played at it, but it was a common diversion for the women and children. *Timo*, or *timotimo*, was another game with the same class. The parties sat on the ground, with a heap of stones by their side, held a small round stone in the right hand, which they threw several feet up into the air, and, before it fell, took up one of the stones from the heap, which they held in the right hand till they caught that which they had thrown up, when they threw down the stone they had taken up, tossed the round stone again, and continued taking up a fresh stone every time they threw the small round one into the air, until the whole heap was removed. The *teatea mata* was a singular play among the children, who stretched open their eyelids by fixing a piece of straw, or stiff grass, perpendicularly across the eye, so as to force open the lids in a most frightful manner. Tupaurupauru, a kind of blind-man's-buff, was also a favourite juvenile sport.

They were very fond of the tahoro, or swing, and frequently suspended a rope from the branch of a lofty tree, and spent hours in swinging backwards and forwards. They used the rope singly, and at the lower end fastened a short stick, which was thus suspended in a horizontal position; upon this stick they sat, and, holding by the rope, were drawn or pushed backwards and forwards by their companions. Walking

The Tahitian Swing.

in stilts was also a favourite amusement with the youth of both sexes. The stilts were formed by nature, and generally consisted of the straight branches of a tree, with a smaller branch projecting on one side. Their naked feet were placed on this short branch, and thus, elevated about three feet from the ground, they pursued their pastime.

The boys were very fond of the uo, or kite, which they raised to a great height. The Tahitian kite was different in shape to the kites of the English boys. It was made of light native cloth, instead of paper, and formed in shape according to the fancy of its owner.

These are only some of the principal games, or amusements, of the natives; others might be added, but these are sufficient to shew that they were not destitute of sources of entertainment, either in their juvenile or more advanced periods of life. With the exception of one or two, they have all, however, been discontinued, especially among the adults; and the number of those followed by the children is greatly diminished. This is, on no account, matter of regret. When we consider the debasing tendency of many, and the inutility of others, we shall rather rejoice that much of the time of the adults is passed in more rational and beneficial pursuits. Few, if any of them, are so sedentary in their habits, as to need these amusements as a means of exercise; and they are not accustomed to apply so closely to any of their avocations, as to require them merely for relaxation.

CHAP. XII.

An account of the Areois, the institution peculiar to the inhabitants of the Pacific—Antiquity of the Areoi society—Tradition of its origin—Account of its founders—Infanticide enjoined with its establishment—General character of the Areois—Their voyages—Public dances—Buildings for their accommodation—Marine exhibitions—Oppression and injury occasioned by their visits—Distinction of rank among them—Estimation in which they are held—Mode of admission—Ceremonies attending advancement to the higher orders—Demoralizing nature of their usages—Singular rites at their death and interment—Description of Rohutunaunoa, the Areois heaven—Reflections on the baneful tendency of the Areoi society—Its dissolution—Conversion of some of the principal Areois—Character and death of Manu—Infanticide connected with the Areoi society—Numbers destroyed—Universality of the crime—Mode of its perpetration—Reasons assigned by the people for its continuance—Disproportion it occasioned between the sexes—Its abolition on the reception of Christianity—Influence of Christian principles. Maternal tenderness—Former treatment of children.

The greatest source of amusement to the people, as a nation, was most probably the existence of a society, peculiar to the islands of the Pacific, if not to the inhabitants of the southern groups. This was an institution called the Areoi society. Many of the regulations of this body, and the practices to which they were addicted, cannot be made public, without violence to every feeling of propriety; but, so far as it can be consistently done, it seems desirable to give some particulars respecting this most singular institution. Although I never met with an account of any institution analogous to this, among the barbarous nations in any parts of the world, I have reason to believe it was not confined to the Society group, and neighbouring islands. It does not appear to have existed in the Marquesas or Sandwich Islands; but the Jesuit Missionaries found an institution, bearing a striking resemblance to it, among the inhabitants of the Caroline or Ladrone Islands; a privileged fraternity, whose practices were, in many respects, similar to those of the Areois of the southern islands. They were called *uritoy;* which, omitting the *t*, would not be much unlike *areoi*. A greater difference exists in the pronunciation of words known to be radically the same.

How long this association has existed in the South Sea Islands, we have no means of ascertaining with correctness. According to the traditions of the people, its antiquity is equal to that of the system of pollution and error with which it was so intimately allied; and, by the same authority, we are informed that there have been Areois almost as long as there have been men. These, however, were all so fabulous, that we can only infer from them that the institution is of ancient origin. According to the traditions of the people, Taaroa created, and, by means of Hina, brought forth, when full-grown, Orotetefa and Urutetefa. They were not his sons; *oriori* is the term employed by the people, which seems to mean *create*. They were called the brothers of Oro, and were numbered among the inferior divinities. They remained in a state of celibacy; and hence the devotees were required to destroy their offspring. The origin of the Areois institution is as follows.

Oro, the son of Taaroa, desired a wife from the daughters of Taata, the first man; he sent two of his brothers, Tufarapainuu and Tufarapairai, to seek among the daughters of man a suitable companion for him; they searched through the whole of the islands, from Tahiti to Borabora, but saw no one that they supposed fit to become the wife of Oro, till they came to Borabora. Here, residing near the foot of Mouatahuhuura, *red-ridged mountain*, they saw Vairaumati. When they beheld her, they said one to the other, This is the excellent woman for our brother. Returning to the skies, they hastened to Oro, and informed him of their success; told him they had found among the daughters of man a wife for him, described the place of her abode, and represented her as a *vahine purotu aiai*, a female possessed of every charm. The god fixed the rainbow in the heavens, one end of it resting in the valley at the foot of the red-ridged mountain, the other penetrating the skies, and thus formed his pathway to the earth.

When he emerged from the vapour which, like a cloud, had encircled the rainbow, he discovered the dwelling of Vairaumati, the fair mistress of the cottage, who became his wife. Every evening he descended on the rainbow, and returned by the same pathway on the following morning to the heavenly regions. His wife bore a son, whom he called *Hoa-tabu-i-te-rai*, friend, sacred to the heavens. This son became a powerful ruler among men.

The absence of Oro from his celestial companions, during the frequent visits he made to the cottage of Vairaumati in the valley of Borabora, induced two of his younger brothers, Orotetefa and Urutetefa, to leave their abode in the skies, and commence a search after him. Descending by the rainbow in the position in which he had placed it, they alighted on the world near the base of the red-ridged mountain, and soon perceived their brother and his wife in their terrestrial habitation. Ashamed to offer their salutations to him and his bride without a present, one of them was transformed on the spot into a pig, and a bunch of *uru*, or red feathers. These acceptable presents the other offered to the inmates of the dwelling, as a gift of congratulation. Oro and his wife expressed their satisfaction at the present; the pig and the feathers remained the same, but the brother of the god assumed his original form.

Such a mark of attention, on such an occasion, was considered by Oro to require some expression of his commendation. He accordingly made them gods, and constituted them Areois, saying, *Ei Areoi orua i ie ao nei, ia noaa ta orua tuhaa:* Be you two Areois in this world, that you may love your portion, (in the government, &c.) In the commemoration of this ludicrous fable of the pig and the feathers, the Areois, in all the taupiti, and public festivals, carried a young pig to the temple; strangled it, bound it in the *ahu haio,* (a loose open kind of cloth,) and placed it on the altar. They also offered the red feathers, which they called the *uru maru no te Areoi;* the shadowy uru of the Areoi, or the red feathers of the party of the Areoi.

It has been already stated that the brothers, who were made gods and kings of the Areois, lived in celibacy; consequently they had no descendants. On this account, although they did not enjoin celibacy upon their devotees, they prohibited their having any offspring. Hence, one of the standing regulations of this institution was, the murder of their children. The first company, the legend states, were nominated, according to Oro's direction, by Urutetefa and Orotetefa, and comprised the following individuals: Huatua, of Tahiti; Tauraatua, of Moorea, or Eimeo; Temaiatea, of Sir Charles Sander's Island; Tetoa and Atae, of Huahine; Taramanini and Airipa, of Raiatea; Mutahaa, of Tahaa; Bunaruu, of Borabora; and Marore, of Maurua. These individuals, selected from the different islands, constituted the first Areoi society. To them, also, the gods whom Oro had placed over them delegated authority, and gave permission to admit to their order all such as were desirous to unite with them, and consented to murder their infants.* These were always the names of the principal Areois in each of the islands; and were borne by them in the several islands at the time of their renouncing idolatry; when the Areois name, and Areois customs, were simultaneously discontinued.

It is a most gratifying fact, that some of those who bore these names, and were ringleaders in all the vice and cruelty connected with the system, have since been distinguished for their active benevolence, and moral and exemplary lives. Auna, one of the first deacons in the church at Huahine, one of the first native teachers sent out by that church to the heathen, and who has been the minister of the church in Sir Charles Sander's Island, an indefatigable, upright, intelligent, and useful man, as a Christian Missionary in the South Sea Islands, was the principal Areoi of Raiatea. He was the Taramanini of that island, until he embraced Christianity.

They were a sort of strolling players, and privi-

* The above is one of the most regular accounts of the origin of the Areois institution, extant among the people. Mr. Barff, to whom I am indebted for it, received it from Auna, and Mahine the king of Huahine.

leged libertines, who spent their days in travelling from island to island, and from one district to another, exhibiting their pantomimes, and spreading a moral contagion throughout society. Great preparation was necessary before the *mareva*, or company, set out. Numbers of pigs were killed, and presented to Oro; large quantities of plantains and bananas, with other fruits, were also offered upon his altars. Several weeks were necessary, to complete the preliminary ceremonies. The concluding parts of these consisted in erecting on board their canoes, two temporary *maraes*, or temples, for the worship of Orotetefa and his brother, the tutelar deities of the society. This was merely a symbol of the presence of the gods; and consisted principally in a stone for each, from Oro's marae, and a few red feathers from the inside of the sacred image. Into these symbols the gods were supposed to enter when the priest pronounced a short *ubu*, or prayer, immediately before the sailing of the fleet. The numbers connected with this fraternity, and the magnitude of some of their expeditions, will appear from the fact of Cook's witnessing, on one occasion, in Huahine, the departure of seventy canoes filled with Areois.

On landing at the place of destination, they proceeded to the residence of the king or chief, and presented their *marotai*, or present; a similar offering was also sent to the temple and to the gods, as an acknowledgment for the preservation they had experienced at sea. If they remained in the neighbourhood, preparations were made for their dances and other performances.

On public occasions, their appearance was, in some respects, such as it is not proper to describe. Their bodies were painted with charcoal, and their faces, especially, stained with the *mati*, or scarlet dye. Sometimes they wore a girdle of the yellow *ti* leaves; which, in appearance, resembled the feather girdles of the Peruvians, or other South American tribes. At other times they wore a vest of ripe yellow plantain leaves, and ornamented their heads with wreaths of the bright yellow and scarlet leaves of the *hutu*, or Barringtonia; but, in general, their appearance was far more repulsive than when they wore these partial coverings.

Upaupa was the name of many of their exhibitions. In performing these, they sometimes sat in a circle on the ground, and recited, in concert, a legend or song in honour of their gods, or some distinguished Areoi. The leader of the party stood in the centre, and introduced the recitation with a sort of prologue, when, with a number of fantastic movements and attitudes, those that sat around began their song in a low and measured tone and voice; which increased as they proceeded, till it became vociferous and unintelligibly rapid. It was also accompanied by movements of the arms and hands, in exact keeping with the tones of the voice, until they were wrought to the highest pitch of excitement. This they continued, until, becoming breathless and exhausted, they were obliged to suspend the performance.

Their public entertainments frequently consisted in delivering speeches, accompanied by every variety of gesture and action; and their representations, on these occasions, assumed something of the histrionic character. The priests, and others, were fearlessly ridiculed in these performances, in which allusion was ludicrously made to public events. In the taupiti, or oroa, they sometimes engaged in wrestling, but never in boxing;

that would have been considered too degrading for them. Dancing, however, appears to have been their favourite and most frequent performance. In this they were always led by the manager or chief. Their bodies, blackened with charcoal, and stained with mati, rendered the exhibition of their persons on these occasions most disgusting. They often maintained their dance through the greater part of the night, accompanied by their voices, and the music of the flute and the drum. These amusements frequently continued for a number of days and nights successively at the same place. The upaupa was then *hui*, or closed, and they journeyed to the next district, or principal chieftain's abode, where the same train of dances, wrestlings, and pantomimic exhibitions, was repeated.

Several other gods were supposed to preside over the upaupa, as well as the two brothers who were the guardian deities of the Areois. The gods of these diversions, according to the ideas of the people, were monsters in vice, and of course patronized every evil practice perpetrated during such seasons of public festivity.

Substantial, spacious, and sometimes highly ornamented houses, were erected in several districts throughout most of the islands, principally for their accommodation, and the exhibition of their performances. The house erected for this purpose, which we saw at Tiataepuaa, was one of the best in Eimeo. Sometimes they performed in their canoes, as they approached the shore; especially if they had the king of the island, or any principal chief, on board their fleet. When one of these companies thus advanced towards the land, with their streamers floating in the wind, their drums and flutes sounding, and the Areois, attended by their chief, who acted as their prompter, appeared on a stage erected for the purpose, with their wild distortions of person, antic gestures, painted bodies, and vociferated songs, mingling with the sound of the drum and the flute, the dashing of the sea, and the rolling and breaking of the surf, on the adjacent reef; the whole must have presented a ludicrous imposing spectacle, accompanied with a confusion of sight and sound, of which it is not very easy to form an adequate idea.

The above were the principal occupations of the Areois; and in the constant repetition of these, often obscene exhibitions, they passed their lives, strolling from the habitation of one chief to that of another, or sailing among the different islands of the group. The farmers did not in general much respect them; but the chiefs, and those addicted to pleasure, held them in high estimation, furnishing them with liberal entertainment, and sparing no property to detain them. This often proved the cause of most unjust and cruel oppression to the poor cultivators. When a party of Areois arrived in a district, in order to provide a daily sumptuous entertainment for them, the chief would send his servants to the best plantations in the neighbourhood; and these grounds, without any ceremony, they plundered of whatever was fit for use. Such lawless acts of robbery were repeated every day, so long as the Areois continued in the district; and when they departed, the gardens often exhibited a scene of desolation and ruin, that, but for the influence of the chiefs, would have brought fearful vengeance upon those who had occasioned it.

A number of distinct classes prevailed among the

Areois, each of which was distinguished by the kind or situation of the tatauing on their bodies. The first or highest class was called *Avae parai*, painted leg; the leg being completely blackened from the foot to the knee. The second class was called *Otiore*, both arms being marked, from the fingers to the shoulders. The third class was denominated *Harotea*, both sides of the body, from the arm-pits downwards, being marked with tatau. The fourth class, called *Hua*, had only two or three small figures, impressed with the same material, on each shoulder. The fifth class, called *Atoro*, had one small stripe, tataued on the left side. Every individual in the sixth class, designated *Ohemara*, had a small circle marked round each ankle. The seventh class, or *Poo*, which included all who were in their noviciate, was usually denominated the *Poo faarearea*, or pleasure-making class, and by them the most laborious part of the pantomimes, dances, &c. was performed; the principal or higher orders of Areois, though plastered over with charcoal, and stained with scarlet dye, were generally careful not to exhaust themselves too much by physical effort, for the amusement of others.

In addition to the seven regular classes of Areois, there were a number of individuals, of both sexes, who attached themselves to the dissipated and wandering fraternity, prepared their food and their dresses, performed a variety of servile occupations, and attended them on their journeys, for the purpose of witnessing their dances, or sharing in their banquets. These were called Fauaunau, because they did not destroy their offspring, which was indispensable with the regular members.

Although addicted to every kind of licentiousness themselves, each Areoi had his own wife, who was also a member of the society; and so jealous were they in this respect, that improper conduct towards the wife of one of their own number, was sometimes punished with death. This summary and fatal punishment was not confined to their society, but was sometimes inflicted, for the same crime, among other classes of the community.

Singular as it may appear, the Areoi institution was held in the greatest repute by the chiefs and higher classes; and, monsters of iniquity as they were, the grand masters, or members of the first order, were regarded as a sort of superhuman beings: they were treated with a corresponding degree of veneration by many of the vulgar and ignorant. The fraternity was not confined to any particular rank or grade in society, but was composed of individuals from every class. But although thus accessible to all, the admission was attended with a variety of ceremonies; a protracted noviciate followed; and it was only by progressive advancement, that any were admitted to the superior distinctions.

It was imagined that those who became Areois were generally prompted or inspired to adopt this course by the gods. When any individual therefore wished to be admitted to their society, he repaired to some public exhibition, in a state of apparent *neneva*, or derangement. He generally wore a girdle of yellow plantain or *ti* leaves round his loins; his face was stained with *mati*, or scarlet dye; his brow decorated with a shade of curiously platted yellow cocoa-nut leaves; his hair perfumed with powerfully scented oil, and ornamented with a profusion of fragrant flowers. Thus arrayed, disfigured, and adorned, he rushed through the crowd assembled round the house in which the actors or dancers were performing, and, leaping into the circle, joined with seeming frantic wildness in the dance or pantomime. He continued in the midst of the performers until the exhibition closed. This was considered an indication of his desire to join their company; and if approved, he was appointed to wait, as a servant, on the principal Areois. After a considerable trial of his natural disposition, docility, and devotedness, in this occupation, if he persevered in his determination to join himself with them, he was inaugurated with all the attendant rites and observances.

This ceremony took place at some taupiti, or other great meeting of the body, when the principal Areoi brought him forth arrayed in the *ahu haio*, a curiously stained sort of native cloth, the badge of their order, and presented him to the members who were convened in full assembly. The Areois, as such, had distinct names, and, at his introduction, the candidate received from the chief of the body, the name by which in future he was to be known among them. He was now directed, in the first instance, to murder his children; a deed of horrid barbarity, which he was in general too ready to perpetrate. He was then instructed to bend his left arm, and strike his right hand upon the bend of the left elbow, which at the same time he struck against his side, whilst he repeated the song or invocation for the occasion; of which the following is a translation.

"The mountain above, *mouna tabu,** sacred mountain. The floor beneath *Tamapua,*† projecting point of

* The conical mountain near the lake of Maeva.

† The central district on the borders of the lake, lying at the foot of the mountain.

the sea. *Manunu,* of majestic or kingly bearing forehead. *Teariitarai,** the splendour in the sky. I am such a one, (pronouncing his new Areoi name,) of the mountain huruhuru." He was then commanded to seize the cloth worn by the chief woman present, and by this act he completed his initiation, and became a member, or one of the seventh class.

The lowest members of the society were the principal actors in all their exhibitions, and on them chiefly devolved the labour and drudgery of dancing and performing, for the amusement of the spectators. The superior classes led a life of dissipation and luxurious indolence. On this account, those who were novices continued a long time in the lower class; and were only admitted to the higher order, at the discretion of the leaders or grand masters.

The advancement of an Areoi from the lower classes, took place also at some public festival, when all the members of the fraternity in the island were expected to be present. Each individual appointed to receive this high honour, attended in the full costume of the order. The ceremonies were commenced by the principal Areoi, who arose, and uttered an invocation to *Te buaa ra,* (which, I presume, must mean the sacred pig,) to the sacred company of *Tabutabuatea,* (the name of all the principal national temples,) belonging to Taramanini, the chief Areoi of Raiatea. He then paused, and another exclaimed, Give us such an individual, or individuals, mentioning the names of the party nominated for the intended elevation.

When the gods had been thus required to sanction their advancment, they were taken to the temple. Here,

* The hereditary name of the king or highest chief of Huahine.

in the presence of the gods, they were solemnly anointed, the forehead of each person being sprinkled with fragrant oil. The sacred pig, clothed or wrapped in the *haio* or cloth of the order, was next put into his hand, and offered to the god. Each individual was then declared, by the person officiating on the occasion, to be an Areoi of the order to which he was thus raised. If the pig wrapped in the sacred cloth was killed, which was sometimes done, it was buried in the temple; but if alive, its ears were ornamented with the *orooro*, or sacred braid and tassel, of cocoa-nut fibre. It was then liberated, and being regarded as sacred, or considered as belonging to the god to whom it had been offered, it was allowed to range the district uncontrolled till it died.

The artist or priest of the tatau was now employed to imprint, in his unfading marks, the distinctive badges of the rank or class to which the individuals had been raised. As this operation was attended with considerable suffering to the parties invested with these insignia of rank, it was usually deferred till the termination of the festival which followed the ceremony. This was generally furnished with an extravagant profusion: every kind of food was prepared, and large bales of native cloth were also provided, as presents to the Areois, among whom it was divided. The greatest peculiarity, however, connected with this entertainment was, that the restrictions of tabu, which prohibited females, on pain of death, from eating the flesh of the animals offered in sacrifice to the gods, were removed, and they partook, with the men, of the pigs, and other kinds of food considered sacred, which had been provided for the occasion. Music, dancing, and pantomime exhibitions, followed, and were sometimes continued for several days.

These, though the general amusements of the Areois, were not the only purposes for which they assembled. They included

"All monstrous, all prodigious things."

And these were abominable, unutterable; in some of their meetings, they appear to have placed their invention on the rack, to discover the worst pollutions of which it was possible for man to be guilty, and to have striven to outdo each other in the most revolting practices. The mysteries of iniquity, and acts of more than bestial degradation, to which they were at times addicted, must remain in the darkness in which even they felt it sometimes expedient to conceal them. I will not do violence to my own feelings, or offend those of my readers, by details of conduct, which the mind cannot contemplate without pollution and pain. I should not have alluded to them, but for the purpose of shewing the affecting debasement, and humiliating demoralization, to which ignorance, idolatry, and the evil propensities of the human heart, when uncontrolled or unrestrained by the institutions and relations of civilized society and sacred truth, are capable of reducing mankind, even under circumstances highly favourable to the culture of virtue, purity, and happiness.

In these pastimes, in their accompanying abominations, and the often-repeated practices of the most unrelenting, murderous cruelty, these wandering Areois passed their lives, esteemed by the people as a superior order of beings, closely allied to the gods, and deriving from them direct sanction, not only for their abominations, but even for their heartless murders. Free from labour or care, they roved from island to island, supported by the chiefs and the priests; and were often feasted with provision plundered from the industrious husbandman, whose gardens were spoiled by the hands of lawless violence, to provide their entertainments, while his own family was not unfrequently deprived thereby, for a time, of the means of subsistence. Such was their life of luxurious and licentious indolence and crime. And such was the character of their delusive system of superstition, that, for them, too, was reserved the Elysium which their fabulous mythology taught them to believe, was provided in a future state of existence, for those so preeminently favoured by the gods.

A number of singular ceremonies were, on this account, performed at the death of an Areoi. The *otohaa*, or general lamentation, was continued for two or three days. During this time the body remained at the place of its decease, surrounded by the relatives and friends of the departed. It was then taken by the Areois to the grand temple, where the bones of the kings were deposited. Soon after the body had been brought within the precincts of the marae, the priest of Oro came, and, standing over the corpse, offered a long prayer to his god. This prayer, and the ceremonies connected therewith, were designed to divest the body of all sacred and mysterious influence the individual was supposed to have received from the god, when, in the presence of the idol, the perfumed oil had been sprinkled upon him, and he had been raised to the order or rank in which he died. By this act it was imagined they were all returned to Oro, by whom they had been originally imparted. The body was then buried as the body of a common man, within the precincts of the temple, in which the bodies of chiefs were interred. This ceremony was not much unlike certain portions of the degrading rites performed on the person of a heretic, in connexion with an auto de fé, in the Romish church.

The resources of the Areois were ample. They were, therefore, always enabled to employ the priest of Roma tane, who was supposed to have the keys of Rohutu noa-noa, the Tahitian's paradise. This priest consequently succeeded the priest of Oro, in the funeral ceremonies: he stood by the dead body, and offered his petitions to Urutaetae, who was not altogether the Charon of their mythology, but the god whose office it was to conduct the spirits of Areois and others, for whom the priest of Romatane was employed, to the place of happiness.

This Rohutu noanoa, (literally, perfumed or fragrant Rohutu,) was altogether a Mahomedan paradise. It was supposed to be near a lofty and stupendous mountain in Raiatea, situated in the vicinity of Hamaniino harbour, and called *Temehani unauna*, splendid or glorious Temehani. It was, however, said to be invisible to mortal eyes, being in the *reva*, or aerial regions. The country was described as most lovely and enchanting in appearance, adorned with flowers of every form and hue, and perfumed with odours of every fragrance. The air was free from every noxious vapour, pure, and most salubrious. Every species of enjoyment, to which the Areois and other favoured classes had been accustomed on earth, was to be participated there. Rich viands and delicious fruits were supposed to be furnished in abundance, for the frequent and sumptuous festivals celebrated there. Handsome youths and women, *purotu anae*, all perfection, thronged the place. These honours and gratifications were only for the privileged orders, the Areois, and the chiefs, who could afford to pay the priests for the passport thither: the charges were so great, that the common people seldom or never thought of attempting to procure it for their relatives; besides, it is probable that the high distinction kept up between the chiefs and people here, would be expected to exist in a future state, and to exclude every individual of the lower ranks, from the society of his superiors.

Those who had been kings or Areois in this world, were the same there for ever. They were supposed to be employed in a succession of amusements and indulgences similar to those to which they had been addicted on earth, often perpetrating the most unnatural crimes, which their tutelar gods were represented as sanctioning by their own example.

These are some of the principal traditions and particulars relative to this singular and demoralizing institution, which, if not confined to the Georgian and Society Islands, appears to have been patronized and carried to a greater extent there than among any other islands of the Pacific. Considering the imagined source in which it originated, the express appointment of Oro, their powerful god, the antiquity it claimed, its remarkable adaptation to the indolent habits and depraved uncontrolled passions of the people, the sanction it received here, and the prospect it presented to its members of the perpetuity, in a future state, of gratifications most congenial to those to whom they were exhibited, the Areoi institution appears a master-piece of satanic delusion and deadly infatuation, exerting an influence over the minds of an ignorant, indolent, and demoralized people, which no human power, and nothing less than a Divine agency, could counteract or destroy.

The entire dissolution of this association, and the abolition of its cruel and abominable practices, on the introduction of Christianity, is one of the most powerful demonstrations God has given to his church and to the world, of the irresistible operation of those means which he has appointed for the complete demolition of the very strong holds of paganism, and the universal extension of virtue, purity, and happiness among the most profligate and debased of mankind. It is a matter of devout acknowledgment to the Almighty, by whose power alone the means employed have been rendered efficacious in its annihilation, and furnishes a cause of hallowed triumph to the friends of moral order, humanity, and religion.

No sooner did these deluded, polluted, and cruel people, receive the gospel of Christ, the elevated sentiments, sacred purity, and humane tendency of which, convinced them that it must have originated in a source as opposite to that whence idolatry had sprung, as light is to darkness, than the spell in which they had been for ages bound was dissolved, and the chains of their captivity were burst asunder. They were astonished at themselves, and were a wonder to all who beheld them. The fabled legends by which, as by enchantment, they had been deceived, were banished from their recollections; the abominations and the bloodshed to which they had been addicted, ceased; and they became moral, virtuous, affectionate, devout, and upright members of a Christian community. There is reason to believe that many, even of the Areois, have been purified from their moral defilement, in that blood which cleanseth from all sin, and that the language addressed by the apostle to the Corinthians may with propriety be applied to them.

The astonishing and gratifying change which has taken place among them, nothing but Christian principles could have effected. Numbers of the Areois early embraced Christianity, and some from the highest orders were amongst the first converts. With few exceptions, they have been distinguished by ardour of zeal, and steady adherence to the religion of the Bible. Many of them have been the most regular and laborious teachers in our schools, and the most efficient and successful native Missionaries. Among this class, also, as might naturally be expected, have been experienced the most distressing apprehensions as to the consequences of sin, and the greatest compunction of mind on account of it. Many of them immediately changed their names, and others would be happy to obliterate every mark of that fraternity, the badges of which they once considered an honourable distinction. I have heard several wish they could remove from their bodies the marks tataued upon them, but these figures remain too deeply fixed to be obliterated, and perpetually remind them of what they once were. It is satisfactory to know, that not a few have enjoyed a sense of the pardoning mercy of God, and though some have been distressed in the prospect of death, others have been happy in the cheering hope, not of a pagan elysium, or a sensual sort of Turkish paradise, but of a holy and peaceful rest in the regions of blessedness.

One of these, whose name was *Manu*, a bird, resided at Bunaauïa, in the district of Atehuru. His age and bodily infirmities were such as to prevent his learning to read, yet he constantly attended the school, and, from listening to others, was able to repeat with correctness large portions of the scriptures, which were regularly read by the pupils. From meditation on these, he derived the highest consolation and support. He was an early convert to Christianity; his deportment was uniformly upright; his character respected by all who knew him; and for several years before his death, he was a member of the Christian church at Burder's Point. The recollection of the abominations and iniquity of which he had been guilty while a member of the Areoi institution, though not greater than those of his companions in crime, often filled his mind with horror and dismay. Whenever he alluded to that society, or to the crimes committed by its members, it was always with evident feelings of the deepest distress. From these it was his mercy to find relief, through faith in the atonement of Christ. This was his only ground of hope for pardon from God; and when, by thus looking to the great means of purity and peace, he was enabled to rest in hope, and his mind became calm, and peaceful, tears of contrition were often seen, while he gratefully remembered the amazing love of God. Towards the latter part of his life, his pastor had the pleasure of observing the greatest circumspection and moral purity in his whole conduct, with a high and increasing degree of spirituality of mind and tranquil joy. How striking the contrast which the evening of his days must have presented, to the early part of his life, spent among the impure, degraded, and wretched members of that infamous association to which he belonged! It is not surprising that his own mind should have been so deeply affected; but from all the moral pollution and guilt then contracted, he was washed and renewed, and prepared for the society of the blessed in the abode of purity and happiness. He died suddenly on the 5th of March, 1823; and, to use the language of the Missionary who watched his progress and his end with the deepest interest, we doubt not he is gone to be with that Saviour, "whom he loved with all his heart."

Infanticide, the most revolting and unnatural crime that prevails, even amongst the habitations of cruelty which fill the dark places of the earth, was intimately connected with this execrable institution. This affecting species of murder was not peculiar to the inhabitants of the Pacific. It has prevailed in different parts of the world, in ancient and modern times, among civilized as well as barbarous nations: but, until the introduction of Christianity, it was probably practised to a greater extent, and with more heartless barbarity, by the South Sea Islanders, than by any other people with whose history we are acquainted. Although we have been unable accurately to ascertain the date of its introduction to Tahiti and the adjacent isles, the traditions of the people warrant the inference, that it is of no very recent origin. I am, however, inclined to think it was practised less extensively in former times than during the fifty years immediately preceding the subversion of their ancient system of idolatry. There is every reason to suppose that, had the inhabitants murdered their infants during the early periods of their history, in any great degree, much less to the extent to which they have carried this crime in subsequent years, the population would never have become so numerous, as it evidently was, not many generations prior to their discovery.

It is difficult to learn to what extent infanticide was practised at the time Wallis discovered Tahiti, or the subsequent visits the islanders received from Cook; but its frequency and avowed perpetration was such as to attract the attention of the latter. Captain Cook's general conduct among the natives, notwithstanding the harsh measures he deemed it expedient to pursue towards the inhabitants of Eimeo, was humane; he took every opportunity of remonstrating with the king and chiefs, against a usage so truly merciless and savage.

When the Missionaries arrived in the Duff, this was one of the first and most affecting appendages of idolatry that awakened their sympathies, and called forth their expostulation and interference. Adult murder sometimes occurred; many were slain in war; and during the first years of their residence in Tahiti, human victims were frequently immolated. Yet the amount of all these and other murders did not equal that of infanticide alone. No sense of irresolution or horror appeared to exist in the bosoms of those parents who deliberately resolved on the deed before the child was born. They often visited the dwellings of the foreigners, and spoke with perfect complacency of their cruel purpose. On these occasions, the Missionaries employed every inducement to dissuade them from executing their intention, warning them, in the name of the living God, urging them also by every consideration of maternal tenderness, and always offering to provide the little stranger with a home, and the means of education. The only answer they generally received was, that it was the custom of the country; and the only result of their efforts, was the distressing conviction of the inefficacy of their humane endeavours. The murderous parents often came to their houses almost before their hands were cleansed from their children's blood, and spoke of the deed with worse than brutal insensibility, or with vaunting satisfaction at the triumph of their customs over the persuasions of their teachers.

In their earliest public negociations with the king and the chiefs, who constituted the government of the island, the Missionaries had enjoined, from motives of policy, as well as humanity and a regard to the law of God, the abolition of this cruel practice. The king Pomare acknowledged that he believed it was not right; that Captain Cook, for whom they entertained the highest respect, had told him it ought not to be allowed; and that for his part he was willing to discontinue it. These, however, were bare professions; for his own children were afterwards murdered, as well as those of his subjects.

In point of number, the disproportion between the infants spared and those destroyed, was truly distressing. It was not easy to learn exactly what this disproportion was; but the first Missionaries published it as their opinion, that not less than two-thirds of the children were murdered by their own parents. Subsequent intercourse with the people, and the affecting details many have given since their reception of Christianity, authorize the adoption of the opinion as correct. The first three infants, they observed, were frequently killed; and in the event of twins being born, both were rarely permitted to live. In the largest families more than two or three children were seldom spared, while the numbers that were killed were incredible. The very circumstance of their destroying, instead of nursing their children, rendered their offspring more numerous than it would otherwise have been. We have been acquainted with a number of parents, who, according to their own confessions, or the united testimony of their friends and neighbours, had inhumanly consigned to an untimely grave, four, or six, or eight, or ten children, and some even a greater number. I feel hence the painful and humiliating conviction which I have ever been reluctant to admit, forced upon me from the testimony of the natives themselves, the proportion of children found by the first Missionaries, and existing in the population at the time of our arrival, that during the generations immediately preceding the subversion of paganism, not less than two-thirds of the children were massacred. A female, who was frequently accustomed to wash the linen for our family, had thus cruelly destroyed five or six. Another, who resided very near us, had been the mother of eight, of which only one had been spared. But I will not multiply instances, which are numerous in every island, and of the accounts of which the recollection is most distinct. I am desirous to establish beyond doubt the belief of the practice, as it is one which, from every consideration, is adapted to awaken in the Christian mind liveliest gratitude to the Father of mercies, strongest convictions of the miseries inseparable from idolatry, tenderest commiseration for the heathen, and most vigorous efforts for the amelioration of their wretchedness.

The universality of the crime was no less painful and astonishing than its repeated perpetration by the same individuals. It does not appear to have been confined to any rank or class in the community; and though it was one of the indispensable regulations of the Areoi society, enforced on the authority of those gods whom they were accustomed to consider as the founders of their order, it was not peculiar to them.

It was perhaps less practised by the raatiras, or farmers, than any other class, yet they were not innocent. I do not recollect having met with a female in the islands, during the whole period of my residence there, who had been a mother while idolatry prevailed, who had not imbrued her hands in the blood of her offspring. I conversed more than once on the subject with Mr. Nott, during his recent visit to his native country. On one occasion, in answer to my inquiry, he stated, that he did not recollect having, in the course of the thirty years he had spent in the South Sea Islands, known a female, who was a mother under the former system of superstition, that had not been guilty of this unnatural crime. Startling and affecting as the inference is, it is perhaps not too much to suppose, that few, if any, became mothers, in those later periods of the existence of idolatry, who did not also commit infanticide. Without reference to other deeds of barbarism, they were in this respect a nation of murderers; and, in connexion with the Areoi institution, murder was sanctioned by their laws.

The various methods by which it was effected are most of them of such a nature, as to prohibit their publication. It does not appear that they ever buried them alive, as the Sandwich Islanders were accustomed to do, by digging a hole, sometimes in the floor of the dwelling, laying a piece of native cloth upon the infant's mouth, and treading down the earth upon the helpless child. Neither were the children as liable to be destroyed, after having been suffered to live for any length of time. The horrid deed was always perpetrated before the victim had seen the light, or in a hurried manner, and immediately after birth. The infants thus disposed of, were called *tamarii huihia, uumihea, or tahihia,* children stabbed or pierced with a sharp-pointed strip of bamboo cane, strangled by placing the thumbs on the throat, or *tahihia,* trodden or stamped upon. These were the mildest methods; others, sometimes employed, were too barbarous to be mentioned.

The parents themselves, or their nearest relatives, who often attended on the occasion for this express purpose, were the executioners. Often, almost before the new-born babe could breathe the vital air, gaze upon the light of heaven, or experience the sensations of its new existence, that existence has been extinguished by its cruel mother's hand; and the "felon sire," instead of welcoming with all a father's joy, a daughter or a son, has dug its grave upon the spot, or among the thick grown bushes a few yards distant. On receiving the warm palpitating body from its mother's hand, he has, with awful unconcern, deposited the precious charge, not in a father's arms, but in its early sepulchre; and instead of gazing, with all that thrilling rapture which a father only knows, upon the tender babe, has concealed it from his view, by covering its mangled form with the unconscious earth; and, to obliterate all traces of the deed, has trodden down the yielding soil, and strewed it over with green boughs, or covered it with verdant turf. This is not an exaggerated description, but the narrative of actual fact; other details, more touching and acute, have been repeatedly given to me in the islands, by individuals, who had been themselves employed in these unnatural deeds.

The horrid act, if not committed at the time the infant entered the world, was not perpetrated at any subsequent period. Whether this was a kind of law among the people, or whether it was the power of paternal affection, by which they were influenced, it is not necessary now to inquire; but the fact is consolatory. If the little stranger was, from irresolution, the mingled emotions that struggled for mastery in its mother's bosom, or any other cause, suffered to live ten minutes or half an hour, it was safe; instead of a monster's grasp, it received a mother's caresses, and a mother's smile, and was afterwards treated with the greatest tenderness. The cruel act was indeed often committed by the mother's hand; but there were times, when a mother's love, a mother's feelings, overcame the iron force of pagan custom, and all the mother's influence and endeavours have been used to preserve her child. Most affecting instances, which I forbear reciting, have been detailed by some, who now perhaps are childless, of the struggles between the mother to preserve, and the father and relatives to destroy, the infant. This has arisen from the motives of false pride, by which they were on some occasions influenced.

The reasons assigned for this practice, though varied, were uniformly shameful and criminal. The first is, the regulation of the Areoi institution, in order to be a member of which, it was necessary, in obedience to the express injunction of the tutelar gods of the order, that no child should be permitted to live. Another cause was the weakness and transient duration of the conjugal bond, whereby, although the marriage contract was formed by individuals in the higher ranks of society, with persons of corresponding rank, fidelity was seldom maintained.

The marriage tie was dissolved whenever either of the parties desired it; and though amongst their principal chiefs it was allowed nominally to remain, the husband took other wives, and the wife other husbands. These were mostly individuals of personal attractions, but of inferior rank in society. The progeny of such a union was almost invariably destroyed, if not by the parents themselves, by the relatives of those superior in rank, lest the dignity of the family, or their standing in society, should be injured by being blended with those of an inferior class. More infant murders have probably been committed under these circumstances, from notions of family pride, than from any other cause. One of my Missionary companions* states, that by the murder of such children, the party of inferior birth has been progressively elevated in rank, and that the degree of distinction attained, was according to the number destroyed,—that by this means, parties, before unequal, were considered as corresponding in rank, and their offspring allowed to live.

The *raatiras*, or secondary class of chiefs, and others by whom it was practised, appear to have been influenced by the example of their superiors, or the shameless love of idleness. The spontaneous productions of the soil were so abundant, that little care or labour was necessary to provide the means of subsistence: the climate was so warm, that the clothing required, as well as the food, could be procured with the greatest facility; yet they considered the little trouble required as an irksome task. A man with three or four children, and this was a rare occurrence, was said to be a *taata taubuubuu*, a man with an unwieldy or cumbrous burden; and there is reason to believe that, simply to avoid the trifling care and effort necessary

* Mr. Williams.

to provide for their offspring during the helpless periods of infancy and childhood, multitudes were consigned to an untimely grave. The females were subject to the most abasing degradation during the whole of their lives; and their sex was often, at their birth, the cause of their destruction: if the purpose of the unnatural parents had not been fully matured before, the circumstances of its being a female child, was often sufficient to fix their determination on its death. Whenever we have asked them, what could induce them to make a distinction so invidious, they have generally answered,—that the fisheries, the service of the temple, and especially war, were the only purposes for which they thought it desirable to rear children; that in these pursuits women were comparatively useless; and therefore female children were frequently not suffered to live. Facts fully confirm these statements.

In the adult population of the islands at the time of our arrival, the disproportion between the sexes was very great. There were, probably, four or five men to one woman. In all the schools established on the first reception of Christianity, the same disproportion prevailed. In more recent years the sexes are nearly equal. In addition to this cruel practice, others, equally unnatural, prevailed, for which the people had not only the sanction of their priests, but the direct example of their respective deities.

Without pursuing this painful subject any further, or inquiring into its antiquity or its origin, which is probably co-eval with that of the monstrous Areoi institutions; these details are of a kind that must impress every mind, susceptible of the common sympathies of humanity, with the greatest abhorrence of paganism, under the sanction of which such cruelties were perpetrated. They are also adapted to convey a most powerful conviction of the true character of heathenism, and the miseries which its victims endure.

The abolition of this practice, with the subversion of idolatry, is a grateful reward to those who have sent the mild and humane principles of true religion to those islands. This single fact demands the gratitude of every Christian parent, especially of every Christian female, and affords the most cheering encouragement to those engaged in spreading the gospel throughout the world.

The elevating, mild, and humanizing influence of Christianity, has not only effected its entire abolition, but it has revived and cherished those emotions of parental tenderness and affection originally implanted in the human bosom. A change of feeling and of conduct, in this respect, has taken place, as delightful as it is astonishing. The most civilized and Christian parts of the world do not furnish more affectionate parents than the Society Islanders now are. In general, they are too tender towards their children, and do not exercise that discipline and control over them which the well-being of the child, and the happiness of the parent, requires.

The most decisive instances of parental affection are every day presented; sometimes, in reference to the present enjoyment and temporal advantage of the children, and not unfrequently in regard to their spiritual benefit. I have often been deeply affected, on beholding a mother, whom I have known to have been guilty of the destruction of four, five, or six children, come into the place of public worship, with a little babe in her arms, often gazing, with evident tenderness and delight, on its smiling countenance. Sometimes a mother has been seen conducting her child to school, or applying for a book at our dwelling; and, occasionally, we have beheld a mother reading and explaining the scriptures to her children, or joining with them in prayer to the Most High. These changes of feeling and practice have taken place, not by a gradual process during successive generations, but among the same people, and in regard to the same individuals, who were subject to all the cruel insensibility, and addicted to all the barbarities, of infant murder.

In the treatment of those children formerly spared, a number of singular customs were observed, and several ceremonies performed. The mother bathed in the sea immediately after a profuse perspiration had been induced, and the infant was taken to the water almost as soon as it entered the world. It was also taken to the marae, where a variety of ceremonies were celebrated. In some of the islands, a number of these were attended to before its birth. When the mother repaired to the temple, the priest, after presenting the costly and numerous offerings, caught the god in a kind of snare, or loop made with human hair, and also offered up his prayer that the child might be an honour to his family, a benefit to the nation, and be more famous than any of his ancestors had been. This ceremony prevailed in the Hervey Islands. A number of *amoa* were performed in the Society Islands, which were a kind of religious ceremony, referring as much to the relatives as to the child. If it was of high rank, the inhabitants of the district were prohibited from kindling fires, or from burning torches, for several days. The child was, soon after its birth, invested with the name and office of its father, who was henceforward considered its inferior. This, however, during the minority of the child, was merely nominal; the father exercised all authority, though in the name of the child. The children were frequently nursed at the breast till they were able to walk, although they were fed with other food.

As soon as the child was able to eat, a basket was provided, and its food was kept distinct from that of the parent. During the period of infancy, the children were seldom clothed, and were generally laid or carried in a horizontal position. They were never confined in bandages, or wrapped in tight clothing, but though remarkably plump and healthy in appearance, they were generally very weak until nearly twelve months old.

The Tahitian parents and nurses were careful in observing the features of the countenance, and the shape of the child's head, during the period of infancy, and often pressed or spread out the nostrils of the females, as a flat nose was considered by them a mark of beauty. The forehead and the back of the head, of the boys, were pressed upwards, so that the upper part of the skull appeared in the shape of a wedge. This, they said, was done, to add to the terror of their aspect, when they should become warriors. They were very careful to *haune*, or shave, the child's head with a shark's tooth. This must have been a tedious, and sometimes a painful operation, yet it was frequently repeated; and although every idolatrous ceremony, connected with the treatment of their children, has been discontinued for a number of years, the mothers are still very fond of shaving the heads, or cutting the hair of their infants as close as possible. This often gives them a very singular appearance. The children are in general large, and finely formed; and, but for the prevalence of the disease which produces such a distortion of the spine, there is reason to believe that a deformed person would be very rarely seen among the inhabitants of the South Sea Islands.

Having suspended the narrative of my personal proceedings and observations during several preceding chapters, for the purpose of introducing an account of the remarkable change which had taken place in the islands, prior to our arrival, but which had been a very frequent topic of conversation during our residence with our predecessors,—having, also, given a brief account of some of the principal institutions and usages, belonging to the system which the nation had abolished,—I propose to resume the narrative with the commencement of the succeeding chapter.

CHAP. XIII.

Voyage to A-fa-re-ai-tu—Means of subsistence among the islands—Pigs—Dogs—Fowls—Different varieties of fish eaten by the people—Methods of dressing animal food—Edible vegetables and fruits—Description of the bread-fruit tree and fruit—Various methods of preparing it—Arum, or Ta-ro—U-hi, or yam—U-ma-ra, or sweet potato—Culture, preparation, and method of dressing arrow-root—Growth, appearance, and value of the cocoa-nut tree—Several stages of growth in which the fruit is used by the people—Process of manufacturing cocoa-nut oil—Varieties of plantain, or banana—Vi, or Brazilian plum—A-hia, or jambo—Inocarpus, or native chestnut—Varieties of Dracanæ—Combinations of native fruits, &c.—Foreign fruits and vegetables that flourish in the islands.

It was soon after sunrise on the 25th of March, 1817, that we left Papetoai in a *tipairua*, or large double canoe. The wind was contrary when we started; and, after proceeding only five miles, we landed at Tiataepuaa, the usual residence of the chiefs of Eimeo. Here we found Mr. Crook and his family waiting our arrival, to join in partaking of the breakfast they had prepared.

As soon as our men had refreshed themselves, we embarked in our respective canoes, and, resuming our voyage, proceeded along the smooth surface of the sea between the reefs and the shore. The wind died away, and a perfect calm succeeded. The heat of the sun was intense, and its scorching effect on our faces was increased by the reflection of the sea. This considerably diminished the pleasure we derived from watching through the perfectly transparent waters, the playful movements of the shoals of small and variegated rock-fish, of every rich and glowing hue, which often shone in brilliant contrast with the novel and beautiful groves of many-coloured coral, that rendered the sandy bottom of the sea, though frequently several fathoms beneath us, in appearance at least, an extensive and charming submarine shrubbery, or flower-garden. The corallines were spread out with all the endless variety and wild independence exhibited in the verdant landscape of the adjacent shore.

The heat of the sun, and the oppressiveness of the atmosphere, with the labour of rowing with their paddles our heavily laden canoes every inch of the way, had so fatigued our men, that when we reached A-ti-ma-ha, fifteen miles from the place whence we started in the morning, we deemed it expedient to land for the night.

I took a ramble through the district a short time before sunset, and was delighted with the wild and romantic beauty of the surrounding scenery,—the luxuriant groves of trees, and the shrubs, that now covered the fertile parts of this almost uninhabited district. In every part I met with sections of pavement, and other vestiges of former inhabitants; and was deeply affected in witnessing the depopulation thus indicated, and which is found to have taken place throughout the island.

Notwithstanding the total absence of every thing resembling accommodation in our lodging, where we spread our bed upon the ground, we should probably have enjoyed a night of refreshing sleep, but for the musquitoes. In these thinly peopled, damp, and woody districts, they are exceedingly numerous and annoying, especially to those who have recently arrived; and although during my subsequent residence in the island, I was less incommoded by them, I was on this occasion glad to escape their noise, &c. by leaving the house soon after midnight, and walking along the shore, or sitting on the beach until day-break.

Heavy showers detained us at Atimaha until ten o'clock in the forenoon, when we pursued our voyage. At Maatea I landed about twelve o'clock, and walked through the district of Haume to Afareaitu. The wind was contrary throughout the day, and it was near sunset before Mrs. Ellis and our little girl, with her nurse, arrived in the canoe. We had suffered much from exposure to the sun, and from the fatigue of our tedious voyage; we were, however, thankful to have reached our destination in safety. The natives cheerfully gave up a large oval-shaped house for our accommodation: Mr. and Mrs Crook occupied one end of it, and we took up our abode in the other. The floor was of earth; upon this we spread some clean white sand, which was covered over with plaited leaves of the cocoa-nut tree. There were no partitions; but by hanging up some mats and native cloth, we soon succeeded in partitioning off a comfortable bed-room, sitting-room, and store-room. Our kitchen was the open yard behind our dwelling; and its only fixtures were a couple of large stones placed in the ground, parallel to each other, and about six inches apart. This was our stove, or fire-place, and during the dry season, answered tolerably well.

With the study of the language, the erection of a printing-office and a dwelling house now demanded my attention. A spot near the principal stream was selected for their site; the inhabitants of the district undertook to build the printing-office, while the king's people, and the inhabitants of Maatea, agreed to put up the frame of my dwelling-house. The acquisition of the language I commenced with Mr. Crook, and was happy to avail myself of the aid of Mr. Davies, who was well acquainted with it, and willing to render us every assistance which his other avocations would admit.

The natives of Afareaitu, and of the neighbouring districts, were rejoiced at our coming among them; they seemed a people predisposed to receive instruction. A spacious chapel was erected prior to our arrival, and a large school was subsequently built; multitudes from other parts of the island took up their abode in the settlement, the school was filled with scholars, and the chapel well attended.

The indigenous productions of the island were abundant in the neighbourhood, and were comparatively cheap, as this part of the island had been but little visited by foreigners. When the flour, and other foreign articles of provision which we had brought from Port Jackson, were nearly expended, we subsisted almost entirely on native food; and though most of it was rather unsavoury at first, it afterwards became tolerably palatable. Wheat is not grown in any of the islands; it has often been tried, but, either from the heat of the climate, the exceeding fertility of the soil, or the absence of regular seasons, it has always failed. No other kind of grain, with the exception of a small quantity of maize, or Indian corn, is cultivated. Flour is, consequently, now only to be obtained from the vessels visiting the islands. It is, however, frequently brought from New South Wales, and from South or North America, and a tolerably good supply may, in general, be regularly obtained. The European is thus enabled to procure bread; which, amid all the varied productions of the country, is still to him "the staff of life."

The islands are certainly well stocked with all that the natives need for subsistence, in greater abundance than is, perhaps, to be found in any other part of the world, and, with a very small degree of care and industry, the inhabitants may, at all seasons, secure whatever is necessary to their comfortable maintenance. They have, it is true, neither beef nor mutton, nor any great variety of animal food; and, considering the heat of the climate, and the indolence of their habits, a vegetable diet is probably the most conducive to health. On public occasions, however, a considerable quantity of meat is dressed, and the chiefs seldom take a meal without it; but the generality of the people use, comparatively, but little animal food.

The flesh of swine, called by the natives *buaa*, constituted their principal meat. Pigs, which the natives say were brought by the first inhabitants, were found in the island by Wallis and Cook. Those originally found there differed considerably from the present breed, which is a mixture of English and Spanish. They are described as having been smaller than the generality of hogs now are, with long legs, long noses, curly or almost woolly hair, and short erect ears. An animal of this kind is now and then seen, and the people say such were the only hogs formerly in Tahiti. It was also said, that they, unlike all other swine, were wholly averse to the mire; and a phenomenon so novel among the habits of their species, produced a poetical effusion, which appeared in a periodical publication about five or six and twenty years ago. If such were the cleanly habits of the swine in Tahiti at that time, they have degenerated very much since, for I have often seen them stretched out at ease in a miry slough, apparently as much at home as the greatest pig would be in such a situation, in any other part of the world.

The hogs now reared are large, and often well fed; they are never confined in sties, but range about in search of food. Those that feed in the heads of the valleys live chiefly upon fruit, while those kept about the houses of the natives are fed occasionally with bread-fruit or cocoa-nuts. Unless well fed, they are very destructive to the fences and the native gardens, and will bite through a stick one or two inches in diameter, with very little effort: sometimes the natives break their teeth, or put a kind of yoke upon them; which, in some of the islands of the Pacific, is rather a singular one. A circular piece, as large as a shilling or a half-crown, is cut out of each ear, and when the wound has healed, a single stick, eighteen inches or two feet long, is passed through the apertures. This wooden bar lies horizontally across the upper part of the pig's head, and, coming in contact with the upright sticks of a fence, arrests his progress, even when he has succeeded in forcing his head through. The flesh of the pig is in general soft, rich, and sweet; it is not so fine as English-fed pork, neither has it the peculiarly agreeable taste by which the latter is distinguished. This is probably caused by the Tahitian swine feeding so much upon cocoa-nuts, and other sweet fruit. For the kind, however, native pork is very good; but, having little meat besides, we soon became tired of it. Although capable, when all the bones are taken out, of being preserved by salt, the natives never, till lately, thought of sitting down to less than a hog baked whole. Several of the chiefs, however, now only dress as much as is necessary for the immediate consumption of their families, and salt the remainder.

Next to the flesh of swine, that of the dog was formerly prized by the Tahitians, as an article of food. Nevertheless, dogs do not appear to have been reared for food so generally as among the Sandwich Islanders; here they were fed rather as an article of luxury, and principally eaten by the chiefs. They were usually of a small or middling size, and appear a kind of terrier breed, but were by no means ferocious; and, excepting their shape and habits, they have few of the characteristics of the English dog: this probably arises from their different food. The hog and the dog were the only quadrupeds whose flesh was eaten by the Tahitians. Rats were occasionally eaten uncooked by the Friendly Islanders; but, although numerous, they do not appear to have been used for that purpose here. The Tahitians have no kinds of game, but the common domestic fowls are reared in great numbers; these were used as food: they had also a few wild ducks, together with pigeons, in the mountains, and several kinds of aquatic birds. The fowls, although good, are not now much eaten by the natives, but are usually reared for the purpose of supplying the ships, which touch at the islands for refreshments.

Fish abound on the coasts, among the coral reefs that surround the islands, and in their extensive lagoons. The islanders are usually expert fishermen, and fish is a principal means of support for those who reside near the shore. The albicore, bonito, ray, sword-fish, and shark, are among the larger sea-fish that are eaten by them; in addition to which, they have an almost endless variety of rock-fish, which are remarkably sweet and good.

In the rivers they find prawns and eels, and in their lakes, where there is an opening to the sea, multitudes of excellent fish are always found; among others is a salmon, which, at certain seasons of the year, is taken in great abundance. It exactly resembles the northern salmon in size, shape, and structure, but the flesh is much whiter than that of the salmon of Europe, or those taken on the northern coasts of America; the taste is also the same, excepting that the Tahitian salmon is rather drier than the other. In the sand they find muscles and cockles, and on the coral reef a great variety of shell-fish; among which, the principal are crabs, lobsters, welks, a large species of cham, and several varieties of echinis, or sea-egg. Numbers of turtle are also found among the reefs and low coralline or sandy islands. The turtle was formerly considered sacred; a part of every one taken was offered to the gods, and the rest dressed with sacred fire, and eaten only by the king and chiefs; and then, I think, either within the precincts of the temple, or in its immediate vicinity; now they are eaten by whomsoever they are caught. Most of their fish is very good, and furnishes a dish of which we were never tired.

Formerly, the natives had but two methods of dressing their meat, fowl, and fish; these were, by wrapping it in leaves, and placing it in an oven of heated stones, or broiling it over the fire. Cooking utensils are now, however, introduced among them, and are generally used, where the natives have the means of purchasing them.

Edible fruits, roots, and vegetables, are found in plenty

and variety. The bread-fruit, *artocarpus*, is the principal, being produced in greater abundance, and used more generally, than any other. The tree on which it grows is large and umbrageous; the bark is light-coloured and rough; the trunk of the tree is sometimes two or three feet in diameter, and rises from twelve to twenty feet without a branch. The outline of the tree is remarkably beautiful, the leaves are broad, and indented somewhat like those of the fig-tree, frequently twelve or eighteen inches long, and rather thick, of a dark green colour, with a surface glossy as that of the richest evergreens.

The fruit is generally circular or oval, and is, on an average, six inches in diameter; it is covered with a roughish rind, which is marked with small square or lozenge-shaped divisions, having each a small elevation in the centre, and is at first of a light pea-green colour, subsequently it changes to brown, and when fully ripe assumes a rich yellowish tinge. It is attached to the small branches of the tree by a short thick stalk, and hangs either singly, or in clusters of two or three together.

There is nothing very attractive or pleasing in the blossom; but a fine stately tree, clothed with dark shining leaves, and loaded with many hundreds of large light green or yellowish coloured fruit, is one of the most splendid and beautiful objects to be met with, among the rich and diversified scenery of a Tahitian landscape. Two or three of these trees are often seen growing around the rustic native cottage, and embowering it with their interwoven and prolific branches. The tree is propagated by shoots from the root, it bears in about five years, and will probably continue bearing for fifty.

The bread-fruit is never eaten raw, except by pigs; the natives, however, have several methods of dressing it. When travelling on a journey, they often roast it in the flame or embers of a wood-fire; and, peeling off the rind, eat the pulp of the fruit: this mode of dressing is called *tunu pa,* crust or shell roasting. Sometimes, when thus dressed, it is immersed in a stream of water, and, when completely saturated, forms a soft, sweet, spongy pulp, or sort of paste; of which the natives are exceedingly fond.

The general and the best way of dressing the bread-fruit, is by baking it in an oven of heated stones. The rind is scraped off, each fruit is cut in three or four pieces, and the core carefully taken out; heated stones are then spread over the bottom of the cavity forming the oven, and covered with leaves, upon which the pieces of bread-fruit are laid; a layer of green leaves is placed over the fruit, and other heated stones are laid on the top; the whole is then covered in with earth and leaves, several inches in depth. In this state, the oven remains half an hour or longer, when the earth is cleared away, the leaves are removed, and the pieces of bread-fruit taken out; the outsides are in general nicely browned, and the inner part presents a white or yellowish, cellular, pulpy substance, in appearance, slightly resembling the crumb of a small wheaten loaf. Its colour, size, and structure are, however, the only resemblance it has to bread. It has but little taste, and that is frequently rather sweet; it is somewhat farinaceous, but by no means so much so as several other vegetables, and probably less so than the English potato, to which in flavour it is also inferior. It is slightly astringent, and, as a vegetable, it is very good, but is a very indifferent substitute for English bread.

To the natives of the South Sea Islands it is the principal article of diet, and may indeed be called their staff of life. They are exceedingly fond of it, and it is evidently adapted to their constitutions, and highly nutritive, as a very perceptible improvement is often witnessed in the appearance of many of the people, a few weeks after the bread-fruit season has commenced. For the chiefs, it is usually dressed two or three times a day; but the peasantry, &c. seldom prepare more than one oven during the same period; and frequently *tihana*, or bake it again, on the second day.

During the bread-fruit season, the inhabitants of a district sometimes join, to prepare a quantity of *opio*. This is generally baked in an immense oven. A large pit, twenty or thirty feet in circumference, is dug out; the bottom is filled with large stones, logs of firewood are piled upon them, and the whole is covered with other large stones. The wood is then kindled, and the heat is often so intense, as to reduce the stones to a state of liquefaction. When thoroughly heated, the stones are removed to the sides; many hundred ripe bread-fruit are then thrown in, just as they have been gathered from the trees, and are piled up in the centre of the pit; a few leaves are spread upon them, the remaining hot stones built up like an arch over the heap, and the whole is covered, a foot or eighteen inches thick, with leaves and earth. In this state it remains a day or two; a hole is then dug on one side, and the parties to whom it belongs take out what they want, till the whole is consumed. Bread-fruit baked in this manner, will keep good several weeks after the oven is opened.

Although the general or district ovens of opio were in their tendency less injurious than the public stills, often erected in the different districts, they were usually attended with debauchery and excess, highly injurious to the health, and debasing to the morals of the people, who generally relinquished their ordinary employment, and devoted their nights and days to mere animal existence, of the lowest kind—rioting, feasting, and sleeping, until the opio was consumed. Within the last ten years, very few ovens of opio have been prepared, those have been comparatively small, and they are now almost entirely discontinued.

Another mode of preserving the bread-fruit is by submitting it to a slight degree of fermentation, and reducing it to a soft substance, which they call *mahi*. When the fruit is ripe, a large quantity is gathered, the rind scraped off, the core taken out, and the whole thrown in a heap. In this state it remains until it has undergone the process of fermentation, when it is beaten into a kind of paste. A hole is now dug in the ground, the bottom and sides of which are lined with green *ti* leaves; the mahi is put into the pit, covered over with *ti* leaves, and then with earth or large stones. In this state it might be preserved several months; and, although rather sour and indigestible, it is generally esteemed by the natives as a good article of food during the scarce season. Previous to its being eaten, it is rolled up in small portions, enclosed in bread-fruit leaves, and baked in the native ovens.

The tree on which the bread-fruit grows, besides producing three, and in some cases four crops in a year, of so excellent an article of food, furnishes a valuable gum, or resin, which exudes from the bark, when punctured, in a thick mucilaginous fluid, which is hardened by exposure to the sun, and is very serviceable in rendering water-tight the seams of their canoes. The bark of the young branches is used in making several varieties of native cloth. The trunk of the tree also furnishes one of the most valuable kinds of timber which the natives possess, it being used in building their canoes and houses, and in the manufacture of their articles of furniture. It is of a rich yellow colour, and assumes, from the effects of the air, the appearance of mahogany; it is not tough, but durable when not exposed to the weather.

It is very probable, that in no group of the Pacific Islands is there a greater variety in the kinds of this valuable fruit, than in the South Sea Islands. The several varieties ripen at different seasons, and the same kinds also come to perfection at an earlier period in one part of Tahiti than in another; so that there are but few months in the year in which ripe fruit is not to be found in the several parts of this island. The Missionaries are acquainted with nearly fifty varieties, for which the natives have distinct names—these I have by me, but it is unnecessary to insert them—the principal are, the *paea,* artocarpus incisa, and the *uru maohe,* artocarpus integrifolia.

Next to the bread-fruit, the *taro,* or *arum,* is the most serviceable article of food the natives possess, and its culture receives a considerable share of their attention. It has a large, solid, tuberous root, of an oblong shape, sometimes nine or twelve inches in length, and five or six in diameter. The plant has no stalk; the broad heart-shaped leaves rise from the upper end of the root, and the flower is contained in a sheath or spathe. There are several varieties; for thirty-three of which the natives have distinct names; and it is cultivated in low marshy parts, as the plant is found to thrive best in moist situations. A large kind, called ape, *arum costatum*, which is frequently planted in the dry grounds, is also used in some seasons, but is considered inferior to the taro.

All the varieties are so exceedingly acrid and pungent in their raw state, as to cause the greatest pain, if not excoriation, should they be applied to the tongue or palate. They are always baked in the same manner as bread-fruit is dressed; the rind, or skin, being first scraped off with a shell. The roots are solid, and generally of a mottled green or gray colour; and when baked, are palatable, farinaceous, and nutritive, resembling the Irish potato more than any other root in the islands.

The different varieties of arum are propagated either by transplanting the small tubers, which they call *pohiri*, that grow round the principal root, or setting the top or crown of those roots used for food. When destitute of foreign supplies, we have attempted to make flour with both the bread-fruit and the taro, by employing the natives to scrape the root and fruit into a kind of pulpy paste, then drying it in the sun, and grinding it in a hand-mill. The taro in this state was sometimes rather improved, but the bread-fruit seldom is so good as when dressed immediately after it has been gathered.

The *uhi*, or yam, *dioscoria alata*, a most valuable root, appears to be indigenous in most of the South Sea Islands, and flourishes remarkably well. Several kinds grow in the mountains; their shape is generally long and round, and the substance rather fibrous, but remarkably farinaceous and sweet. The kind most in use is generally of a dark brown colour, with a roughish skin; it is called by the natives *obura*.

The yam is cultivated with much care, though to no very great extent, on account of the labour and attention required. The sides of the inferior hills, and the sunny banks occasionally met with in the bottoms of the valleys, are selected for its growth. Here, a number of small terraces are formed one above another, covered with a mixture of rich earth and decayed leaves. The roots intended for planting are kept in baskets till they begin to sprout; a yam is then taken, and each eye, or sprout, cut off, with a part of the outside of the root, an inch long and a quarter of an inch thick, attached to it; these pieces, sometimes containing two eyes each, are spread upon a board, and left in some part of the house to dry; the remainder of the root is baked and eaten. This mode of preparing the parts for planting does not appear to result from motives of economy, as is the case in some parts, where the Irish potato is prepared for planting in a similar manner; but because the natives imagine it is better thus to plant the eyes when they first begin to open, or germinate, with only a small part of the root, than to plant the whole yam, which they say is likely to rot. Whether the same plan might be adopted in planting the sweet potato, and other roots, I am not prepared to say, as it is only in raising the yam that it is practised in the horticulture of the natives. When the pieces are sufficiently dry, they are carefully put in the ground with the sprouts uppermost, a small portion of dried leaves is laid upon each, and the whole lightly covered with mould. When the roots begin to swell, they watch their enlargement, and keep them covered with light rich earth, which is generally spread over them about an inch in thickness.

The yam is one of the best flavoured and most nutritive roots which the islands produce. The natives usually bake them; they are, however, equally good when boiled; and, as they may be preserved longer out of the ground than any other, they are the most valuable sea-stock to be procured; and it is to be regretted that they are not more generally cultivated. Few are reared in the Georgian Islands; more perhaps in the Society cluster; but Sir Charles Sander's Island is more celebrated for its yams than any other of the group.

The *umara*, or sweet potato, *convolvulus batatus*, or *chrysorizus*, is grown by the natives as an article of food. The richest black mould is chosen for its culture; and the earth is raised in mounds nine or ten feet in diameter, and about three feet high. They do not plant the roots; but in the top of these mounds insert a small bunch of the vines, which germinating, produce the tuberous roots eaten by the natives. In the Sandwich Islands, the sweet potato is one of the principal means of subsistence; here it is only partially cultivated, and is greatly inferior to those grown in the northern islands, probably from the difference of soil and climate. The roots are large, and covered with a thin smooth skin. In size, shape, and structure, they resemble several kinds of the Irish potato. The umara is very sweet, seldom mealy, and sometimes quite soft, and altogether less palatable than the taro or the yam. It is dressed by the natives in their stone ovens, and is only used when the bread-fruit is scarce.

Patara, is a root growing wild in the valleys, in shape and taste resembling a potato more than any other root found in Tahiti. It is highly farinaceous, though less nutritive than the yam; the stem resembles the woodbine or convolvulus. The natives say the flower is small and white; I never saw one, for it is not cultivated, and but seldom sought, as the tuberous root is small, and more than two are seldom found attached to the same vine or stalk.

The natives are acquainted with rice; but, although both the soil and climate would probably favour its growth, it has not yet been added to the edibles of Tahiti. We have not been very anxious to introduce it, as the quantity of water required for its culture, would, we have supposed, induce in such a climate a state of atmosphere by no means conducive to health. But though they have not rice, they have a plant which they call *hoi,* the shape and growth of which resembles the Patara; but in taste and appearance it is so much like rice, that the natives call the latter by the native designation of the former. It is very insipid, and only sought in seasons of scarcity.

The *pia,* or arrow-root, *chailea tacca,* is indigenous and abundant. It is sometimes cultivated; but in most of the islands it grows spontaneously on the high sandy banks near the sea, or on the sides of the lower mountains, and appears to thrive in a light soil and dry situation. Though evidently of a superior quality, and capable of being procured in any quantity, it requires some labour to render it fit for food, and on this account was not extensively used by the natives, but formed rather a variety in their dishes at public feastings, than an article of general consumption.

The growth of the arrow-root resembles that of the potato. Although indigenous, and growing spontaneously, it is occasionally cultivated in the native gardens, by which means much finer roots are procured. When it is raised in this manner, a single root uncut is planted; a number of tuberous roots, about the size of large new potatoes, are formed at the extremities of fibres, proceeding from the root which had been planted. The leaves are of a light green colour, and deeply indented; they are not attached to one common stem, but the stalk of each distinct leaf proceeds from the root. The stalk, bearing the flower, rises in a single shaft, resembling a reed, or arrow, three or four feet high, crowned with a tuft of light pea-green petalled flowers. These are succeeded by a bunch of green berries, resembling the berries of the potato.

When the leaves from the stalk dry or decay, the roots are dug up and washed; after which the rind is scraped off with a cowrie shell. The root is then grated on a piece of coral, and the pulp pressed through a sieve made with the wiry fibrous matting of the cocoa-nut husk. This is designed to remove the fibres and other woody matter which the root may contain. The pulp, or powder, is received in a large trough of water, placed beneath the rustic sieve. Here, after having been repeatedly stirred, it is allowed to subside to the bottom, and the water is poured off. Fresh water is applied and removed, until it flows from the pulp, tasteless and colourless; the arrow-root is then taken out, dried in the sun, and is fit for use.

The process is simple, but it requires considerable care to dry it properly. When partially dry, the natives were formerly accustomed to knead or roll it up in circular masses, containing six or seven pounds each, and in this state expose it to the sun till sufficiently dry to be preserved for use. By this process they prepared much that has been exported from the islands, which may account for its inferior colour, as the whole mass was seldom sufficiently dry to prevent its turning mouldy, and assuming a brown or unfavourable colour.

They had no means of boiling it, but were accustomed to put a quantity of the arrow-root powder with the expressed milk from the kernel of the cocoa-nut into a large wooden tray, or dish; and, having mixed them well together, to throw in a number of red-hot stones, which being moved about by thin white sticks, heated the whole mass nearly to boiling, and occasioned it to assume a thick, broken, jellied appearance. In this state it is served up in baskets of cocoa-nut leaves, and is a very rich sweet kind of food, usually forming a part of every public entertainment.

Arrow-root has recently been prepared in large quantities, as an article of exportation to England; but although it is by no means inferior to that brought from the West Indies, it has not been so well cleaned, dried, or packed, and has consequently appeared very inferior when it has been brought into the market. There is reason, however, to believe, that when the natives shall have acquired better methods of preparing their arrow-root, it may become a valuable article of commerce.

There is a very large and beautiful species of fern, called by the natives *nahe*; the leaves of which are fragrant, and, in seasons of scarcity, the large tuberous kind of root is baked and eaten. It is insipid, affords but little nutriment, and is only resorted to when other supplies fail. It is altogether a different plant from the fern, the root of which is eaten by the natives of New

Zealand. The berries, or apples, of the *nono, morindo citrifolia,* and the stalks of the *pohue, convolvulus Brasiliensis,* are also eaten in times of famine.

The fruits of the islands are not so numerous as in some continental countries of similar temperature, but they are valuable; and, next to the bread-fruit, the *haari,* or cocoa-nut, *coccos nucifera,* is the most serviceable. The tree on which it grows is also one of the most useful and ornamental in the islands, imparting to the landscape, in which it fails not to form a conspicuous object, all the richness and elegance of intertropical verdure.

The stem is perfectly cylindrical, three or four feet in diameter at the root, very gradually tapering to the top, where it is probably not more than eighteen inches round. It is one single stem from the root to the crown, composed apparently of a vast number of small hollow reeds, united by a kind of resinous pith, and enclosed in a rough, brittle, and exceedingly hard kind of bark. The stem is without branch or leaf, excepting at the top, where a beautiful crown or tuft of long green leaves appears like a graceful plume waving in the fitful breeze, or nodding over the spreading wood, or the humble shrubbery. The nut begins to grow in a few months after it is planted; in about five or six years, the stem is seven or eight feet high, and the tree begins to bear. It continues to grow and bear fifty or sixty years, or perhaps longer, as there are many groves of trees, apparently in their highest perfection, which were planted by Pomare nearly forty years ago. While the plants are young, they require fencing, in order to protect them from the pigs; but after the crown has reached a few feet above the ground, the plants require no further care.

The bread-fruit, the plantain, and almost every other tree furnishing any valuable fruit, arrives at perfection only in the most fertile soil; but the cocoa-nut, although it will grow in the rich bottoms of the valleys, and by the side of the streams that flow through them, yet flourishes equally on the barren sea-beach, amid fragments of coral and sand, where its roots are washed by every rising tide; and on the sun-burnt sides of the mountains, where the soil is shallow, and remote from the streams so favourable to vegetation. The trunk of the tree is used for a variety of purposes: their best spears were made with cocoa-nut wood; wall plates, rafters, and pillars for their larger houses, were often of the same material; their instrument for splitting bread-fruit, their rollers for their canoes, and also their most durable fences, were made with its trunk. It is also a valuable kind of fuel, and makes excellent charcoal.

The timber is not the only valuable article the cocoa-nut tree furnishes. The leaves, called *niau,* are composed of strong stalks twelve or fifteen feet long. A number of long narrow pointed leaflets are ranged alternately on opposite sides. The leaflets are often plaited, when the whole leaf is called *paua,* and forms an excellent skreen for the sides of their houses, or covering for their floors. Several kinds of baskets are also made with the leaves, one of which, called *arairi,* is neat, convenient, and durable. They were also plaited for bonnets or shades for the forehead and eyes, and were worn by both sexes. In many of their religious ceremonies they were used, and the *niau,* or leaf, was also an emblem of authority, and was sent by the chief to his dependents, when any requisition was made: bunches or strings of the leaflets were also suspended in the temple on certain occasions, and answered the same purpose as beads in Roman Catholic worship, reminding the priest or the worshipper of the order of his prayers. On the tough and stiff stalks of the leaflets, the candle-nuts, employed for lighting their houses, were strung when used.

Round that part of the stem of the leaf which is attached to the trunk of the tree, there is a singular provision of nature, for the security of the long leaves against the violence of the winds. A remarkably fine, strong, fibrous matting, attached to the bark under the bottom of the stalk, extending half way round the trunk, and reaching perhaps two or three feet up the leaf, acting like a bracing of network to each side of the stalk, keeps it steadily fixed to the trunk. While the leaves are young, this substance is remarkably white, transparent, and as fine in texture as silver paper. In this state it is occasionally cut into long narrow slips, tied up in bunches, and used by the natives to ornament their hair. Its remarkable flexibility, beautiful whiteness, and glossy surface, render it a singularly novel, light, and elegant plume; the effect of which is heightened by its contrast with the black and shining ringlets of the native hair it surmounts. As the leaf increases in size, and the matting is exposed to the air, it becomes coarser and stronger, assuming a yellowish colour, and is called *Aa.*

There is a kind of seam along the centre, exactly under the stem of the leaf, from both sides of which long and tough fibres, about the size of a bristle, regularly diverge in an oblique direction. Sometimes there appear to be two layers of fibres, which cross each other, and the whole is cemented with a still finer, fibrous, and adhesive substance. The length and evenness of the threads or fibres, the regular manner in which they cross each other at oblique angles; the extent of surface, and the thickness of the piece, corresponding with that of coarse cotton cloth; the singular manner in which the fibres are attached to each other—cause this curious substance, woven in the loom of nature, to present to the eye a remarkable resemblance to cloth spun and woven by human ingenuity.

This singular fibrous matting is sometimes taken off by the natives in pieces two or three feet wide, and used as wrapping for their arrow-root, or made up into bags. It is also occasionally employed in preparing articles of clothing. Jackets, coats, and even shirts, are made with the *aa*, though the coarsest linen cloth would be much more soft and flexible. To these shirts the natives generally fix a cotton collar and wristbands, and seem susceptible of but little irritation from its wiry texture and surface. It is a favourite dress with the fishermen, and others occupied on the sea.

The fruit, however, is the most valuable part of this serviceable, hardy, and beautiful plant. The flowers are small and white, insignificant when compared with the size of the tree or the fruit. They are ranged along the sides of a tough, succulent, branching stalk, surrounded by a sheath, which the natives call *aroe*, and are fixed to the trunk of the tree, immediately above the bottom of the leaf. Fruit in every stage, from the first formation after the falling of the blossom, to the hard, dry, ripe, and full-grown nut, that has almost begun to germinate, may be seen at one time on the same tree, and frequently fruit in several distinct stages on the same bunch, attached to the trunk of the same stalk.

The tree is slow in growth, and the fruit does not, probably, come to perfection in much less than twelve months after the blossoms have fallen. A bunch will sometimes contain twenty or thirty nuts, and there are, perhaps, six or seven bunches on the tree at a time. Each nut is surrounded by a tough fibrous husk, in some parts two inches thick; and when it has reached its full size, it contains, enclosed in a soft white shell, a pint or a pint and a half of the juice usually called cocoa-nut milk.

There is at this time no pulp whatever in the inside. In this stage of its growth the nut is called *oua,* and the liquid is preferred to that found in the nut in any other state. It is perfectly clear, and in taste combines a degree of acidity and sweetness, which renders it equal to the best lemonade. No accurate idea of the consistency and taste of the juice of the cocoa-nut can be formed from that found in the nuts brought to England. These are old and dry, and the fluid comparatively rancid; in this state they are never used by the natives, except for the purpose of planting or extracting oil. The shell of the *oua,* or young cocoa-nut, is often used medicinally.

In a few weeks after the nut has reached its full size, a soft white pulp, remarkably delicate and sweet, resembling, in consistency and appearance, the white of a slightly boiled egg, is formed around the inside of the shell. In this state it is called *niaa,* and is eaten by the chiefs as an article of luxury, and used in preparing many of what may be called the made-dishes of Tahitian banquets. After remaining a month or six weeks longer, the pulp on the inside becomes much firmer, and rather more than half an inch in thickness.

The juice assumes a whitish colour, and a sharper taste. It is now called *omoto,* and is not so much used. If allowed to hang two or three months longer on the tree, the outside skin becomes yellow and brown, the shell hardens, the kernel increases to an inch or an inch and a quarter in thickness, and the liquid is reduced to less than half a pint. It is now called *opau,* and, after hanging some months on the tree, falls to the ground. The hard nut is sometimes broken in two and broiled, or eaten as taken from the tree, but is generally used in making oil.

If the cocoa-nut be kept long after it is fully ripe, a white, sweet, spongy substance is formed in the inside, originating at the inner end of the germ which is enclosed in the kernel, immediately opposite one of the three apertures or eyes, in the sharpest end of the shell. This fibrous sponge ultimately absorbs the water, and fills the concavity, dissolving the hard kernel, and combining it with its own substance, so that the shell, instead of containing a kernel and milk, encloses only a soft cellular substance. While this truly wonderful process is going on within the nut, a single bud or shoot, of a white colour but hard texture, forces its way through one of the holes in the shell, perforates the tough fibrous husk, and, after rising some inches, begins to unfold its pale green leaves to the light and the air; at this time, also, two thick white fibres, originating in the same point, push away the stoppers or covering from the other two holes in the shell, pierce the husk in an opposite direction, and finally penetrate the ground. If allowed to remain, the shell, which no knife would cut, and which a saw would scarcely penetrate, is burst by an expansive power, generated within itself; the husk and the shell gradually decay, and, forming a light manure, facilitate the growth of the young plant, which gradually strikes its roots deeper, elevates its stalk, and expands its leaves, until it becomes a lofty, fruitful, and graceful tree.

There are many varieties of the cocoa-nut tree, in some of which the fruit is rather small and sweet. For each variety the natives have a distinct name, as well as for the same nut in its different stages of perfection. I have the names of six sorts, but it is unnecessary to insert them.

The juice of the nuts growing on the sea-shore does not appear to partake, in any degree, of the saline property of the water that must constantly moisten the roots of the tree. The milk of the nuts from the sandy beach or the rocky mountain, is often as sweet and as rich as that grown in the most fertile parts of the valley.

On first arriving in the islands, we used the cocoa-nut milk freely, but, subsequently, preferred plain water as a beverage; not that the milk became less agreeable, but because we supposed the free use of it predisposed to certain dropsical complaints prevalent among the people. Cocoa-nuts were formerly a considerable article of food among the common people, and were used with profusion on every feast of the chiefs; but, for some years past, they have been preserved, and allowed to ripen on the tree, for the purpose of preparing oil, which has recently become an article of exportation, although the value is so small as to afford them but little encouragement to its extended manufacture.

The cocoa-nut trees are remarkably high, sometimes

sixty or seventy feet, with only a tuft of leaves, and a number of bunches of fruit, on the top; yet the natives gather the fruit with comparative ease. A little boy strips off a piece of bark from a *purau,* branch, and fastens it round his feet, leaving a space of four or five inches between them, and then, clasping the tree, he vaults up its trunk with greater agility and ease than a European could ascend a ladder to an equal elevation. When they gather a bunch at a time, they lower them down by a rope; but when they pluck the fruit singly, they cast them on the ground. In throwing down the nuts, they give them a whirling motion, that they may fall on the point, and not on the side, whereby they would be likely to burst.

The cocoa-nut oil is procured from the pulp, and is prepared by grating the kernel of the old nut, and depositing it in a long wooden trough, usually the trunk of a tree hollowed out. This is placed in the sun every morning, and exposed during the day; after a few days the grated nut is piled up in heaps in the trough, leaving a small space between each heap. As the oil exudes, it drains into the hollows, whence it is scooped in bamboo canes, and preserved for sale or use. After the oil ceases to collect in the trough, the kernel is put into a bag, of the matted fibres, and submitted to the action of a rude lever press; but the additional quantity of oil, thus obtained, is inferior in quality to that produced by the heat of the sun.

In addition to these advantages, the shells of the large old cocoa-nuts are used as water-bottles, the largest of which will hold a quart; they are of a black colour, frequently highly polished, and, with care, last a number of years. All the cups and drinking vessels of the natives are made with cocoa-nut shells, usually of the omoto, which is of a yellow colour. It is scraped very thin, and is often slightly transparent. Their ava cups were generally black, highly polished, and sometimes ingeniously carved with a variety of devices, but the Tahitians did not excel in carving. The fibres of the husk are separated from the pulp by soaking them in water, and are used in making various kinds of cinet and cordage, especially a valuable coiar rope.

It is impossible to contemplate either the bread-fruit or cocoa-nut tree, in their gigantic and spontaneous growth, their majestic appearance, the value and abundance of their fruit, and the varied purposes to which they are subservient, without admiring the wisdom and benevolence of the Creator, and his distinguishing kindness towards the inhabitants of these interesting islands.

More rich and sweet to the taste, though far less serviceable as an article of food, is the *maia*, plantain and banana, *musa paradisaica*, and *musa sapientum*. These are also indigenous, although generally cultivated in the native gardens. They are a rich nutritive fruit, common within the tropics, and so generally known as to need no particular description here. There are not, perhaps, fewer than thirty varieties cultivated by the natives, besides nearly twenty kinds, very large and serviceable, that grow wild in the mountains. The *orea*, or maiden plantain, with the other varieties, comes to the highest perfection in the South Sea Islands, and is a delicious fruit. The stalk, or tree, on which these fruits grow, is seldom above eight or twelve feet high; the leaves are fine broad specimens of the luxuriance of tropical vegetation, being frequently twelve or sixteen feet long, eighteen inches or two feet wide, of a beautiful pea-green colour when fresh, and a rich bright yellow when dry. The fruit is about nine inches long, and in shape somewhat like a cucumber, excepting that the angles are frequently well defined, which gives to the fruit the appearance of a triangular or quadrangular prism when ripe, of a bright delicate yellow colour. Sixty or seventy single fruit are occasionally attached to one stalk. Each plantain stem, or tree, produces only one bunch of fruit, and when the fruit is ripe, it is cut down, and its place supplied by the suckers that rise around the root whence it originally sprung. If the suckers, or offsets, be four or five feet high when the parent stem is cut down, they will bear in about twelve months.

The fruit is not often allowed to ripen on the trees, but it is generally cut down as soon as it has reached its full size, and while yet green; the bunch is then hung up in the native houses to ripen, and eaten as each turns yellow. When they wish to accelerate their ripeness for a public entertainment, they cut them down green, wrap them in leaves, and bury them thirty-six or forty-eight hours in the earth, and on taking them out they are quite soft, and apparently ripe, but much more insipid than those which had gradually ripened on the tree, or even in the house. The kinds growing in the mountains are large, and, though rich and agreeable when baked, are most unpalatable when raw; they have a red skin, and a bright yellow pulp. Their native name is *fei:* their habits of growth are singular; for, while the fruit of all the other varieties hangs pendent from the stem, this rises erect from a short thick stalk in the centre of the crown or tuft of leaves at the top. In several of these islands, the *fei* is the principal support of the inhabitants. The plantain is a fruit that is always acceptable, and resembles in flavour a soft, sweet, but not juicy pear; it is very good in milk, and also in puddings and pies, and, when fermented, makes excellent vinegar.

The *vi*, or Brazilian plum, a variety of *spondias*, (*spondias dulcis* of Parkinson,) is an abundant and excellent fruit, of an oval or oblong shape, and bright yellow colour. In form and taste it somewhat resembles a magnum-bonum plum, but it is larger, and, instead of a stone, has a hard and spiked core, containing a number of seeds. The tree on which it grows is deciduous, and one of the largest found in the islands, the trunk being frequently four or five feet in diameter. The bark is gray and smooth, the leaf pinnatc, of a light green colour; the fruit hangs in bunches, and is often so plentiful, that the ground underneath the trees is covered with ripe fruit, while the satisfied, and almost surfeited pigs, lie sleeping round its roots.

The *ahia*, or jambo, *eugenia Mallaccensis*, is perhaps the most juicy of the indigenous fruits of the Society Islands. It resembles, in shape, a small oblong apple, is of a bright beautiful red colour, and has a white, juicy, but rather insipid pulp. Though grateful in a warm climate like Tahiti, its flavour is by no means so good as that of the ahia growing on the Sandwich Islands. Like the vi, it bears but one crop in the year, and does not continue in season longer than two or three months. Both these trees are propagated by seed.

In certain seasons of the year, if the bread-fruit be scarce, the natives supply the deficiency thus occasioned with the fruit of the mape or rata, a native chestnut, *tuscarpus edulis.* Like other chestnut-trees; the mape is of stately growth and splendid foliage. It is occasionally seen in the high grounds, but flourishes only in the rich bottoms of the valleys, and seldom appears in greater perfection than on the margin of a stream. From the top of a mountain I have often been able to mark the course of a river by the winding and almost unbroken line of chestnuts, that have towered in majesty above the trees of humbler growth. The mape is branching, but the trunk, which is the most singular part of it, usually rises ten or twelve feet without a branch, after which the arms are large and spreading.

During the first seven or eight years of its growth, the stem is tolerably round, but after that period, as it enlarges, instead of continuing cylindrical, it assumes a different shape altogether. In four or five places round the trunk, small projections appear, extending in nearly straight lines from the root to the branches. The centre of the tree seems to remain stationary; while these projections increasing, at length seem like so many planks covered with bark, and fixed round the tree, or like a number of natural buttresses for its support. The centre of the tree often continues many years with perhaps not more than two or three inches of wood round the medula, or pith; while the buttresses, though only about two inches thick, extend two, three, and four feet, being widest at the bottom. I have observed buttresses, not more than two inches in thickness, projecting four feet from the tree, and forming between each buttress natural recesses, in which I have often taken shelter from the rain. When the tree becomes old, its form is still more picturesque, as a number of knots and contortions are formed on the buttress and branches, which render the outlines more broken and fantastic.

The wood of the rata has a fine straight grain, but being remarkably perishable, is seldom used, excepting for fire-wood. Occasionally, however, they cut off one of the buttresses, and thus obtain a good natural plank, with which they make the long paddles for their canoes, or axe-handles. The leaf is large and beautiful, six or eight inches in length, oblong in shape, of a dark green colour, and, though an evergreen, exceedingly light and delicate in its structure. The tree bears a small white racimated panicle flower, esteemed by the natives on account of its fragrance. The fruit, which hangs singly, or in small clusters, from the slender twigs, is flat, and somewhat kidneyshaped. The same term is also used by the natives for this fruit, and the kidney of an animal. The nut is a single kernel, in a hard, tough, fibrous shell, covered with a thin, compact, fibrous husk. It is not eaten in a raw state; but, though rather hard when fully ripe, it is, when roasted in a green state, soft, and pleasant to the taste.

In addition to these, the *ti*-root, *dracanæ terminalis*, resembling exactly that found in the Sandwich Islands, is baked and eaten; and the *to*, or sugar-cane, *saccharum officinarum*, which grows spontaneously, and perhaps in greater perfection than in any other part of the world, was formerly cultivated, and eaten raw. On a journey, the natives often carry a piece of sugar-cane, which furnishes a sweet and nourishing juice, appeasing at once, to a certain degree, both thirst and hunger. Within a few years, they have been taught to extract the juice, and, by boiling it, to prepare a very good sugar.

These valuable indigenous productions are not only eaten when dressed, as taken from the tree, or dug out of the ground, but, by a variety of combinations, several excellent kinds of food are prepared from them, which may be termed the confectionary, or made-dishes, of Tahiti. With the ripe bread-fruit and plantains they make what they call *pepe,* which, when baked, looks not unlike soft gingerbread. A mixture of arrow-root and grated cocoa-nut kernel, wrapped in green bread-fruit leaves, and baked, is called *taota;* with the arrow-root and plantain they also make a number of sweet puddings, which are wrapped in plantain leaves, and baked in the native ovens. A rich sauce, called *taiero,* is made with very finely grated young cocoa-nut; which undergoes, before using, a slight degree of fermentation. It is prepared with much care; and being considered an article of great luxury, is usually thought an essential dish in their public entertainments. The taste of this sauce is not unsavoury; but it is too oily to suit an English palate.

The most general dish in the Southern Islands is what they call *popoi,* nearly resembling the *poe* of the Sandwich Islands. It is made with the ripe mountain plantain, either raw or baked, beaten up to a paste or jelly, and diluted with cocoa-nut milk. Another kind of *popoi* is made with bread-fruit, or *opio,* beaten up and diluted with cocoa-nut or plain water. With the riper cocoa-nuts they make a sauce called *mitiaro,* which is prepared by cutting the kernel of the cocoa-nut in thin slices, and putting it into a calabash with salt water, in which it is shaken every day until the nut is dissolved. This, although a most unpleasant mixture, is eaten by the natives as sauce to their fish, bread-fruit, and almost every other article of food.

The several kinds of animal food, and the varied edible vegetables, are good; and, could flour be procured, and occasionally beef or mutton, Europeans would find the diet every way adapted to their support; but the principal animal food being, with the exception of a few goats, either fowls or pork, (the kind of meat of which we are perhaps likely to tire sooner than of any other,) and wanting bread and all the varied preparations of flour, the native food is generally found not to afford that strength and nourishment which Europeans are accustomed to derive from the diet of their own country. The native fruits are delicious; and their number has been greatly increased by the addition of many of the most valuable tropical fruits. Oranges, shaddocks, limes, and other plants, were introduced by Captains Cook, Bligh, and Vancouver. Vines were originally taken by the Missionaries, but nearly destroyed by the natives in their wars. In 1824 I brought a number of plants from the Sandwich Islands; which, I have since heard, thrive well. Citrons, tamarinds, pine-apples, guavas, Cape mulberries, and figs; custard apples, *annona triloba,* and coffee plants, have at different times been introduced, and successfully cultivated, by the Missionaries. Many foreign vegetables have been tried, yet few of them thrive. The growth of wheat has been more than once attempted. Pumpkins, melons, water-melons, cucumbers, cabbages, and French-beans, flourish better than any other foreign vegetables.

To the list of the edible vegetables, fruits, and roots of the Society Islands, given in the preceding chapter, others might probably be added, but these are sufficient to shew the abundance, diversity, nutritiveness, delicacy, and richness of the provision spontaneously furnished to gratify the palate, and supply the necessities, of their inhabitants. Here man seemed to live only for enjoyment, and appeared to have been placed in circumstances, where every desire was satisfied, and where it might be imagined that even the apprehension of want was a thing unknown. Amid the unrestrained enjoyment of a bounty so diversified and profuse, it is hardly possible to suppose that the divine Author of all should neither be recognized nor acknowledged; or, that his very mercies should foster insensibility, and alienate the hearts of the participants in his bounty. Such, however, was the melancholy fact: Although

—— —— —— —— the soil untill'd
Pour'd forth spontaneous and abundant harvests,
The forests cast their fruits, in husk or rind,
Yielding sweet kernels or delicious pulp,
Smooth oil, cool milk, and unfermented wine,
In rich and exquisite variety;
On these the indolent inhabitants
Fed without care or forethought.

CHAP. XIV.

Times of taking food among the islands—Tradition of the origin of the bread-fruit tree—Tahitian architecture—Materials employed in the erection of native houses—Description of their various kinds of buildings—Usual enclosures—Increased demand for books—Establishment of the printing press—Eager anticipations of the people—First printing in the island done by the king—Printing the Gospel of St. Luke—Liberal aid from the British and Foreign Bible Society—Influence of the process of printing, &c. on the minds of the people—Visit of a party of natives from the eastern archipelago—Desire of the inhabitants for the scriptures—Applicants from different islands—Estimation in which the scriptures are held—Influence of the press in the nation—Number of works printed.

THE natives of the South Sea Islands have no regular times for eating, but arrange their meals, in a great measure, according to their avocations, or the supply of their provision. They usually eat some time in the forenoon; but their principal meal is taken towards the evening. Their food being lighter, and of a less stimulating kind, than that of Europeans, is usually consumed by them in much larger quantities at a time. They do not appear ever to have been very temperate in their diet, excepting from necessity, and many seem to have made the gratification of their appetite the means of shortening their existence.

We have often endeavoured to learn from the natives whether the vegetable productions used as food when the islands were discovered by Captain Wallis, were found on the islands by those who first peopled them; whether these colonists, from whatsoever country they may have come, had brought any seeds or roots with them; or whether they had been, at a more recent period, conveyed thither from any other islands,—but their answers with regard to the origin of most of them, have been so absurd and fabulous, that no correct inference can be drawn from them.

In reference to the origin of the bread-fruit, one of their traditionary legends states, that in the reign of a certain king, when the people ate *araea*, red earth, a husband and wife had an only son, whom they tenderly loved. The youth was weak and delicate; and one day the husband said to the wife, "I compassionate our son, he is unable to eat the red earth. I will die, and become food for our son." The wife said, "How will you become food?" He answered, "I will pray to my god; he has power, and he will enable me to do it." Accordingly, he repaired to the family marae, and presented his petition to the deity. A favourable answer was given to his prayer, and in the evening he called his wife to him, and said, "I am about to die; when I am dead, take my body, separate it, plant my head in one place, my heart and stomach in another, &c. and then come into the house and wait. When you shall hear at first a sound like that of a leaf, then of a flower, afterwards of an unripe fruit, and subsequently of a ripe round fruit falling on the ground, know that it is I, who am become food for our son." He died soon after. His wife obeyed his injunctions, planting the stomach near the house, as directed. After a while, she heard a leaf fall, then the large scales of the flower, then a small unripe fruit, afterwards one full-grown and ripe. By this time it was daylight; she awoke her son, took him out, and they beheld a large and handsome tree, clothed with broad shining leaves, and loaded with bread-fruit. She directed him to gather a number, take the first to the family god and to the king; to eat no more red earth, but to roast and eat the fruit of the tree growing before them.—This is only a brief outline of the tradition which the natives give of the origin of the bread-fruit. The account is much longer, and I wrote it out in detail once or twice from the mouth of the natives, but it is too absurd to demand attention or afford information. It was probably invented by some priest, to uphold the influence of the gods, and the tribute of first-fruits paid to the king. The origin of the cocoa-nut, chestnut, and yam, are derived from similar sources; the cocoa-nut having grown from the head of a man, the chestnut from his kidneys, the yams from his legs,—and other vegetable productions from different parts of his body.

The *fei*, or mountain plantain, beaten into a pulp, and diluted with cocoa-nut milk or water till brought to the consistency of arrow-root, as ordinarily prepared in England, was much used. Large quantities were usually prepared for every festival; a kind of cistern was made, with a framework of wood, and a lining of leaves, which when filled was a sufficient load for six men to carry. Seven or eight of these were sometimes filled, and carried on men's shoulders to one feast. The mode of preparing their made-dishes was seldom, according to our ideas, the most cleanly, and we rarely partook of any of their dressed food, excepting it had been cooked as brought from the garden, or prepared by our own servants.

From this enumeration of the various articles of diet procurable among the islands, it will be evident, that though neither wheat, oats, barley, pease, and beans, nor other pulse and grain, are grown, yet the aborigines with a moderate degree of labour may obtain the necessaries, and many of what are by them esteemed the luxuries, of life. Their diet and modes of living are, however, still very different from those to which a European has been accustomed, and which he finds, even in their altered climate, most conducive to his health. In this respect, the first Missionaries endured far greater privations than those who have at subsequent periods joined them. They were often without tea and sugar, had no other animal food than that which they procured in common with the natives, and but seldom obtained flour. For some years after our arrival in the islands, the supply of this last important article of diet was very inadequate and uncertain; we have been many months at a time without tasting it, either in the form of bread or any other preparation. The supply now procured is, however, more regular, and the introduction of goats furnishing milk, and the flesh of the kid, the feeding of cattle, by which means the residents are able to make butter and occasionally to kill an ox, has greatly improved their circumstances.

In a short time after our arrival at Afareaitu, the people began to erect the printing-office, and the frame of our dwelling. According to the directions of the king, and the arrangements among themselves, the work was divided between several parties. The people of Afareaitu erected the printing-office; and those of Maatea, a neighbouring district, my dwelling. The king wrote a letter to the chief of the district, hastening him in the undertaking, and in a few weeks came over himself, in order to encourage and stimulate the parties engaged in the work.

Fa-re is the term for house in most of the islands, and an account of the erection of those we occupied here will convey a general idea of their plan of building. The timber being prepared, they planted the square posts which support the ridge-pole about three feet deep. The piece forming the ridge was nearly triangular, flat underneath, but raised along the centre on the upper side, and about nine inches wide; the joints were accurately fitted, and square mortises were made, to receive the tenons formed on the top of the posts. As soon as these were firmly secured, it was raised by ropes, and fixed in its proper place. The side-posts were next planted, about three or four feet apart; these were square, and nearly nine inches wide. In the top of each post, a groove, about six inches deep and an inch and a half wide, was cut; in this was fixed a strong board, eight or nine inches broad, bevelled on the upper edge, forming a kind of wall-plate along the side of the house. The rafters, which they call *aho,* were put on next; they are usually straight branches of the *purau,* hibiscus tileaceus, an exceedingly useful tree, growing luxuriantly in every part of the islands. The poles used for rafters are about four inches in diameter at the largest end. As soon as they are cut, the bark is stripped off, and used in the manufacture of cordage, lines, &c. The rafters are then deposited in a stream of water for a number of days, in order to extract the juices with which they are impregnated, and which, the natives suppose, attract a number of insects, that soon destroy them. When taken out, the poles are dried, and considered fit for use. The wood is remarkably light, its growth is rapid, and though the old parts of the tree are exceedingly tough, the young branches or poles, used for rafters and other purposes, are soft and brittle, resembling the texture and strength of branches of the English willow. The foot of the rafter is partially sharpened, and about eighteen inches from the end a deep notch is cut, which receives the bevelled edge of the *ra-pe*, or wall-plate, while the upper extremity rests upon the ridge. The rafters are generally ranged along on one side, three feet apart, with parallel rafters on the opposite side, which cross each other at the top of the ridge, where they are firmly tied together with cinet, or the strong fibres of the *ieie*, a remarkably tough mountain plant. A pole is then fixed along, above the junction of the opposite rafters, and the whole tied down to pegs fastened in the piece of timber forming the ridge. The large wood used in building is of a fine yellow colour, the rafters are beautifully white; and as the house is often left some days in frame, its appearance is at once novel and agreeable.

The buildings are thatched with *rau fara*, (the leaves of the pandanus,) which are prepared with great care: When first gathered from the trees, they are soaked three or four days in the sea, or a stream of water. The sound leaves are then selected, and each leaf, after having been stretched singly on a stiff stick fixed in the ground, is coiled up with the concave side outwards. In this state they remain till they are perfectly flat, when each leaf is doubled about one-third of the way from the stalk, over a strong reed or cane six feet long, and the folded leaf laced together with the stiff stalks of the cocoa-nut leaflets. The thatch, thus prepared, is taken to the building, and a number of lines of cinct are extended above the rafters, and in each of the spaces between, from the lower edge to the ridge. The thatchers now take a reed of leaves, and fasten it to the lower ends of the rafters at the left extremity of the roof, and, placing another reed about an inch above it, pierce the leaves with a long wooden needle, and sew it to the lines fixed on the outer side of the rafters and in the space between them: when six or eight reeds are thus fixed, they pass the cord with which they are sewn two or three times round each of the three rafters over which the reed extends. Placing every successive reed about an inch above the last, they proceed until they reach the ridge. The workmen now descend, and carry up another course of thatch, in the same way inserting the ends of the reeds of the fresh course into the bent part of the leaves on the former. It is singular to see a number of men working underneath the rafters, in thatching a house.

When the roof is finished, the points only of the long palm-leaves are seen hanging on the outside; and the appearance within, from the shining brown colour of the leaves bent over the reeds, and the whiteness of the rafters, is exceedingly neat and ingenious. The inside of the rafters of the chiefs' houses, or public buildings, is frequently ornamented with braided cords of various colours, or finely-fringed white or chequered matting. These are bound or wrapped round the rafters, and the extremities sometimes hanging down twelve or thirteen inches, give to their roof or ceiling a light and elegant appearance. Most of the natives are able to thatch a house, but covering in the ridge is more difficult, and is only understood by those who have been regularly trained for the work. A quantity of large cocoa-nut, or fern leaves, is first laid on the upper part of the thatch, and afterwards a species of long grass, called *aretu*, is curiously fixed or woven from one end to the other, so as to remain attached to the thatch, and yet cover the ridge of the house.

The roof being finished, they generally level the ground within, and enclose the sides. In the erection of my house, this part was allotted to the king's servants. About thirty of them came one morning with a number of bundles of large white purau poles, from two to three inches in diameter. After levelling the floor, they dug a trench a foot deep round the outside, and then, cutting the poles to a proper length, planted them an inch and a half or two inches apart, until the building was completely enclosed, excepting the space left for a door in the front and opposite sides. In order to keep the poles in their proper place, two or three light sticks, called *tea*, were tied horizontally along the outside. Partitions were then erected in the same manner, as we were desirous, contrary to the native practice, to have more than one room. The house was now finished, and in structure resembled a large birdcage. In two of the rooms we laid down boards which we had brought from Port Jackson, and either paved the remainder of the floor with stones, or plastered it with lime. The outside was skreened with platted cocoa-nut leaves, lined with native cloth. This also constituted our curtains, and, hung up before the entrance to some of the apartments, answered the purpose of a door. Thus fitted up, our native house proved a comfortable dwelling during the months we remained at Afareaitu.

The houses of the natives, although varying in size and shape, were all built with the same kind of materials, and in a similar manner. Some of them were exceedingly large, capable of containing two or three thousand people. *Nanu*, a house belonging to the king, on the borders of Pare, was three hundred and ninety-seven feet in length. Others were a hundred, or a hundred and forty feet long. These, however, were erected only for the leading chiefs. As the population has decreased, a diminution has also taken place in the size of the dwellings, yet, for some time after our arrival, several remained an hundred feet in length. The chiefs seem always to have been attended by a numerous retinue of dependants, or Areois, and other idlers. The unemployed inhabitants of the districts where they might be staying, were also accustomed to attend the entertainments given for the amusement of the chiefs, and this probably induced the people to erect such capacious buildings for their accommodation.

Some of the houses were straight at each end, and resembled in shape an English dwelling; this was called *haupape:* but the most common form for the chiefs' houses was what they called *poté*, which was parallel along the sides, and circular at the ends. Houses of this kind have a very neat, light, and yet compact appearance. The above are the usual forms of their permanent habitations, and the durability of the house depends much upon the manner in which it is thatched: if there is much space between the reeds, it soon decays; but if they are placed close together, it will last five or seven years without admitting the rain. Occasionally two or three coverings of thatch are put on the same frame. The Tahitians are a social people, naturally fond of conversation, song, and dance; hence several families often resided under the same roof.

In addition to the oval and the oblong house, they often had the *fare pora*, the *fare rau*, and the *buhapa*, or other temporary dwellings, for encampments during the period of war, or when journeying through the mountains; and their *farau vaa*, or canoe houses, which were large, and built with care; a number of what they call *oa* were planted at unequal distances on both sides of the rafter and post, which being one piece of timber, tended to strengthen the building.

The floor of their dwellings was covered with long dried grass, which, although comfortable when first laid down, was not often changed, and, from the moisture occasioned by the water spilled at meals and other times, was frequently much worse than the naked sand or soil would have been. Their door was an ingenious contrivance, being usually a light trellis-frame of bamboo-cane, suspended by a number of braided thongs, and attached to a long cane in the upper part of the inside of the wall-plate—the thongs sliding backwards and forwards like the rings of a curtain, whenever it was opened or closed. Many of their houses are erected within their enclosures or plantations, but they generally stand on the shore, or by the wayside. Every chief of rank, or person of what in Tahiti would be termed respectability, has an enclosure round his dwelling, leaving a space of ten or twenty feet width withinside. This court is often kept clean, sometimes spread over with dry grass, but generally covered with black basaltic pebbles, or *anaana*, beautifully white fragments of coral. The *aumoa* is a neat and durable fence, about four feet high; the upright pieces are tenoned into a polished rail along the top, or surmounted with the straight and peeled branches of the purau or tamanu.

The size, structure, and conveniency of the Tahitian houses, such as Wallis found, and such as are here described, exhibit no small degree of invention, skill, and attention to comfort, and shew that the natives were even then far removed from a state of barbarism. They also warranted the inference that they were not deficient in capacity for improvement, and that, with better models and tuition, they would improve in the cultivation of every art of civilized life, especially when they should be put in the possession of iron and iron tools, as those they had heretofore used were rude stone adzes, or chisels of bone.

It is, however, proper to remark, that although all were capable of building good native houses, and many erected comfortable dwellings, yet great numbers, from indolence or want of tools, reared only temporary and wretched huts, as unsightly in the midst of the beautiful landscape, as they were unwholesome and comfortless to their abject inhabitants.

When our printing-office was finished, as the purau branches afforded but an indifferent shelter from the rain and wind, the sides of the printing-office were boarded, and one or two glass windows introduced; probably the first ever seen in Eimeo. The floor was covered partly with the trunks of trees split in two, and partly paved with stone. In searching for suitable stones, we pulled down the remaining ruins of one or two maraes in the neighbourhood, and, finding among them a number of smooth and level-surfaced basaltic stones, we were happy to remove them from the temple, and fix them in the pavement of the printing-office floor; thus appropriating them to a purpose very different indeed from that for which they were primarily designed, by those who had evidently prepared them with considerable labour and care.

Numbers of the inhabitants of several parts of Tahiti and Eimeo flocked to Afareaitu, to attend the means of instruction, and the public ordinances of religion, as it was more convenient to many than Papetoai. They were also anxious to see this wonderful machine, the printing-press, in operation, having heard much of the facility with which, when once it should be established, they would be supplied with books; an article at that time more valuable, in their estimation, than any other.

A few copies of the spelling-book printed in England had been taken to the island in the year 1811. Some hundred copies of a smaller spelling-book, and a brief summary of the Old and New Testament, the latter containing about seventy-five 12mo pages, had been printed at Port Jackson, and were in circulation; but many hundreds of the natives who had learned to read, were still destitute of a book. Others could repeat correctly, from memory, the whole of the books, and were anxious for fresh ones. In many families, where all were scholars, there was but one book; while others were totally destitute. The inhabitants of the neighbouring islands were in still greater need. I have seen many who had written out the whole of the spelling-book on sheets of writing paper; and others who, unable to procure paper, had prepared pieces of native cloth with great care, and then, with a reed immersed in red or purple native dye, had written out the alphabet, spelling, and reading lessons, on these pieces of cloth, made with the bark of a tree. It was also truly affecting to see many of them, not with phylacteries, but with portions of scripture, or the texts they had heard preached from, written on scraps of paper, or fragments of cloth, preserved with care, and read till fixed in the memory of their possessors. This state of affairs, together with the earnest desire of the people to increase their knowledge of sacred truth, rendered it exceedingly desirable that the press should be set to work as soon as possible. Within three months after our arrival at Afareaitu, every thing was in readiness, and on the 10th of June, 1817, the operations preparatory to printing were commenced.

Pomare, who was exceedingly delighted when he heard of its arrival, and had furnished every assistance in his power, both in the erection of the building, and the removal of the press, types, &c. from Papetoai, where they had been landed, was not less anxious to see it actually at work. He had for this purpose visited Afareaitu, and, on his return to the other side of the island, requested that he might be sent for whenever we should begin. A letter having been forwarded to inform him that we were nearly ready, he hastened to our settlement, and, in the afternoon of the day appointed, came to the printing-office, accompanied by a few favourite chiefs, and followed by a large concourse of people.

Soon after his arrival, I took the composing-stick in my hand, and observing Pomare looking with curious delight at the new and shining types, I asked him if he would like to put together the first A B or alphabet. His countenance was lighted up with evident satisfaction, as he answered in the affirmative. I then placed the composing-stick in his hand; he took the capital letters, one by one, out of their respective compartments, and, fixing them, concluded the alphabet. He put together the small letters in the same manner, and the few monosyllables composing the first page of the small spelling-book were afterwards added. He was delighted when he saw the first page complete, and appeared desirous to have it struck off at once; but when informed that it would not be printed till as many were composed as would fill a sheet, he requested that he might be sent for whenever it was ready. He visited us almost daily until the 30th, when, having received intimation that it was ready for the press, he came, attended by only two of his favourite chiefs. They were, however, followed by a numerous train of his attendants, &c. who had by some means heard that the work was about to commence. Crowds of the natives were already collected around the door, but they made way for him, and, after he and his two companions had been admitted, the door was closed, and the small window next the sea darkened, as he did not wish to be overlooked by the people on the outside. The king examined, with great minuteness and pleasure, the form as it lay on the press, and prepared to try to take off the first sheet ever printed in his dominions. Having been told how it was to be done, he jocosely charged his companions not to look very particularly at him, and not to laugh if he should not do it right. I put the printer's ink-ball into his hand, and directed him to strike it two or three times upon the face of the letters; this he did, and then placing a sheet of clean paper upon the parchment, I covered it down, and, turning it under the press, directed the king to pull the handle. He did so, and when the paper was removed from beneath the press, and the covering lifted up, the chiefs and attendants rushed towards it, to see what effect the king's pressure had produced. When they beheld the letters black, and large, and well defined, there was one simultaneous expression of wonder and delight.

The king took up the sheet, and having looked first at the paper and then at the types with attentive admiration, handed it to one of his chiefs, and expressed a wish to take another. He printed two more; and, while he was so engaged, the first sheet was shewn to the crowd without, who, when they saw it, raised one general shout of astonishment and joy. When the king had printed three or four sheets, he examined the press in all its parts with great attention. On being asked what he thought of it, he said it was very surprising; but that he had supposed, notwithstanding all the descriptions which had been given of its operation, that the paper was laid down, and the letters by some means pressed upon it, instead of the paper being pressed upon the types. He remained attentively watching the press, and admiring the facility with which, by its mechanism, so many pages were printed at one time, until it was near sunset, when he left us; taking with him the sheets he had printed, to his encampment on the opposite side of the bay.

When the benefits which the Tahitians have already derived from education, and the circulation of books, are considered, with the increasing advantages which it is presumed future generations will derive from the establishment of the press, we cannot but view the introduction of printing as an auspicious event. The 30th of June 1817, was, on this account, an important day in the annals of Tahiti; and there is no act of Pomare's life, excepting his abolition of idolatry, his clemency after the battle of Bunaïna, and his devotedness in visiting every district of the island, inducing the chiefs and people to embrace Christianity, that will be remembered with more grateful feeling than the circumstance of his printing the first page of the first book published in the South Sea Islands.

The spelling-book being most needed, was first put to press, and an edition of 2600 copies soon finished. The king with his attendants passed by the printing-office every afternoon, on their way to his favourite bathing-place, and seldom omitted to call, and spend some time in watching the progress of the work. He engaged in counting several of the letters, and appeared surprised when he found that, in sixteen pages of the spelling-book, there were upwards of five thousand of the letter *a*. An edition of 2300 copies of the Tahitian Catechism, and a Collection of Texts, or Extracts, from Scripture, were next printed; after which, St. Luke's Gospel, which had been translated by Mr. Nott, was put to press.

While the spelling-book was in hand, Mr. and Mrs. Orsmond arrived in the islands, and took up their residence at Afareaitu; increasing thereby the enjoyment of our social hours.

The first sheet of St. Luke's Gospel was nearly printed, when the Active, with six Missionaries from England, arrived. Among them were our fellow-voyagers, Mr. and Mrs. Threlkeld, and our esteemed friends Mr. and Mrs. Barff; we had parted with them in England, and were truly rejoiced to welcome them to the distant shores of our future dwelling-place. By the same vessel, a supply of printing paper was sent from the British and Foreign Bible Society. Its arrival was most providential. The paper sent by the Missionary Society was only sufficient, after the elementary books had been finished, to enable us to print 1500 copies; but the arrival of the liberal grant from the Bible Society enabled us at once to double the number of copies. Although the demand has increased, and larger editions of the subsequent books have been necessary, the British and Foreign Bible Society has generously furnished the paper for every subsequent portion of the Scriptures that has been printed in the islands.

The composition and press-work of the elementary books, and of the greater portion of the edition of nearly 3000 copies of St Luke's Gospel, was performed almost entirely by Mr. Crook and myself. In the mean time, two natives were instructed to perform the most laborious parts; and, before the book was finished, they were able, under proper superintendence, to relieve us from the mechanical labour of press-work,—a department in which, they with others have been ever since employed; receiving regular payment for the same. In all works subsequently published, the Missionaries, on whom the management of printing has devolved, have been in a great measure relieved, by the aid of those instructed in that department of this useful art.

We laboured eight, and sometimes ten, hours daily, yet found that the work advanced but slowly. Notwithstanding all the care that had been exercised in selecting the printing materials and the accompanying apparatus, many things were either deficient or spoiled; here we could procure no proper supply, and the edition was not completed until the beginning of 1818. It was entitled, "*Te Evanelia na Luka, iritihia ei parau Tahiti,*" literally, The Gospel of Luke, taken out to be, or transferred to, the language of Tahiti; *E-parau hae-*

rehia te parau maitai o te hau nei e ati paatoai te ao nei ia ite te mau fenua atoa, was the motto "This good word (or gospel) of the kingdom shall be published in all the world," Matt. xxiv. 14. and the imprint was, *Neneihia i te nenei raa parau a te mau Misionari,* 1818. Pressed at the (paper or book) presser of the Missionaries.—There being no term in the native language answering to the word translated Gospel, the Greek word *Evangelion* was introduced, some of the consonants being omitted in conforming it to the native idiom.

The curiosity awakened in the inhabitants of Afareaitu by the establishment of the press, was not soon satisfied: day after day Pomare visited the printing-office; the chiefs applied to be admitted inside, while the people thronged the windows, doors, and every crevice through which they could peep, often involuntarily exclaiming, *Beri-ta-nie! fenua paari:* Oh Britain! land of skill, or knowledge. The press soon became a matter of universal conversation; and the facility with which books could be multiplied, filled the minds of the people in general with wonderful delight. Multitudes arrived from every district of Eimeo, and even from other islands, to procure books, and to see this astonishing machine. The excitement manifested frequently resembled that with which the people of England would hasten to witness, for the first time, the ascent of a balloon, or the movement of a steam-carriage. So great was the influx of strangers, that for several weeks before the first portion of the Scriptures was finished, the district of Afareaitu resembled a public fair. The beach was lined with canoes from distant parts of Eimeo and other islands; the houses of the inhabitants were thronged, and small parties had erected their temporary encampments in every direction. The school during the week, and chapel on the Sabbath, though capable of containing 600 persons, were found too small for those who sought admittance. The printing-office was daily crowded by the strangers, who thronged the doors, &c. in such numbers, as to climb upon each others backs, or on the sides of the windows, so as frequently to darken the place. The house had been enclosed with a fence five or six feet high; but this, instead of presenting an obstacle to the gratification of their curiosity, was converted into a means of facilitating it: numbers were constantly seen sitting on the top of the railing; whereby they were able to look over the heads of their companions who were round the windows.

Among the various parties in Afareaitu, at this time, were a number of the natives of the Paumotu, or Pearl Islands, which lie to the north-east of Tahiti, and constitute what is called the Dangerous Archipelago. These numerous islands, like those of Tetuaroa to the north, are of coralline formation, and the most elevated parts of many of them are seldom more than two or three feet above high-water mark. The principal, and almost only edible vegetable they produce, is the fruit of the cocoa-nut. On these, with the numerous kinds of fishes resorting to their shores or found among the coral reefs, the inhabitants entirely subsist. They appear a hardy and industrious race, capable of enduring great privations. The Tahitians believe them to be cannibals; but as to the evidence or extent of this charge, we cannot speak confidently. They are in general firm and muscular, but of a more spare habit of body than the Tahitians. Their limbs are well formed, their stature generally tall. The expression of their countenance, and the outline of their features, greatly resemble those of the Society Islanders; their manners are, however, more rude and uncourteous. The greater part of the body is tataued, sometimes in broad stripes, at others in large masses of black, and always without any of the taste and elegance frequently exhibited in the figures marked on the persons of the Tahitians. By the latter, the natives of the Pearl Islands were formerly regarded with the greatest contempt, as *teehae* and *maua,* savages and barbarians. It was some months since they had arrived from their native islands, which they had left for the purpose of procuring books and teachers for their countrymen. From the time of their landing, Pomare had taken them under his protection; and when he came over to Eimeo, they followed in his train.

A considerable party of the Aura tribe came one day to the printing-office, to see the press. When they were admitted, and beheld the native printer at work, their surprise and astonishment were truly affecting. They were some time before they would approach very near, and appeared at a loss whether to consider it as an animal or a machine. As their language is strikingly analogous to that spoken in the Society Islands, I entered into conversation with them. They were very urgent to be supplied with spelling-books, which I regretted my inability to effect to any extent, as our edition was nearly expended. Learning that they had discontinued idol-worship, I asked why they had abandoned their gods. They replied, that they were evil spirits, and had never done them any good, but had caused frequent and desolating wars. Moorea,* they

* He had been a professor of Christianity, and a pupil in the Mission-school some time before our arrival.

said, was their teacher, and had instructed them concerning the true God, for whose worship in the island of Anaa,* whence most of them came, they had already erected three chapels.

But little time was allowed for the drying of the printed sheets. The natives were in want of books, and most eager for them: the first inquiry of every party that arrived, usually was, "When will the books be ready?" The presses were therefore fixed, and, having acquired some knowledge of bookbinding as well as printing, before leaving England, I proceeded, as soon as the printing was finished, to binding, though but inadequately furnished with materials.

The first bound copy was sent to Papetoai, and is still, I believe, in Mr. Nott's possession; the second, half-bound in red morocco, was presented to the king, who evidently received it with high satisfaction. The queen and chiefs were next supplied, and preparations made for meeting the demands of the people. In order to preserve the books, it was deemed inexpedient to give them into the hands of the natives, either unbound, or merely covered as pamphlets. We had only a small quantity of mill-boards, and it was necessary to increase them on the spot; a large quantity of native cloth, made with the bark of a tree, was therefore purchased, and females employed to beat a number of layers or folds together, usually from seven to ten. These were afterwards submitted to the action of a powerful upright screw-press, and, when gradually dried, formed a good stiff pasteboard. For their covers, the few sheep-skins brought from England were cut into slips for the backs and corners, and a large bundle of old

* Prince of Wales's Island

newspapers dyed, for covers to the sides. In staining these papers, they were covered over with the juice of the stems of the mountain plantain, or fei. The young plants brought from the mountains were generally two or three inches in diameter at the lower end. The root was cut off above the part that had been in the ground, and the stem being then fixed over a vessel, half a pint sometimes of thick purple juice exuded from it. This was immediately spread upon the paper, imparting to the sheet, when dried in the sun, a rich glossy purple colour, which remained as long as the paper lasted. If lime-juice was sprinkled upon it, a beautiful and delicate pink was produced. When the juice of the *fei* was allowed to remain till the next day, the liquor became much thinner, assumed a brownish red tinge, and imparted only a slight colour to the paper.

The process of binding appeared to the natives much more simple than that of printing; yet, in addition to those whom we were endeavouring to instruct, each of the principal chiefs sent one of his most clever men, to learn how to put a book together. For some time we bound every book that was given to the natives; but our materials being expended long before they were supplied, and the people continuing impatient for the books, even in sheets—rather than keep them destitute of the Scripture already printed, they were thus distributed.

Those among the natives who had learned to bind were now overwhelmed with business, and derived no inconsiderable emolument from their trade, as they required each person to bring the pasteboard necessary for his own books, and also a piece of skin or leather for the back, or for the whole cover. Many soon learned to sew the sheets together, others cut pieces of wood very thin, instead of pasteboard, which were fastened to the sides; the edges of the leaves were then cut with a knife; and the book used in this state daily, while the owner was searching for a skin or a piece of leather, with which to cover it for more effectual preservation. This was the most difficult article to procure, and many books were used in this state for many months.

Leather was now the article in greatest requisition among all classes; and the poor animals, that had heretofore lived in undisturbed ease and freedom, were hunted solely for their skins. The printing-office was converted into a tanyard; old canoes, filled with lime-water, were prepared; and all kinds of skins brought to have the hair extracted, and the oily matters dissipated. It was quite amusing to see goats' dogs' and cats' skins collected to be prepared for book-covers. Sometimes they procured the tough skin of a large dog, or an old goat, with long shaggy matted hair and beard attached to it, or the thin skin of a wild kitten taken in the mountains. As soon as the natives had seen how they were prepared, which was simply by extracting the hair and the oil, they did this at their own houses; and in walking through the district at this period, no object was more common than a skin stretched on a frame, and suspended on the branch of a tree, to dry in the sun.

All the books, hitherto in circulation among the people, had been gratuitously distributed; but when the first portion of Scripture was finished, as it was a larger book than had yet been published, it was thought best to receive a small equivalent for it, lest the people should expect that books afterwards printed would be given also, and lest, from the circumstance of their receiving them without payment, they should be induced to undervalue them. A small quantity of cocoa-nut oil, the article they could most easily procure, was therefore demanded for each book, and cheerfully paid by every native. This was not done with a view of deriving any profit from the sale of the books, but merely to teach the people their value; as no higher price was required than what it was supposed would cover the expense of paper and printing materials,—and we still continued to distribute elementary books gratuitously.

The season occupied in the printing and binding of these books was one of incessant labour, which, in a tropical climate, and a season when the sun was vertical, was often found exceedingly oppressive; yet it was one of the happiest periods of my life. It was cheering to behold the people so prepared to receive the sacred volume, and anxious to possess it. I have frequently seen thirty or forty canoes from distant parts of Eimeo, or from some other island, lying along the beach; in each of which, five or six persons had arrived—whose only errand was to procure copies of the Scriptures. For these many waited five or six weeks, while they were printing. Sometimes I have seen a canoe arrive with six or ten persons for books; who when they have landed, have brought a large bundle of letters, perhaps thirty or forty, written on plantain leaves, and rolled up like a scroll. These letters had been written by individuals, who were unable to come and apply personally for a book, and had therefore thus sent, in order to procure a copy. Often, when standing at my door, which was but a short distance from the sea-beach, as I have gazed on the varied beauties of the rich and glowing landscape, and the truly picturesque appearance of the island of Tahiti, fourteen or eighteen miles distant, the scene has been enlivened by the light and nautilus-like sail of the buoyant canoe, first seen in the distant horizon as a small white speck, sometimes scarcely distinguishable from the crest of the waters, at others brilliantly reflecting the last rays of the retiring sun, and appearing in bold and beautiful relief before

> "The impassioned splendour of those clouds
> That wait upon the sun at his departure."

The effect of this magnificent scene has often been heightened by the impression that the voyagers, whose approaching bark became every moment more conspicuous among the surrounding objects, were not coming in search of pearls or gems, but the more valuable treasure contained in the sacred Scriptures, deemed by them "more precious than gold, yea, than much fine gold." One evening, about sunset, a canoe from Tahiti with five men arrived on this errand. They landed on the beach, lowered their sail, and, drawing their canoes on the sand, hastened to my native dwelling. I met them at the door, and asked them their errand. *Luka,* or *Te parau na Luka,* "Luke, or, The word of Luke," was the simultaneous reply, accompanied with the exhibition of the bamboo-canes filled with cocoa-nut oil which they held up in their hands, and had brought as payment for the copies required. I told them I had none ready that night, but that if they would come on the morrow, I would give them as many as they needed; recommending them, in the mean time, to go and lodge with some friend in the village. Twilight in the tropics is always short, it soon grew dark; I wished them good night, and afterwards retired to rest, supposing they had gone to sleep at the house of some friend; but, on looking out of my window about daybreak, I saw these five men lying along on the ground on the outside of my house, their only bed being some platted cocoa-nut leaves, and their only covering the large native cloth they usually wear over their shoulders. I hastened out, and asked them if they had been there all night: they said they had: I then inquired why they did not, as I had directed them, go and lodge at some house, and come again. Their answer surprised and delighted me: they said, "We were afraid that, had we gone away, some one might have come before us this morning, and have taken what books you had to spare, and then we should have been obliged to return without any; therefore, after you left us last night, we determined not to go away till we had procured the books." I called them into the printing-office, and, as soon as I could put the sheets together, gave them each a copy; they then requested two copies more, one for a mother, the other for a sister; for which they had brought payment. I gave these also. Each wrapped his book up in a piece of white native cloth, put it in his bosom, wished me good morning, and without, I believe, eating or drinking, or calling on any person in the settlement, hastened to the beach, launched their canoe, hoisted their matting sail, and steered rejoicing to their native island. This is only one instance among many that occurred at the time, both at Afareaitu and Papetoai, exhibiting the ardent desire of the people in general to possess the Scriptures as soon as they could be prepared for them They frequently expressed their apprehensions lest the number of the books should not be sufficient for those who were waiting; and have more than once told us, that the fear of being disappointed has often deprived them of sleep.

Many were doubtless influenced by motives of curiosity, others by a desire to possess an article of property now so highly esteemed by all parties, but many were certainly influenced by a desire to become more fully acquainted with the revelation God had made to man, and to read for themselves, in their own language, those truths that were able to make them "wise unto salvation." By some, after the first emotion of curiosity had subsided, the books were probably neglected; but by most they were carefully and regularly read, becoming at once the constant companion of their possessors, and the source of their highest enjoyment.

When the Gospel of Luke was finished, an edition of Hymns in the native language was printed, partly original and partly translations from our most approved English compositions; and although the book was but small, it was acceptable to the people, who are exceedingly fond of metrical compositions, their history and traditions having been preserved in a metrical kind of ballad. This circumstance rendered the Hymn-book which was completed at Huahine, quite a favourite, and afforded the means, not only of assisting them in the matter of their praises to Almighty God, but enabled them to convey the most important truths of revelation in the manner most attractive and familiar to the native mind.

While engaged in these labours, the principal object besides, that occupied our attention, was the study of the language. Several hours every day were devoted to its acquisition, and twice a week we met, when we were assisted by the instructions of Mr. Davies, who favoured us with the use of his manuscript vocabulary, and the outlines of a grammar, which he had prepared several years before. In addition to these means, I found the composing, or setting, of the types for the Tahitian books, the best method of acquiring all that was printed in the language. Every letter in every word passing repeatedly, not only under my eye, but through my hand, I acquired almost mechanically the orthography. The number of natives by whom we were always surrounded, afforded the best opportunities for learning the meaning of those words which we did not understand. The structure of many sentences was also acquired by the same means; and, in much less than twelve months, I could converse familiarly on any common subject.

My acquisition of the language was thus facilitated by attention to the printing in the native language. The use of the press in the different islands, we naturally regard as one of the most powerful human agencies that can be employed in forming the mental and moral character of the inhabitants, imparting to their pursuits a salutary direction, and elevating the whole community. It is not easy to estimate correctly the advantages already derived from this important engine of improvement. The sacred scriptures, and the codes of laws, are the only standard works of importance yet printed. The whole of the New, and detached portions of the Old Testament, have been finished, and the remaining parts are in progress.

In the native language, they also possess Old and New Testament histories—several large editions of spelling-books, reading lessons, and different catechisms—a short system of arithmetic—the codes of laws for the different islands—regulations for barter, and their intercourse with shipping. Numerous addresses on the subject of Christian practice—several editions of the native hymn-book—the reports of their different Societies—and, lastly, they have commenced a periodical publication called the Repository. I have received the first number, and most earnestly hope they will be able to carry it on. Every work yet printed has been prepared by the Missionaries, with the assistance of the most intelligent among the people. But we look forward, with pleasing anticipation, to the time when the natives themselves shall become writers. In the investigation and illustration of many things connected with the peculiar genius and character of their own countrymen, they will have advantages which no individual, who is a foreigner, can ever possess; and we may hope that the time is not far distant, when they will not only have standard works by native authors, but that their periodical literature will circulate widely, and spread knowledge and piety among all classes of the people.

CHAP. XV.

Arrival of Missionaries from England—Retrospect of labour at Afareaitu—Honesty of the people—Departure from Eimeo—Voyage to the Society Islands—Appearance of Huahine—Fa-re harbour and surrounding country—Accommodations on shore—Building and launching of the Haweis—Re-occupation of Matavai—New stations in Tahiti—Journeys across the interior of Eimeo—Village of Tamai—State of the inhabitants of Huahine—Commencement of Missionary labours—Influence of presents to the people.

About a month after our departure from Papetoai, Mr. Orsmond, who had sailed from England about July, 1816, arrived at Eimeo, and, after residing some time with the Missionaries at Papetoai, he removed to Afareaitu, pursued harmoniously with us the study of the language, assisted in preparing books for the people, and in other duties of the station, and subsequently accompanied us to the the Leeward Islands. On the 17th of November, in the same year, Messrs. Bourne, Darling, Platt, and Williams, with their wives, who sailed from England 17th November, 1816, reached the islands. Mr. and Mrs. Threlkeld, who had sailed with us from England, but had been obliged by domestic affliction to remain at Rio Janeiro, and Mr. and Mrs. Barff, who had originally left England with Mr. Orsmond, joined us by the same conveyance. This event was truly cheering to their predecessors, conveying the strongest evidence of the desire, on the part of the

Society at home, to relieve them from every distressing anxiety as to their successors, and to afford every aid in the prosecution of their important and extending work. To us it was a matter of gratitude and satisfaction. With some who had now arrived, we had parted nearly two years before in our native land; others we had left among strangers on a foreign shore; but we were now, in the providence of God, brought together under circumstances peculiarly encouraging; and not only permitted to enjoy each others' society, but to combine our energies for the advancement of that cause to which our lives were devoted.

The arrival of so large a reinforcement enabled the Missionaries to make arrangements for re-occupying their original station in Tahiti, and establishing a Mission in the Society, or, as they are usually termed, when spoken of in connexion with Tahiti and Eimeo, the Leeward Islands. It was, however, thought desirable that no division of their numbers should take place until the vessel, which had been commenced building soon after the return from Port Jackson, should be finished, and the works prepared for the press were printed.

As soon as these objects were accomplished, we prepared to remove to the island of Huahine, the most windward of the group properly called the Society Islands. This name was given to the cluster (which includes Sir Charles Sander's Island, Huahine, Raiatea, Tahaa, Borabora, Maupiti or Maurua, Tubai, and the small islets surrounding them,) by Captain Cook, in honour of the Royal Society, at whose recommendation the voyages to the South Seas, which led to their discovery, were undertaken.

The king, and many of the chiefs of Tahiti and Eimeo, appeared to lament the removal of the press; but as Mr. Bourne, who was acquainted with the art of printing, had a small press and types, and others had been requested from England, it was the less to be regretted. The principal object attempted in the establishment of a station at Afareaitu having been accomplished, we left our houses and gardens, and took a most affectionate leave of our friends, who evinced great regret at our departure.

The season we had spent with them had been to us a period of no ordinary activity and excitement, and would probably be regarded by them as forming an era in their history. We trust some advantage was derived from the instructions they had received; and we have every reason to remember, with gratitude and satisfaction, the hospitality and kindness we experienced. Once a week, the people of Maatea, a neighbouring district, brought our family a present of bread-fruit, and other articles of food; the inhabitants of Afareaitu, and the district of Teavaro, took a similar one to our companions. We reposed the most entire confidence in the people, and had no reason to regret even the exposure of our property. We were robbed by an English servant, whom we had taken from Port Jackson, of linen and clothing; but, although we had no lock, and for a long time no bolt on our door, (which, when fastened, a native could at any time have opened, by putting his hand through the sticks and pushing back the bolt,) and though sometimes the door was left open all night,—yet we do not know that one single article was stolen from us by the natives, during the eighteen months we resided among them.

I have visited the district only once since; and although welcomed with every expression of pleasure by the people, I experienced a sensation of melancholy interest, in walking over the garden, the fences of which had been taken down, and a few flourishing shrubs only remained, to mark its original situation. Most of the valuable plants had been removed by the people to their own gardens, as the spot selected by me was not one which they would have preferred. A few cocoa-nuts which I had planted near the printing-office appeared to thrive, as they were protected by a light fence round each of the trees.

The vessel came round, took our goods, and the articles belonging to the printing-office, &c., on board, and proceeded to Papetoai, where we shipped our cattle. On the eighteenth of June, 1818, Mr. Davies, Mr. and Mrs. Williams, Mr. and Mrs. Orsmond, Mrs. Ellis, and myself, accompanied by a number of the principal chiefs, sailed from Eimeo to the Leeward Islands. We arrived at Huahine late on the evening of the following day, and some of our party went on shore, but it was not till the morning of the 20th that we reached the anchorage in Fare harbour.

Here I looked abroad with new and mingled emotions on the scene in which I was to commence my labours, and probably to spend the remainder of my life. The clear sky was reflected in the unruffled waters of the bay, which was bordered with a fine beach strewed with various shells. The luxuriant convolvulus, presenting its broad and shining leaves in striking contrast with the white coral and sand, spread its vines across the beach, even to the margin of the water, over which the slender shrub or the flowering tree often extended their verdant branches, while the groves of stately bread-fruit, and the clumps of umbrageous *callophyllum*, or tamanu trees, and the tall and gracefully waving cocoa-nuts, shaded the different parts of the shore.

The district of *Fa-re*, bordering the harbour of the same name, is about a mile and a half, or two miles, in length, and reaches from the shore to the centre of the island. It is bounded on the south by a range of mountains separating it from the district of Haapape, and on the north by the small district of Buaoa, whence a long, bleak point of land, called the *Faaao*, extending a considerable distance into the sea, and covered with tall cocoa-nut trees, adds much to the beauty of the shore, and the security of the harbour. A ridge of inferior hills divides the district in the centre, and it greatly increases the picturesque appearance of its scenery. A small river rises on the northern side of this ridge, and, flowing along the boundary between the two districts, meets the sea exactly opposite the northern entrance. Another stream, more broad and rapid, rises at the head of the principal valley, and flows in a circuitous course to the southern part of the bay. The district is well watered and wooded. The lower hills, at the time of our arrival, were clothed with verdure, and the mountains in the centre of the island, whose summits appeared to penetrate the clouds, were often entirely covered with trees. All was rich and luxuriant in vegetation, but it was the richness and the luxuriance of a wilderness; scarcely a trace of human culture could be seen, yet I could but think the scene

"How fair,
Were it but from sin refined:
Man how free, how happy here,
Were he pure as God is kind."

A few native houses were visible: there were not probably more than ten or twelve in the district, and the inhabitants might be occasionally seen guiding the light canoe across the bay, or leisurely walking beneath the grateful shade of the spreading trees. They were the rude untutored tenants of the place; their appearance and their actions were in perfect keeping with the scenes of wildness by which they were surrounded. The only clothing most of them wore was a girdle of cloth bound loosely round the waist, and a shade of cocoa-nut leaves over their foreheads. Notwithstanding this, it was impossible to behold without emotion either the scenery or the inhabitants.

The accompanying Plate exhibits an accurate representation of the outline and character of the scenery in the north-eastern parts of the district and harbour, although it was taken at a period subsequent to our arrival, when the landscape had been improved by partially clearing the ground near the shore, and erecting a number of houses.

In the forenoon of the day after we came to anchor, accompanied by Matapuupuu, we walked through the district, in search of a house for Mr. Orsmond and myself, and at length selected one on the southern side of the bay; belonging to Taaroarii, the young chief of the island, while Mr. and Mrs. Williams were accommodated with another belonging to Maau, a raatira, who resided near the anchorage. Towards noon, our goods were most of them landed, and taken into our new habitation. It was a large oval building, standing within ten or twelve yards of the sea, without either partitions or even sides, consisting simply of a large roof, supported by three pillars along the centre, and a number round the sides.

Drawn by Capt.[n] Rob.[t] Elliot. R.N. — Engraved by W. Le Pettit.

NORTH-EAST VIEW OF THE DISTRICT OF FA-RE, IN HUAHINE.

The floor was composed of stones, sand, and clay. Mr. and Mrs. Orsmond occupied one end, and we took up our abode in the other.

When our goods, &c. were all brought under its cover, and the boats had returned to the ship, we sat down to rest, and could not avoid gazing on the scenes around us, before we began to adjust our luggage. Large fragments of rock were scattered at the base of the mountains that rose on one side of our dwelling, the sea rolled within a few yards on the other, and in each direction along the shore there was one wild and uncultivated wilderness. A pair of cattle that we had brought from New South Wales, with a young calf, all of which had been landed from the ship during the morning, were tied to an adjacent bread-fruit tree; two or three milch goats from Eimeo, fastened together by bands of hibiscus bark tied round their horns, had already taken their station on the craggy projections at the foot of the mountain, and were cropping the herbage that grew in the fissures of the rocks. One of our little ones was smiling in the lap of its native nurse, while the other was playing on the dried grass lying by the side of the boxes on which we were sitting, and the natives, under the full influence of highly excited curiosity thronged around us in such numbers as to impede the circulation of the air.

Our first effort was to prepare some refreshment. The chiefs had sent us a present of bread-fruit and fish. A native youth, fourteen or fifteen years of age, leaving the crowd, came forward, and asked if he should cook us some bread-fruit. We accepted his offer; he became a faithful servant, and continued with us till we removed from the islands. He fixed two large stones in the ground for a fire-place, and, bringing a bundle of dry sticks from the adjacent bushes, lighted a fire between the stones, upon which he placed the tea-kettle. While he was employed in dressing our bread-fruit, &c. we removed some of the boxes, piled up our luggage as compactly as we could, and, when the food was prepared, sat down to a pleasant repast of fried fish, bread-fruit, and plantains, cocoa-nut milk, and tea. As a beverage, we always preferred the latter, although the former is exceedingly pleasant.

The large island of Raiatea lies immediately to the west of Fare harbour, and, by the time we had finished our meal, the sun was partly hid behind the high and broken summits of its mountains. This admonished us to prepare our sleeping-place, as the twilight is short, and we were not sure of procuring lights for the evening. The natives cut down four stout sticks from the neighbouring trees; these we fixed in the earthen floor, and fastening sheets and native cloth from one to the other, enclosed our bed-room; a couple of chests were carried into it, upon which we spread our bed, making up one for the children, by the side of our own, on some packages that lay on the floor. We procured cocoa-nut oil, and when it grew dark, breaking a cocoa-nut in half, took one end, and winding a little cotton-wool round the thin stalk of the leaflet of the tree, fixed it erect in the kernel of the nut. This we filled with the oil, and thus our lamp and oil were entirely the production of the cocoa-nut tree; the small piece of cotton-wick gathered from the garden in Eimeo, being the only article it had not supplied. These were the only kind of lamps we had for some years, and, though rude in appearance, gave a good light, when kept steady, and sheltered from the wind. Shortly, however, after sunset this evening, the land-breeze came down from the mountains. As we had no shelter for our lamp, we found it difficult to keep it burning, and at an early hour retired to rest, tying our screen down with strips of bark, to prevent its being blown aside by the wind. Notwithstanding the novelty of our situation, the exposure to the air from the mountains, the roaring of the heavy surf on the reefs, the inroads of dogs, pigs, and natives, with no other shelter than a pile of boxes; we passed a comfortable night, and rose refreshed in the morning, thankful for the kind protection we had experienced, gratified also to find that no article of our property had been stolen, though all was unavoidably exposed.

The vessel that had conveyed us from Eimeo to Huahine, and in the building of which the Missionaries were engaged when we arrived, had been undertaken jointly by them and the king, at the recommendation of the Governor of New South Wales, and the Rev S. Marsden. The king proposed to find materials, and the Missionaries labour. By this means they hoped they might be enabled to instil into the minds of the natives a spirit of enterprise, and induce them to desire to build ships for themselves. It was intended to employ the vessel in the pearl-fishery, among the Paumotu Islands to the eastward; to work her with native seamen; to take the pearls, and mother-of-pearl shell, to Port Jackson; bringing from that settlement tools, cutlery, and manufactured goods for the natives, and supplies for the Mission; thus proving, at the same time, a means of stimulating the people to habits of industry, and defraying to a certain degree the expenses of the Mission. Such were the views with which the building of the vessel was undertaken; but circumstances had arisen since that time, which left but little hope that these ends would ever be answered. The work was, however, already so far advanced, that all parties were unwilling to abandon it.

The vessel was about seventy tons burden, and the hull was nearly completed. The Missionaries who had arrived undertook to complete what their predecessors had commenced; and although it was an undertaking of great labour, it was ready to be launched in a few weeks after they had landed.

The 7th of December, 1817, being the day fixed for the launch, crowds of the inhabitants assembled to witness the spectacle: when the preparations were completed, the wedges were removed; but as the vessel did not move, strong ropes were passed round her stern, and a number of the natives on each side began pulling her towards the water. Pomare was present, and exerted all his influence to stimulate the natives employed in launching the ship. One of the king's orators, a short, plump, round-faced man, about fifty years of age, was perched upon a projecting rock by the sea-side, vociferating one of their *ubus*, or songs, on the launching of their own large canoes, suiting the action to the word, and using at times the most violent gesticulations, as if he imagined his own muscular powers alone were to move the vessel. They have a number of these kind of songs, some of considerable length, which I have at different times written down. They were designed to stimulate the men who were drawing the canoes into the water.

The natives employed in this work generally laid down on the beach short logs of the cylindrical trunk of the cocoa-nut tree, and drew the canoes over these natural rollers into the sea. Some of these songs were very short, as "*Iriti i mua, iriti i muri, e to, e to tau vaa ie:*" Lift up the stem, lift up the stern, and pull, and pull, my strong canoe. The song employed on the present occasion appeared rather a long one: I tried to comprehend its import, but, notwithstanding all the vociferation of the orator, it was recited with such rapidity, and there was so much din and clamour among the people, who on such occasions only put forth their strength in proportion to the noise which they make, that I could only now and then distinguish the word *pahi*, a large canoe or ship. Had I been able to hear more distinctly, it is probable that at that time I should not have understood the bard, as many words not in common use are found in their songs.

At length the vessel moved towards the sea, amid the shouts of the assembled multitudes. Before, however, she fairly floated, an accident occurred, which threw a damp over the spirits of all present. As she glided smoothly along towards the water, Pomare, who had stationed himself by the sea-side for the purpose, gave the vessel her name, by throwing a bottle of wine at her, and exclaiming, *Ia ora na oe e Haweis*, Prosperity to you, O Haweis. It having been agreed to designate the first vessel of any size built in the islands The Haweis, in honour of the late Dr. Haweis, who was the steady friend of the South Sea Mission, and in some respects may be said to have been its founder.

The circumstance of the king's throwing the wine at the ship, the breaking of the bottle, the red wine spreading abroad, and pieces of glass bottle flying in every direction, startled the natives who were pulling the ropes on that side of the vessel. They immediately left hold of the ropes, and stood gazing in astonishment alternately at the king, and the place against which the bottle had been thrown. Those on the opposite side continued pulling with all their might, and soon drew the vessel on one side, till she fell. One simultaneous cry, *Aue te pahi e,* Alas, the ship! or Oh, the ship! resounded in every direction, and the king seemed to think she would never be launched. With great effort she was replaced, during the same afternoon, in an upright position, and subsequently launched upon the bosom of the Pacific, amid the exulting shouts of the multitudes who thronged the shores.

The Haweis was afterwards rigged, and employed in conveying the Missionary families to their respective stations; after which she made one or two very profitless voyages to New South Wales. On account of the heavy expenses attending every voyage, although it was of great importance to maintain regular intercourse between the respective stations, and between the islands and the colony, it was found necessary to dispose of the ship, which had been built with so much cost and labour; she was sold in New South Wales, and is now employed in trading between Port Jackson and Van Diemen's Land.

Although finishing the vessel, and printing, required the greater number of the Missionaries to continue in Eimeo, these duties did not detain the whole, but left several at liberty to extend, in some degree, their efforts. Matavai, the original Missionary station, was the first that was re-occupied. Mr. Wilson, one of the Missionaries who first landed from the Royal Admiral in 1801, resumed his labours here in the early part of 1818, within a quarter of a mile of the spot from which he had been obliged to fly when the Mission was broken up in the close of the year 1809, and not far from the place where Mr. Lewis was murdered.

Mr. Bicknell, accompanied by Mr. Tessier, formed a station under the auspices of Tati, in the populous district of Papara. A new station was also commenced by Mr. Crook and Mr. Bourne at Papaoa, in the district of Faa; and when the Haweis was finished, Mr. Darling joined Mr. Wilson at Matavai. At the urgent request of Utami, the chief of the populous district of Atehuru, he subsequently commenced a Mission among his people at Bunaauïa or Burder's Point, whither Mr. Bourne also repaired.

The two stations at Eimeo being on opposite sides of the island, occasioned us frequent journeys from Afareaitu to Papetoai. These excursions, although they gave us an opportunity of examining more extensively the aspect of the country and the state of its inhabitants, often proved fatiguing. Sometimes we walked along the beach to Papeare, several miles to the north of our abode—ascended a low ridge of mountains, extending nearly to the sea—crossed the elevated eastern range—and extending through the defiles and ravines of the interior mountains, descended on the opposite side of the island, and approached the shore near the inland boundary of Opunohu bay. At other times, we travelled round in the neighbourhood of the shore, alternately walking on the beach, or, proceeding in a light canoe, paddled along the shallow water near the shore. Occasionally we passed through the inland village of

Tamae; and although, whenever we took this route, we had to walk three-quarters of a mile along the margin of the lake, up to our knees in water, yet we have always been amply repaid, by beholding the neatness of the gardens, and the sequestered peace of the village, by experiencing the generous hospitality, and receiving unequivocal proofs of the simple piety of its inhabitants. Once or twice, when approaching Tamae about sunrise, we have met the natives returning from the bushes, whither, by the break of day, they had retired for meditation and secret prayer. Their countenance beamed with peace and delight; and, *Ia ora oe ia Iesu, Ia ora oe i te Atua*—Peace to you from Jesus, Blessing on you from God—was the general strain of their salutation.

More than once we had to take our little boy, even before he was three months old, from Afareaitu, where he was born, to Papetoai, for medical advice.

These journeys were exceedingly wearisome: returning from one of them, night overtook us many miles before we reached our home; we travelled part of the way in a single canoe, but for several miles, where there was no passage between the reef and the shore, and the fragile bark was exposed without shelter to the long heavy billows of the Pacific, we proceeded along the beach, while the natives rowed the canoe upon the open sea. Two native female attendants alternately carried the child, while Mrs. Ellis and I walked on the shore, occasionally climbing over the rocks, or sinking up to our ankles in fragments of coral and sand. Wearied with our walk, we were obliged to rest before we reached the place where we expected to embark again. Mrs. Ellis, unable to walk any further, sat down upon a rock of coral, and gave our infant the breast, while I hailed the natives, and directed them to bring the canoe over the reef, and take us on board. Happily for us, the evening was fair, the moon shone brightly, and her mild beams, silvering the foliage of the shrubs that grew near the shore, and playing on the rippled and undulating wave of the ocean, added a charm to the singularity of the prospect, and enlivened the loneliness of our situation. The scene was unusually impressive. I remember distinctly my feelings as I stood, wearied with my walk, leaning on a light staff by the side of the rock on which Mrs. Ellis with our infant was sitting, and behind which our female attendants stood. On one side the mountains of the interior, having their outline edged, as it were with silver, from the rays of the moon, rose in lofty magnificence, while the indistinct form, rich and diversified verdure, of the shrubs and trees, increased the effect of the whole. On the other hand was the illimitable sea, rolling in solemn majesty its swelling waves over the rocks which defended the spot on which we stood. The most profound silence pervaded the whole scene, and we might have fancied that we were the only beings in existence, for no sound was heard, excepting the gentle rustling of the leaves of the cocoa-nut tree, as the light breeze from the mountain swept through them, or the loud hollow roar of the surf, and the rolling of the foaming wave, as it broke over the distant reef, and the splashing of the paddle of our canoe, as it approached the shore. It was impossible, at such a season, to behold this scene, exhibiting impressively the grandeur of creation, and the insignificance of man, without experiencing emotions of adoring wonder and elevated devotion, and exclaiming with the psalmist,

"When I consider thy heavens, the work of thy fingers, the moon and the stars which thou hast ordained; what is man, that thou art mindful of him, and the son of man, that thou visitest him?"

The canoe at length reached the shore; we seated ourselves in its stern, and, advancing pleasantly along for seven or eight miles, reached our habitation about midnight.

The island of Huahine had, in common with the others forming the leeward group, been visited by Mr. Nott, who had travelled round it, preaching to the inhabitants of the principal villages. The Missionaries who had been expelled from Tahiti, had remained here some months prior to their final departure for Port Jackson; but at these periods only a temporary impression had been made upon the minds of the people, which had in a great degree, if not altogether, subsided. After the abolition of idolatry in Tahiti and Eimeo, and the subsequent adoption of Christianity by their inhabitants, Mahine, the king of Huahine, had sent down Vahaivi, one of his principal men, with directions to the chiefs to burn the idols, demolish the temples, and discontinue the ceremonies and worship connected therewith. This commission was executed, and not only were their objects of worship destroyed, their temples thrown down, the houses of their idols consumed, and idol-worship no longer practised; but the rude stills employed in preparing ardent spirits from the sugarcane, and other indigenous productions, were either broken, or hid under ground. Intoxication, infant murder, and some of the more degrading vices, indulged under the sanction of their superstition, were also discontinued.

This change, although approved and effected by the principal chiefs on the islands, in conjunction with the messenger of the king, was nevertheless opposed. Several chiefs, of inferior influence, collecting their dependents, encamped on the borders of the lake near Maeva, and threatened to avenge the insult to the gods, by attacking the chiefs who had sanctioned their destruction. Both parties, however, after assuming a hostile attitude for some time, adjusted their differences, and returned in peace to their respective districts, mutually agreeing to embrace Christianity, and wait the arrival of the Missionaries, whose residence among them they had been led to expect. In this state we found them when we landed; they had, with the exception of one or two individuals, forsaken idolatry, and, in profession at least, had become Christians; probably without understanding the nature of Christianity, or feeling in any great degree its moral restraints or its sacred influence. A few, including two or three who had been to Eimeo, had acquired the elements of reading, or had learned to repeat the lessons in the spelling-book, more from memory than acquaintance with spelling and reading; the rest were nearly in the same state in which they were when visited in 1808 and 1809, excepting that their superstitious ceremonies were discontinued, and they had a building for the worship of the true God.

For a number of Sabbaths after our arrival, but few of the inhabitants assembled for public worship, and the schools were very thinly attended. Those who came were so little acquainted with the gospel, that in the lessons given in the school, and the addresses delivered to assemblies met for worship, it was found necessary to begin with the first principles of instruction, and of Christianity. Numbers excused themselves from attending, on account of the wearisomeness of learning their letters, when there was every reason to believe that unwillingness to conform to the precepts inculcated, was the true cause of their disinclination. They usually neglected public worship, because they said they did not know how to read; this being considered a sufficient apology for the non-observance of the Sabbath, or the social duties of religion. Such neglect was also frequently used as a cover for their vices. When spoken to on the impropriety of their conduct, they would sometimes answer, "We are not scholars," or, "We are not praying people;" these being the terms employed to designate those who made a profession of religion. Many were induced to keep back from the schools, and the place of public worship, from a desire to remain free from those restraints on their vicious practices, which such profession of Christianity was considered to impose.

Under these circumstances we acted upon the principles by which our predecessors had invariably regulated their endeavours to teach the inhabitants of Tahiti and Eimeo; and respecting which, after careful observation, I believe we are unanimous in our conviction that they are the true principles upon which any attempts to instruct a rude untutored people can be prosecuted with a prospect of the greatest ultimate success. We made no presents to those who were our scholars, more than to others from whom we had experienced an equal degree of hospitality; we offered no reward to any one for learning, and held out no prospect of personal or temporal advantage to our pupils and hearers; and studiously avoided presenting any other inducements to learn, than the advantages that would be secured to our scholars themselves, by the possession of that knowledge, which we were not only willing but desirous to impart. At the same time we were most anxious, distinctly and powerfully to impress on their minds the desirableness and necessity of their possessing correct ideas of the true God—the means of seeking his favour—the happiness that would result therefrom in the present life, and in that state of existence after death, to which this was but preparative—together with the vast increase of knowledge and enjoyment that would attend their being able to read the printed books,—preserve whatever they heard that was valuable, by making it fast upon the paper,—and corresponding by letter with their friends at a distance, as familiarly and distinctly as if they were present. By representations such as these, we endeavoured to excite in their minds a desire to hear the Scriptures read, and the Gospel preached, in the chapels, and to attend our instructions in the schools.

Had our means been ample, and had we, on landing, or when inviting the attention of the chiefs and people to the objects of our proposed residence among them, liberally distributed presents of cloth, ironmongery, &c. or even engaged in part to support the children that would receive our lessons, the chapel would undoubtedly have been well attended, and the scholars proportionably multiplied; but then it would have been only from the desire to receive a constant supply of such presents—a motive highly prejudicial to the individuals by whom it would have been indulged, destructive of the comfort, and disastrous to the future labours, of the Missionary among them. So long as our distributions had been frequent and increasingly valuable, the expressions of attachment would have been ardent, and the attendance regular; but when these had failed, their zeal, &c. would have declined, and the chapel and the school would have been deserted. In addition to this, whenever a fresh supply of articles, for our own maintenance or use, might have arrived, if we had not been equally liberal in the distribution of our presents, we should have been unhesitatingly charged with keeping for ourselves that which was designed for them, and thus have been involved in most unpleasant altercation.

The plans of procedure, in the commencement of a new Mission, must necessarily be regulated in a great degree by the circumstances of the people among whom it may be established; and the extreme poverty, or fugitive habits, of the parents, may render it desirable for the teachers either wholly or in part to maintain the scholars, in order to secure attendance. These instances are I believe very rare, and absolute necessity alone can warrant recourse to such a plan. Instruction itself will be undervalued; it can never be attempted but on a very limited scale, and will be always liable to vexatious interruptions. A system of maintenance should only be adopted in regard to such pupils as it is hoped are under religious impressions, or are training with a view to their becoming monitors or schoolmasters themselves. In those parts of the world where the scholars could not be supported while at the schools, it would be better for them to devote a portion of their time to such employment as would enable them to procure the means of subsistence themselves, than that they should receive their maintenance from the Mission.

These remarks apply principally to the commencement of a Mission among an unenlightened people, where a school will be an essential part of such establishment; at subsequent periods, rewards to those who have excelled, consisting of books, penknives, inkstands, slates, or other articles connected with the pursuits of the school, may be given with a good effect, tending rather to stimulate to diligent enterprise, than to cherish a spirit of dependent indolence.

In reference to presents made by Missionaries to chiefs, on their first settlement among an unenlightened people, I am disposed to think they are always injurious, when given with a view of gaining influence, or inducing their recipients to attend to religious instruction. Self-interest, or a desire for property, is the principle upon which the intercourse uncivilized persons have with foreigners visiting their country for purposes of commerce, &c. is regulated; the estimation in which such individuals are usually held, and the influence they exercise, is generally proportioned to the extent of their property, or the portions of it which the natives receive. Not a few instances have occurred among the islands of the Pacific, in which individuals, who, while their presents were unsparingly lavished upon the people, were regarded as kings and chiefs among them, but who, when they have experienced a reverse in their circumstances, have been treated with marked and contemptuous neglect. An equal degree of *this* kind of influence, the means of the Missionary will never enable him to gain among the people, nor ought he for a moment to desire it. Discouraging indeed will be his prospects, if the estimation in which he is held by those among whom he labours be only that which arises from their expectation of the presents he may make them. His influence must be of another and a higher order, if he desires to succeed.

The effect of a present on the mind of a rude or partially civilized chieftain is instantaneous, but requires constant repetition, or increase, to prevent its decline. The influence which a Missionary will aim to possess is more difficult to attain; but when once possessed, is of exceedingly greater value. It is the result of a conviction in the minds of the people, that his ultimate aim is their welfare; that he comes among them to promote, not his own, but their interest; and that his efforts tend to increase their knowledge and their enjoyments, and are adapted to put them in possession of the means of multiplying their comforts in this life, and leading them to future blessedness.

To produce and sustain this conviction in the minds of the people around him, should be one of the first and the constant endeavours of a Missionary. Until he has effected this, he can expect but little success; and when once, under the blessing of God, it is attained, one of the greatest difficulties in his way will be removed. This influence is not to be obtained by presents; these, the most rude and untutored heathen know, are seldom given unless an equivalent is expected in return; but it is to be gained by a full, plain, and explicit statement of his objects in the commencement of his work, and a uniform reference, in all his subsequent conduct, to the advancement of these objects. Uncivilized communities are often most shrewd observers of the conduct of those who enter their society, and pay far more regard to the actions and dispositions, than the mere declarations, of strangers. Singleness of aim, and purity of motive, embodied, before such observers, in undeviating and disinterested efforts, will in general be appreciated, although they may not soon yield themselves up to the influence of those efforts.

One of the most effectual means of implanting and preserving this impression, is the exhibition of uniform benevolence. The office and the aim of every Missionary require the exercise of this disposition in the highest degree; and he who would be successful, should by this identify himself, as far as possible, with the objects of his regard. Without officiously interfering with their individual or family affairs, he should interest himself in their welfare, and strive to share and alleviate their distress. Besides the deep commiseration, which their spiritual wretchedness will excite, he will often find their temporal afflictions and sorrows such, as to claim his tenderest sympathy. "Kindness is the key to the human heart," when the spirit is softened or subdued under the influence of sufferings, it is often most susceptible of salutary impression; and the exercise of Christian sympathy and kindness, in such a season, will seldom fail to produce, even among the most barbarous tribes, highly favourable results.

In mere casual visits, or journeys through the countries of uncivilized tribes, presents to their chiefs are necessary, and often desirable, even where a Missionary is a permanent resident; but they should always be given as a token of friendship and personal respect from the Missionary, or of good-will from some friends by whom they may have been sent, and not as a means of obtaining influence, or inducing the people to attend to instruction.

CHAP. XVI.

Arrivals in Huahine—Support of the Mission—Formation of the Tahitian Missionary Society—Place of meeting—Speech of the king—Formation of a Society in Huahine—Establishment of the Mission in Raiatea—Description of the district of Fare—Erection of dwellings—Preaching in the native language—Indolence of the South Sea Islanders—Means adopted for the encouragement of industry—Cotton plantation—Disappointment in returns—Arrival of Mr. Gyles—Introduction of the art of making sugar, &c.—Visit to Tahiti—Sugar plantations and mills in the Leeward Islands—Introduction of coffee from Norfolk Island—Culture and preparation of tobacco for exportation.

Shortly after our arrival in Huahine, a large boat belonging to Mahine, the chief of the island, two others belonging to Messrs. Orsmond and Williams, and a fleet of canoes, brought down from Eimeo a number of chiefs and people belonging to Huahine, Raiatea, &c. They had gone to Tahiti some years before, for the purpose of assisting Pomare in the resumption of his authority, had witnessed and participated the change that had taken place, and had afterwards prolonged their residence, in order to enjoy the advantages of instruction, until a Mission should be established in their native islands. Their arrival was welcomed with the liveliest satisfaction, and we were happy to receive their countenance and cooperation in the prosecution of our work. An excitement, highly beneficial in its tendency, was awakened in the minds of the people; who, influenced by the example and advice of their friends from Eimeo, attended in great numbers daily at the schools, and were seen in the chapel, not only on the Sabbath, but whenever it was open for public worship. Numerous applications were also made for spelling-books, of which, with others of an elementary kind, a supply had been printed in Eimeo.

When the whole of the Missionaries reached Huahine, it was proposed in the first instance to form only one station in the Leeward Islands; and that those of us who had but recently arrived from England, should unitedly prosecute the study of the language, with such assistance as Messrs. Davies and Nott could render us, until we should be able to perform divine service among the people, and conduct the affairs of a distinct station. The acquisition of the language engaged our constant attention; and we not only devoted some hours every day to its study, but met together two or three times a week to receive instruction, and facilitate our improvement.

We had not been many weeks at Fare before Tamatoa the king of Raiatea, with his brother, and a number of chiefs from Raiatea, Tahaa, and Borabora arrived. They were exceedingly anxious that some of our number should at once remove to their islands. Mai, the king or chief of Borabora, who was also at Huahine, had before written to the Missionaries, reminding them that Jesus Christ and his apostles did not confine themselves to one place, but visited different parts, that as many as could might receive their instructions. The necessities of the people were so very obvious, the prospects of usefulness so extensive, and the request of the chiefs so urgent, that, although unwilling to be deprived of the assistance of their seniors, in the acquisition of the language, Mr. Williams and Mr. Threlkeld felt it to be their duty to accompany Tamatoa, and the chiefs who were with him, to the island of Raiatea. They purposed to attempt their civilization, the establishment of schools, and, with the assistance of pious and intelligent natives, their instruction in the use of letters, and the first principles of religion; while they were cultivating such an acquaintance with the language as would enable them more fully to unfold the great objects of their Mission. They represented distinctly the disadvantages under which they should commence public instruction, from their very partial knowledge of the language; but the chiefs always replied, "Never mind that, you possess enough now to teach us more than we know, and we will make it our business to teach you our language." The visitors from Raiatea were supported in their application by a number of chiefs belonging to the same island; who, after residing some years in Eimeo, had now removed to Huahine, and were desirous of returning to their own possessions in Raiatea and Tahaa, yet did not wish to go unaccompanied by some of those, from whose instruction they had derived so much advantage.

It was always a matter of regret with the Missionaries, that the expenses of the establishment in the islands should be sustained altogether by the parent Society; and in order to diminish this, they had from time to time disposed of the fruits of their own industry, to the captains of vessels touching at Tahiti; or they had sent small quantities to New South Wales, receiving, in return, such articles as they were most in need of. The greater portion of the inhabitants having now embraced Christianity, they availed themselves of what appeared to them the most suitable means for impressing the minds of the converts with the principle laid down in the Scriptures, that it is the duty of those who enjoy the gospel, not only to maintain, but also to extend it. It appeared to them that both these ends might be answered most appropriately and effectually, by establishing among the natives a Missionary Society, auxiliary to the London Society, rather than by calling upon them, immediately after their conversion, to support the teachers labouring among them. Such a measure might, while they were but partially acquainted with the true nature and design of Christianity, have induced some, who were perhaps halting between two opinions, to infer that the Missionaries were influenced by motives of pecuniary advantage, in their endeavours to induce them to receive Christian instruction.

The inhabitants of the islands knew that many of the supplies which the families from time to time received, were sent by their friends in England, and procured by the voluntary contributions of those there, who had first sent, and subsequently maintained the Mission; and it was thought that it would be better that their contributions towards the support of Christianity, should be combined with those of the contributors to the Missionary Society; that the supplies for the teachers might still be drawn from this source, while at the same time the natives would be contributing towards the support of their own instructors, and yet identifying themselves with British Christians in their efforts to propagate Christianity throughout the world.

The plan was proposed to the king, and at once

approved by him; it was also mentioned to several of the leading chiefs, by whom it was favourably received. Auna told me that the king one day said to him, "Auna, do you think you could collect five bamboo canes of oil in a year?" He answered, Yes; and the king said, "Do you think you could appropriate so much towards sending the word of God to the heathens?" Again he answered in the affirmative; and the king again said, "Do you think those that value the gospel would think it a great labour to collect so much yearly for this purpose?" Auna answered, that he did not think they would. "Then," said the king, "think about it, and perhaps we can have a combination, or society, for this purpose." The king found several chiefs favourably disposed; the Missionaries also proposed it to others; and, as it met with general approbation, the approaching month of May was appointed for the formation of the association.

Mr. Nott came over to Afareaitu for the purpose of completing the plan. On the 23d of April, in the same year, Messrs. Nott, Davies, Orsmond, and myself, held a meeting with the king, at our house; when the principles upon which the society should be formed, and the rules by which it was proposed to regulate its proceedings, were considered, and, on the following day, finally adjusted.

The 13th of May, 1818, being the anniversary of the parent institution in England, was fixed for the establishment and organization of the native society. The king and chiefs met at Papetoai, and it was a delightful and interesting day to all who were present.

At sunrise we held a prayer-meeting in the English language. The natives held one among themselves at the same hour. The forenoon was appropriated to worship in English; at which time a sermon was preached by Mr. Henry, one of the senior Missionaries; and in the afternoon the services were entirely in the native language.

The chiefs and people assembled from most of the districts of Eimeo, and a number of strangers from Tahiti, residing at Papetoai, were also present. The extension of the Redeemer's kingdom had been the topic of discourses in the native congregation on the preceding Sabbath, and had in some degree prepared the minds of the people for entering more fully into the subject. The public services on this occasion were to commence at three o'clock in the afternoon; but long before the appointed hour, the chapel was crowded, and a far greater number than had gained admission, still remained on the outside.

Three or four hundred yards distant from the chapel, there was a beautiful and extensive grove. To this spot it was proposed to adjourn, and thither the natives immediately repaired, seating themselves on the ground under the cocoa-nut trees. At three o'clock we walked to the grove, and on entering it we beheld one of the most imposing and delightful spectacles I think I ever witnessed in the islands. The sky was clear, the smooth surface of the ocean rippled with the cool and stirring breeze. The grove, stately and rich in all the luxuriance of tropical verdure, extended from the white beach of coral and shells to the very base of the mountains, whose gradual ascent, and rocky projections, led to the interior. The long-winged and interwoven leaves of the trees formed a spreading canopy, through which a straggling sunbeam occasionally found its way, and among whose long and graceful leaflets the breeze from the ocean, sweeping softly, gave even a degree of animation to the whole. The grass that grew underneath appeared like a rich carpet, spread by nature for the interesting ceremony; pendulous plants, some verdant in foliage, others rich and variegated in blossom, hung from the projections of the rocks, while several species of convolvulus and climbing plants were twined round the trunks of the trees, or hung in gay festoons among the gigantic and wide-spread leaves of the grove, ornamenting the whole with their large and splendid pink blossoms. Near one of the large cocoa-nut trees whose cylindrical trunk appeared like a natural pillar supporting the roof, there was a rustic sort of stand, four or five feet above the ground, on which Mr. Nott took his station. Before him, in a large arm-chair provided for the occasion, sat Pomare, supported on the right by Tati, chief of Papara, and on the left by Upaparu, the king's secretary. A number of chiefs, with the queen and chief women of the islands, sat around; while thousands of the natives, attired in their gay and many-coloured native or European dresses, composed the vast assemblage, each one having come as to a public festival, in his best apparel. Pomare was dressed in a fine yellow tiputa, stamped on that part which covered his left breast with a rich and elegant scarlet flower, instead of a star. Most of the chiefs wore the native costume, and the females were arrayed in beautifully white native cloth, and yellow cocoa-nut-leaf shades, or bonnets with wreaths of sweet-scented flowers round their necks, or garlands of the same in their black and glossy hair. The services commenced with singing, in which many of the natives joined. A solemn prayer was offered, after which Mr. Nott delivered a short, animated, and suitable discourse, from the Eunuch's answer to Philip, Acts viii. 30, 31. As soon as this was concluded, Pomare addressed the multitude of his subjects around, proposing the formation of a society.

He began by referring them back to the ages that were past, and to the system of false religion by which they had been so long enslaved, reminding them very feelingly of the rigid exactions imposed in the name of their imaginary gods, for they were but pieces of wood, or cocoanut husk. He then alluded to the toil they endured, and the zeal and diligence so often manifested, in the service of these idols. To them the first-fruits of the field, the choicest fish from the sea, with the most valuable productions of their labour and ingenuity, were offered; and to propitiate their favour, avert their displeasure, and death, its dreaded consequence, human victims were so often slain. While referring to these dark and distressing features of their idolatry, the general seriousness of the assembly, and the indications of remorse or horror in the recollection of these cruelties, appeared to accompany and respond confirmation to his statements. In striking contrast with them, he placed the mild and benevolent motives and tendency of the Gospel of Jesus Christ, and the benefits its introduction had conferred: alluding to the very fact of their being assembled for the purpose which had convened them, as a powerful illustration of his remarks. He then stated the vast obligations they were under to God for sending them his word, and the partial manifestation of gratitude they had yet given. After this, he directed their attention to the miserable situation of those whom God had not thus visited, and proposed that, from a sense of the value of the gospel, and a desire for its dissemination, they should form a Tahitian Missionary Society, to aid the London Society in sending the Gospel to the heathen, especially those in the islands of the surrounding ocean; explaining the kind of remuneration given to the proprietors of ships, and the expensiveness even of sending Missionaries. "The people of Africa," said he, "have already done so; for though, like us, they have no money, they have given of their sheep, and other property. Let us also give of the produce of our islands,—pigs, or arrow-root, or cocoa-nut oil. Yet it must be voluntary, let it not be by compulsion. He that desires the word of God to grow where it has been planted, and to be conveyed to countries wretched as ours was before it was brought to us, will contribute freely and liberally to promote its extension: he who is unacquainted with its influence, and insensible to its claims, will not, perhaps, exert himself in this work. So let it be. Let him not be reproved; neither let the chiefs in general, nor his superiors, be angry with him on that account." Pomare on this occasion seemed anxious to impress the minds of the people with his desire that they should act according to the dictates of their own judgment, and not form themselves into a society, simply because he had recommended it. As he drew to the close of his address, he intimated his wish that those who approved of the proposal he had made, should lift up their right hands. Two or three thousand naked arms were simultaneously elevated from the multitude assembled under the cocoa-nut grove, presenting a spectacle no less imposing and affecting, than it was picturesque and new. The regulations of the society were then read, and the treasurer and secretaries chosen. By this time the shades of the evening began to gather round us, and the sun was just hidden by the distant wave of the horizon, when the king rose from his chair, and the chiefs and people retired to their dwellings, under feelings of high excitement and satisfaction. There was so much rural beauty and secluded quietude in the scene, and so much that was novel and striking in the appearance of the people, momentous and delightful in the object for which they had been convened, that it was altogether one of the most interesting meetings I ever attended.

Mahine, and the Leeward or Society Island chiefs, who had been present at the formation of the Tahitian Missionary Society, were desirous that Huahine, although it had not been equally favoured with facilities for receiving the gospel, should not be behind any of the Windward group in the efforts of its inhabitants to sustain and to propagate it. In a few months after their arrival, therefore, they proposed that a society, upon the plan of that established in Eimeo, should be formed in Huahine, in aid of the parent society in London. We were anxious to aid in the accomplishment of their design; and a day was fixed, on which a public meeting was to be held for its formation. In the forenoon of the 6th of October, 1818, Mahine, and the Missionaries of Huahine, Tamatoa, and those of Raiatea, Mai, and numbers from Borabora, repaired to the chapel, followed by crowds of the people. The place was soon filled, and a far greater number remained outside than were assembled under the roof. In order that as many as possible might hear, directions were given to take down one of the ends of the house;

this was soon done, and those who could not gain admission, were enabled to hear.

Temporary verandas or coverings of cocoa-nut leaves had been attached to the side of the house next the sea, widening it five or six feet, and on the other side it was also thrown open. A sermon was preached in the forenoon, and in the afternoon the people were addressed by Mahine, Taua, and other leading chiefs, on the advantages they had derived from the gospel, the destitute state of those who had not received it, and the obligation they were under to send it; proposing, at the same time, that each person, so disposed, should annually prepare a small quantity of cocoa-nut oil, which should be collected, sent to England, and sold, to aid the Society, which had sent teachers to Tahiti, in sending them to other nations.

Those who had been at Eimeo, and many of the inhabitants of Huahine, appeared interested in the details that were given of the condition of other parts of the world, and the efforts that had been made by Christians in England to send them the means of instruction. The presence of the chiefs of the different islands, with numbers of their people, the former devotees of their respective national idols, and the adherents of the different political parties, who had often within the last twenty years met each other in battle on the shores of Huahine or Raiatea, together with the novelty of the object, and the excitement of feeling which such a concourse of people necessarily produced, rendered the meeting exceedingly interesting, though to us it was less so than one subsequently held in Fare, and that which we had attended in Eimeo.

The Haweis having conveyed the Missionaries to

their respective stations, taken in cocoa-nut oil, and such other productions of the islands as were marketable at Port Jackson, left Tahiti, and touched at Huahine, on her way to the colony of New South Wales. Messrs. Williams and Threlkeld had availed themselves of the visit of the Active, in the month of September, to remove with their families to Raiatea, and form a new station in that large and important island. Tamatoa the king, and his brother, accompanied them, while the rest of the chiefs and people of that island followed in their boats and canoes. In the Haweis, which left Huahine early in December, 1818, Mr. Hayward, from Eimeo, proceeded on a voyage to Port Jackson, and Mr. and Mrs. Orsmond to Raiatea, while Messrs. Nott, Davies, Barff, and myself, remained at Huahine.

Our temporary dwelling was scarcely rendered comfortable, by partitioning the different rooms with bamboo-canes, and covering them with Tahitian cloth, when it was necessary to prepare for the erection of a printing-office, the supply of books brought from Eimeo being found unequal to the increasing demand. Mr. Nott was also revising, for the press, the Gospel by John, and Mr. Davies had the Gospel of Matthew ready. This rendered it expedient to examine the district, that we might select the most eligible place for the erection of our permanent dwelling, to which we purposed to attach the printing-office.

We were desirous of securing the advantages of garden-ground and water; but in seeking these, we avoided obliging the natives to remove from any of those spots which they had already appropriated to their own use. In this there was not much difficulty, the whole district was before us, and but few places, except in the vicinity of the shore, had been selected by the people, who were waiting till we had made our choice, that they might build as near our dwelling as would be convenient.

We explored the district carefully, but often found the brushwood, and interlaced branches of the trees, so impervious, that, without a hatchet, we should have penetrated but a short distance from the winding paths trodden by the natives. The soil was good throughout; and, as the natives had chosen the most eligible places along the shore, we fixed upon a small elevation near the junction of two clear and rapid streamlets, about a quarter of a mile from the entrance of the valley of Mahamene. It was at this time a complete wilderness, overgrown with rank weeds and thick brushwood. We commenced preparing it for the site of our dwelling; and when cleared, it was a most delightful spot.

A garden is a valuable acquisition in this part of the world; and, next to our dwellings, we regarded it as an important part of our domestic establishment. As soon as the sites of our houses were fixed, we employed natives to enclose a piece of ground adjoining them. I had received from governor Macquarie in New South Wales, a hundred ears of Egyptian wheat, which being a kind frequently grown in a warm climate, it was supposed might flourish in the islands. The grain was planted with care, and grew remarkably well; the leaf was green, the stalks high and strong, and the ears large; but as they began to turn yellow, it appeared that scarcely one of them contained a single grain of corn, and the few that were found, were shrivelled and dry. Potatoes were also tried, and have been repeatedly planted since, in different situations and seasons; but although, after the first growth, they usually appear like young potatoes,—if planted again, they are invariably soft and sweet, very small, and by no means so palatable as the indigenous sweet potato.

At Afareaitu, I had sown a number of seeds from England, Rio Janeiro, and New South Wales. Coffee and cashew-nuts, *anacardium occidentale*, I had before planted in boxes; they grew well, but the coffee and the cashew-nuts were totally destroyed by the goats, which, leaping the fence one day, in a few minutes ate up the plants, on which I had bestowed much care. I succeeded, however, in preserving the custard-apple, *anona triloba* or *squamosa*, that I had brought from Rio, and plants from it are now bearing fruit in several of the islands. In addition to these, I was enabled to cultivate the papaw apple, *carica papaya*, French-beans, carrots, turnips, cabbages, and Indian corn; while our little flower-garden, in Huahine, was adorned with the convolvulus major and minor, capsicum, helianthus, and amaranthus, with several brilliant native flowers, among which the *gardenia* and *hibiscus rosea chinensis* were always conspicuous. The front of our house was shaded by orange trees, and our garden enclosed with a citron hedge.

The comfort connected with a garden, and the means of support derived therefrom, were not our only inducements to its culture; we were desirous to increase the vegetable productions of the island, and anxious also that our establishments should become models for the natives in the formation of their own, and in this we were not disappointed. Before I left the islands, a neat little garden was considered by numbers as a necessary appendage to their habitation. The natives display a taste for the beautiful, in their fondness of flowers. The gardenia, hibiscus, and amaranthus, were often woven in most graceful wreaths or garlands, and worn on their brows. They were delighted when the helianthus was added to their flowers. The king and queen passed by my garden when the first ever grown in the islands was in flower, and came in, to admire its size and brilliant colours. Soon after their return, I received a note from the king, asking for a flower for the queen, and also one for her sister; I sent them each a small one; and the next time they appeared in public, the large sunflowers were fixed as ornaments in their hair.

A stream rolled at the bottom of a steep bank, about twenty yards from our houses. Two or three aged and stately chestnut-trees growing on the margin of this bank, extended their branches over the stream and the bank, casting around a grateful and an inviting shelter from the noontide sun.

Immediately behind this spot, *Matoereere*, black rock, the loftiest mountain in the island, towered in majesty above the surrounding hills. The lower part of the mountain appears basaltic; the central strata are composed of a vesicular kind of volcanic rock, while the upper parts are a large kind of breccia. It is verdant to its summit, which is of a beautiful conic shape, supported by a perpendicular rock. The inferior hills, on one side, were not only verdant, but to a considerable extent clothed with shrubs or trees, while a degree of sterile whiteness marked the basaltic and volcanic rocks on the other. These gave a richness and picturesque appearance to the landscape, which was greatly heightened by the lofty mountain in the centre. Often have I seen the mists and clouds resting on its sides, or encircling its brow, while the sunbeams have irradiated its summit; and it has appeared, especially when seen from a distance,

> "As some tall cliff that rears its awful form,
> Swells from the vale, and midway leaves the storm."

On the northern side of the valley, and near the foot of Matoereere, we proposed to erect our dwelling and the printing-house. Mr. Davies selected a spot between this place and the sea, on the same side; and Mr. Orsmond fixed upon one near the southern border of the harbour, and on the opposite side of the valley of Mahamene. which was spacious, fertile, well watered, and sufficiently high to be secure from dampness.

The people readily erected the frame of our house and the printing-office, which was put up much in the same manner as that had been which we occupied in Eimeo; but, as it was intended for a more permanent abode, it was finished with greater care. It had but one floor, excepting that over the printing-office there was a kind of loft for drying the paper. The front was boarded with materials brought from Port Jackson. The walls at the ends and the back were plastered with excellent coral lime; and both the printing-house and dwelling were floored with bread-fruit boards, split or sawn by the natives; the windows in the bed-rooms, sitting-rooms, study, and printing-house, were glazed; and what was a new and strange thing to the natives, our kitchen, in which was a stone oven, fire-place, and chimney, was included under the same roof.

Cooking houses were usually detached from the dwellings of the chiefs and foreigners, but we attached it to our house, that Mrs. Ellis might avoid exposure to the sun, and the heat of the middle of the day, whenever it might be necessary to instruct the servant, or superintend the dressing of our food. The partitions separating the different apartments were framed, wattled with thin sticks, and plastered; and although we found the labour of building oppressive, we were amply compensated by the comfort we subsequently enjoyed. The house was finished early in 1819, became our residence shortly afterwards, and continued so until we embarked for the Sandwich Islands.

Building houses, and avocations of a similar kind, were regarded as secondary objects; our main efforts were directed to the acquisition of the language. Whatever besides we had been able to do, we considered ourselves wholly inefficient, until we were capable of delivering our message to the inhabitants in their own tongue. We had many difficulties to encounter, and were obliged to pick up the greater part of the language from the natives, who, unacquainted with our speech, could only explain to us the meaning of words and phrases by their own: thus their explanations often increased our perplexity. My intimate acquaintance with all that had been printed, afforded me great facility in prosecuting the study of Tahitian. In less than a year, I was able to converse with the people on common topics, and preached my first sermon in Tahitian in the month of November, 1818.

I was much affected on giving up myself to Missionary pursuits, on leaving England, and on reaching the islands, but I had never so deeply felt the responsibility of my situation, and my insufficiency for the work, as I did on the day when I delivered my first native discourse. The congregation was large, the chiefs and Missionaries were present; and, at the appointed time, I commenced the services with singing, reading, and prayer, exercises in which I had occasionally engaged before. I had selected for the text what appeared a most suitable passage with which to commence my public ministry: "This is a faithful saying, and worthy of all acceptation, that Christ Jesus came into the world to save sinners." 1 Tim. i. 15. I was enabled to conclude the service with less difficulty than I expected, and was happy to have an opportunity of declaring, though very imperfectly, truths that were able to make those to whom they were delivered, wise unto salvation, through faith in Christ Jesus. In continuing my labours, I found it necessary, on account of the peculiarities of the native language, to write out most of my discourses, and commit them to memory, before I could venture to address them to the people.

The establishment of schools, the reducing to writing, and a regular grammatical system, uncultivated and oral languages, and the translation of the sacred Scriptures, have ever been acknowledged as important, if not essential parts of a Missionary's duty; but the promulgation of the gospel by the living voice has always been considered by us as the primary means of converting the heathen; and though the other departments of labour have not been neglected, this has been regarded as the first great duty of a Missionary, according with his very designation, the principal design of the institution under whose patronage he is engaged—the practice of the apostles and first Missionaries, and the spirit as well as the letter of the Divine commission, whence he derives his highest sanction, and anticipates greatest success. Preparation for this service has therefore been regarded as demanding particular attention.

Since our arrival at Huahine, in addition to the preparation of their dwellings, Messrs. Nott and Davies had been employed in preaching to the people, and preparing the Gospels of Matthew and John for the press. In the schools, Mr. Barff had been much engaged, and Mr. Orsmond, prior to his removal to Raiatea, had assisted in the instruction of the people, not only of Fare, but also of the adjoining districts.

The indolence of the South Sea Islanders has long been proverbial, and our minds were not less affected on beholding it, than those of other visitors had been. We were convinced that it was the parent of many of their crimes, infant-murder not excepted, and was also a perpetual source of much of their misery. The warmth of the climate, the spontaneous abundance with which the earth and the sea furnished, not merely the necessaries of life, but what was to the inhabitants the means of luxurious indulgence, had, no doubt, strengthened their natural love of ease, and nurtured those habits of excessive indolence in which they passed the greater portion of their lives.

These habits, so perfectly congenial to their uncultivated minds, to the fugitive manner of life, mirthful disposition, and rude state of society that prevailed among the islanders, appeared one of the most formidable barriers to their receiving our instructions, imbibing the spirit and exhibiting the moral influence of religion, and advancing in civilization. All classes were alike insensible to the gratification arising from mental improvement, and ignorant of all the enjoyments of social and domestic life, the comforts of home, and the refinements and conveniences which arts and labour add to the bestowments of Providence. The difficulties we encountered resulted not less from the inveteracy of their idle habits, than from the absence of all inducements to labour, that were sufficiently powerful to call into action their dormant energies. Their wants were few, and their desires limited to the means of mere animal existence and enjoyment; these were supplied without much anxiety or effort, and possessing these, they were satisfied.

During the early periods of their residence in the islands, our predecessors often endeavoured to rouse them from their abject and wretched modes of life, by advising them to build more comfortable dwellings, to wear more decent clothing, and to adopt, so far as circumstances would admit, the conveniences and comforts of Europeans; contrasting at times the condition of their own families with that of the natives around them. While the inhabitants continued heathens, their endeavours were altogether unavailing. The people frequently said, "We should like some of these things very well, but we cannot have them without working;—that we do not like, and therefore would rather do without them. The bananas and the plantains, &c. ripen on the trees, and the pigs fatten on the fruits that are strewed beneath them, even while we sleep; these are all we want, why therefore should we work?"

"They knew no higher, sought no happier state,
Had no fine instinct of superior joys.
Why should they toil to make the earth bring forth,
When without toil she gave them all they wanted?
The bread-fruit ripened, while they lay beneath
Its shadows in luxurious indolence;
The cocoa filled its nuts with milk and kernels,

And while they slumbered from their heavy meals,
In dead forgetfulness of life itself,
The fish were spawning in unsounded depths,
The birds were breeding in adjacent trees,
The game was fattening in delicious pastures,
Unplanted roots were thriving under ground,
To spread the tables of their future banquets!"

They furnish a striking illustration of the sentiment, that to civilize a people they must first be christianized. A change in their views and feelings had now taken place, and, learning from the Scriptures, that idleness, and irregular and debasing habits of life, were as opposed to the principles of Christianity, as to their own personal comfort; they were disposed to attend to the recommendations of their teachers in this, as well as other matters.

Industry, however, soon languishes, unless nurtured by more powerful motives than the effects of abstract principles upon partially enlightened and ill-regulated minds. To increase their wants, or to make some of the comforts and decencies of society as desirable as the bare necessaries of life, appeared to us the most probable method of furnishing the best incitements to permanent industry. It was therefore recommended to them to erect for themselves more comfortable dwellings, and cultivate a larger quantity of ground, to meet the exigencies of those seasons of scarcity which they often experienced during the intervals between the bread-fruit crops. We also persuaded them to use such articles of our clothing as were adapted to their climate and habits, and to adopt our social and domestic habits of life. This not only required a considerable addition of personal labour, but a variety of articles that could not be supplied on the islands, and must be obtained through the medium of commerce with Port Jackson and England; and they could only procure these articles, in a degree equal to that in which they multiplied the productions of the soil, so as to be able to exchange them for the manufactured goods of civilized countries.

None of the spontaneous productions of the islands were available for purposes of barter or exportation. The sandal-wood of the Sandwich Islands, and the pine-timber of New Zealand, produced without effort on the part of the inhabitants, being valuable commodities, and given in exchange for the articles conveyed by foreign vessels to their shores, afforded great inducements to commercial adventure, and furnished the natives of those countries with facilities for increasing their resources and their comforts, of which the Tahitians were destitute. Whatever articles of export they could ever expect to furnish, must be the product of their own industry; this we were desirous to direct in channels the most profitable, such as were best suited to their means, and congenial to their previous habits. We therefore recommended them to direct their attention to the culture of cotton, one variety of which appeared to be an indigenous plant in most of the islands. Several valuable kinds of cotton having been at different times introduced, were also growing remarkably well.

Soon after we reached Huahine, a number of those who accompanied us from Eimeo, with some of the chiefs of the island, united in clearing and fencing a large piece of ground, which they planted with the best seeds they could procure, and called *aua vavae*, cotton-garden, The females were the most active in this work. Whether they were more anxious than the other sex to obtain foreign articles of dress, and the conveniences and the comforts of domestic life—or whether, feeling more peculiarly their obligations to Christianity, and desiring to take the lead in the introduction of those habits which they had been taught to consider as the necessary result of its principles, and the accompaniments of a Christian profession—it is unnecessary to determine; but they laboured diligently and perseveringly, cutting down in the mountains wood for the fencing, employing their own servants to transport it to the shore, clearing away the brushwood, enclosing the ground, digging the soil, planting the seed, watching with constancy its growth, and carefully gathering the cotton.

In order to encourage them by our example, and direct them by our own proceedings, Messrs. Barff, Orsmond, and myself, having obtained permission from the owners of the valley in which we resided, employed natives to clear away the trees and bushes with which it was overgrown, for the purpose of planting it with coffee, sugarcane, or cotton. On this we also bestowed personally many an hour, desirous not only to afford those who were inclined to follow our advice, and cultivate the earth for articles of commerce, the encouragement of our counsel and direction, but to demonstrate the practicability of accomplishing, by means within their power, what had been proposed.

The directors of the Missionary Society were fully sensible of the necessity of introducing a regular system of industry among the islanders, in order to the assuming and maintaining a station amongst Christian or civilized nations; and felt that the interesting and peculiar circumstances of the people at this time, required something beyond the inculcation of the principles of Christianity, and instruction in the use of letters. They justly inferred, that, unless habits of regular industry were introduced, and civilization promoted, the people, if they did not absolutely return to all the absurdities, superstition, and cruelty of paganism, would develop but partially the genius and spirit of Christianity, and exercise very imperfectly its practical virtues. The state of feeling, also, that prevailed among the inhabitants at this time, predisposed them readily to attend to any recommendations of the kind; and the great deference they now paid to the counsel of their teachers, presented an opportunity more favourable than had ever occurred before, or was likely to occur again.

Influenced by these considerations, the Directors sent to the South Sea Islands Mr. Gyles, a gentleman who had been many years manager of a plantation in Jamaica, and who, being well acquainted with the culture of the cane and the manufacture of sugar, was furnished by the Missionary Society with the necessary machinery and apparatus for introducing this branch of industry. Mr. Gyles was engaged for four years, during which time it was supposed he would be able, not only to commence his operations, but to proceed so as to convince the king and chiefs what might be done, and also to improve the natives in the art of cultivating cane, instruct them in the process of boiling, &c., and leave them capable of carrying it on by themselves. He reached Tahiti in August, 1818, and shortly afterwards removed to Eimeo, where he began to erect the machinery, and enclosed a considerable tract of ground in the fertile and extensive valley at the head of the beautiful bay of Opunohu, usually called Taloo Harbour. Circumstances detained the king at Tahiti for many months after Mr. Gyles's arrival in Eimeo, and retarded very materially the progress of the undertaking. Sugarcane was, however, procured from the gardens of the adjacent districts, and sugar made in the presence of the natives, who were delighted on discovering that an article, so highly esteemed, could be made on their own shores, from the spontaneous product of their soil.

The advantageous and expensive arrangements of the Directors, for the purpose of introducing these important branches of commerce and productive labour, although not entirely frustrated, were in the first instance rendered to a great degree unavailing, by the unfounded and injurious reports of unprincipled and interested individuals, who beheld the advancement of the people in knowledge and civilization with any other feelings than those of satisfaction.

Early in the year 1819, the captain of a vessel, the Indus, whom purposes of commerce led to Tahiti, informed the king that Mr. Gyles's errand to Tahiti was merely experimental; and that, should the attempt to manufacture sugar succeed, individuals from distant countries, possessing influence and large resources, would establish themselves in the islands, and, with an armed force, which he would in vain attempt to oppose, would either destroy the inhabitants, or reduce them to slavery. These alarming statements were strengthened by allusion to the present state of the West Indies, where Mr. Gyles had been engaged in the manufacture of sugar and the culture of coffee. This device was employed for a short time with success against the establishment of the Mission among the Sandwich Islands; where the king and chiefs were told, that though foreigners first went in a peaceable and friendly manner to the West Indies, they subsequently went with all the apparatus of war, attacked and defeated the inhabitants, hunted the fugitives with blood-hounds, finally exterminated them, and remained masters of the islands.

Though the inconsistency of this statement with the defenceless manner in which the Missionaries had come amongst them, would have been self-evident to an enlightened mind,—being supported by an incontrovertible historical fact, it was remarkably adapted to operate powerfully upon an individual but partially informed, and exceedingly suspicious of every measure that might permanently alienate the smallest portion of territory, or lead to the establishment of foreign proprietorship, and consequent influence, in the islands.

This view of the enterprise led Pomare to decline rendering that assistance which it was expected would have been readily imparted, and the want of which retarded considerably the progress of the work. The necessary labour required from the natives was paid for at a remarkably high price, and often difficult to obtain on any terms.

Matters continued in this state until the month of May, 1819, when a national assembly of chiefs and people from Tahiti and Eimeo met at Papaoa, in the district of Pare. The Missionaries from the several stations assembled at the same period, for the purpose of commemorating the anniversary of the Tahitian Auxiliary Missionary Society.

Before they returned, the king informed them, that. apprehensive of unfavourable results from the reports already in circulation among the chiefs and people, he could not consent to the prosecution of the manufacture of sugar, &c., excepting on a very limited scale. Pomare was not hasty in forming his decision on any matter of importance, and by no means precipitate in his measures; but on this occasion he appears to have been altogether uninfluenced by that temperate deliberation, and judicious policy, which he generally manifested in matters tending to improve the condition of the people, and increase the national resources.

The Missionaries also appear to have been so strongly influenced by the king's communication, that, instead of endeavouring to remove his objections, by persuading him to allow the trial to be fairly made, and then act accordingly, they deemed it expedient, that so far as they, or the Society by which the machinery had, at great expense, been sent out, were concerned, it should be at once discontinued. Accordingly, on the 14th of May, "in order to satisfy the king, and quiet the minds of the people," they advised Mr. Gyles "to return to New South Wales by the first conveyance."

Shortly after this decision, communications from England required a general meeting of the Missionaries from the several stations; and Messrs. Williams, Barff, and myself, went up from the Leeward Islands to Tahiti and Eimeo. By the same conveyance Mr. and Mrs. Nott removed to Tahiti, where Mr. Nott has since laboured regularly in Matavai, or the adjacent district of Pare. We were detained there about a fortnight; during which period we received from Mr. Gyles much information on the culture of the plant, and the manufacture of sugar. Before we left, Mr Gyles very obligingly had a quantity of cane bruised and boiled, that we might not only understand the theory, but witness the process of grinding canes, boiling the juice, and granulating the sirup, so as to introduce it among the inhabitants of the Leeward Islands.

Our business at Tahiti being finished, Messrs. Barff, Williams, and myself, with a number of natives, sailed from Eimeo about noon, on the 12th of August, in an open boat belonging to Mr. Hayward. Before the sun had set, we had nearly lost sight of the island; and when the night gathered round us, we found ourselves in the midst of the vast Pacific, in a very small and fragile bark, without compass or nautical instrument, or any other means of directing our way than the luminaries of heaven. The night, however, was cloudless, and

> "Star after star, from some unseen abyss,
> Came through the sky, till all the firmament
> Was thronged with constellations, and the sea
> Strown with their images."

The interval between the close of the evening and the dawn of the following day was pleasantly spent; and soon after sun-rise, on the morning of the 13th, we were gladdened by the sight of the lofty mountains in Huahine, which were seen above the line of clouds that rested on the western horizon. About five in the afternoon of the same day, Mr. Barff and myself were restored to the bosom of our families; thankful for the guidance and protection we had enjoyed on the voyage, and the merciful care which those we left had experienced.

The facility with which the manufacture of sugar might be carried on by the people, and the certain market it would always find in Port Jackson should they be able to furnish more than their own necessities required, induced us not only to recommend it to the natives, but also to plant with sugarcane the ground already cleared and enclosed.

The proprietors of the cotton garden watched the progress of the plants with care and anxiety, accompanied probably with some of those golden dreams of future emolument which frequently operate very powerfully on the minds of individuals commencing an enterprise, which, although in some degree uncertain as to its results, yet promises, upon the whole, an increase of wealth or enjoyment. Unhappily for them, the ground they had chosen was unsuitable, and many of the plants were not productive. The first crop, however, was gathered, the seeds carefully picked out, and the cotton packed in baskets. When a ship arrived, they were eager to dispose of it, expecting far more in return than the warmest encouragement in its culture had ever warranted. Their estimate of its value had been formed according to its bulk; and when it was weighed, and they saw a large basket-full weigh only two or three pounds, and a proportionate price offered, they were greatly disappointed. They brought back their cotton, and hung it up in their houses till another ship arrived, when it was again presented for sale; but being again estimated by weight, little if any more was offered for it. Some sold what they had collected, others were so disappointed, that they seemed hardly to care what became of it. This circumstance, together with the length of time and the constant attention that a cotton plantation required, before any return could be received, greatly discouraged them, and prevented their continuing its culture. They chose rather to feed a number of pigs, or cultivate the vegetables in demand by the shipping, dispose of them when vessels might put in for refreshments, and receive at once in exchange, articles of cloth, &c. than wait till the crops should be gathered, and experience so much uncertainty, or meet with such annoying disappointments in the amount of their returns.

Mr. Gyles, on his way to the colony of New South Wales in the month of August 1819, spent some time at Huahine and Raiatea; and we gladly availed ourselves of his visit, to make further inquiries relative to the object for which he had come to the islands. Some spare machinery and boilers, sent out by the Society, were also left at Huahine. Assisted by the natives, we subsequently erected a rustic mill; and, when the cane in our plantation was ripe, commenced our endeavours to convert it into sugar. The cylinders for crushing the cane were perpendicular: an ox was trained to draw in the mill. He was yoked to a lever on one side of the central roller; a number of natives, pushing at another on the opposite side, turned the mill, and pressed the juice from the cane. The natives were surprised at the quantity of juice from a single cane, as they had never been accustomed to see it thus collected, but had generally broken it in small pieces, and, by masticating the cane, extracted the juice.

After boiling it some time, we added the *temper* or mixture of lime and water; and when we supposed the quantity had been sufficiently reduced, directed the natives to remove it to a suitable vessel for cooling, the progress of which we watched very anxiously, and, ultimately, had the satisfaction of beholding fine-grained crystals of sugar formed from the liquid. The natives were delighted and astonished; and although our surprise was not less than theirs, our satisfaction was more chastened; for, notwithstanding we had succeeded so well in our first attempt, we considered it more the result of accident than skill, and were by no means confident that, in a second effort, we should be equally successful.

We were, however, sufficiently encouraged to recommend the people, notwithstanding their disappointment in regard to the cotton, to direct their attention to the culture of sugar, since they had no longer any cause to doubt the practicability of procuring, from their respective plantations, sugar for their own use, or for barter with shipping. Our advice was not unheeded; several of the chiefs were induced to cultivate the cane; the mill we had erected became a kind of public machine, to which they brought their produce; and although, in some instances, we failed in procuring good sugar, in time the people were so well acquainted with the process, as to be able to boil it themselves. The Missionaries in Raiatea also erected a mill, more efficient than the one we had constructed in Huahine, cultivated a quantity of cane, made sugar themselves, and taught the inhabitants of the island to do the same.

Sugarcane grows spontaneously in all the South Sea islands, and more than ten varieties are indigenous. It has been stated, that the best canes now cultivated in the West Indies, are the kinds taken thither by Captain Bligh. In their native islands they grow remarkably fine. I have frequently seen canes as thick as a man's wrist, and ten or twelve feet between the root and the leaves. The *irimotu*, a large yellow cane, and the *to-ura*, of a dark red colour, grow very large, and yield an abundance of juice, but the *patu*, a small light red, long-jointed cane, with a thin husk, or skin, contains the greatest quantity of saccharine matter. Some of the sugar manufactured by Mr. Gyles was of a very superior quality; and if hired labour were less expensive, or the people more industrious, it might be raised with facility in considerable quantities. The return, however, is distant, and the crops are less productive than many other articles that might be cultivated in the islands, especially unconnected with the distillation of rum from the refuse of the juice, or the molasses of the sugar. This is probably the only plan that would render it in any degree profitable; but to the use of rum, the present chiefs, particularly those of the Leeward or Society Islands, are averse; its introduction since embracing Christianity, they have been able to prevent; and it will be matter of deepest regret, if either they or their successors should favour its distillation on the islands, or its importation from abroad. Next to idolatry, and the diseases introduced by foreigners, it is the greatest scourge that has ever spread its desolations through their country.

But although these circumstances have hitherto operated against the general culture of the cane, the chiefs and some of the people make sugar for their own consumption, and have occasionally supplied captains of ships, who have wished to replenish their sea-stock. In this respect, although the attempt of the Directors to introduce extensively its cultivation, has failed in the first instance; the natives have nevertheless acquired, from Mr. Gyles's transient residence among them, an acquaintance with the process of manufacturing this valuable article of commerce, which, it is presumed, will prove to the nation an important and a permanent advantage.

The Haweis, in returning to the islands in the spring of 1819, touched at Norfolk Island, formerly an appendage to the colony of New South Wales, and I believe re-occupied since that period. From this island the captain brought away a number of young coffee plants, which, on his arrival in the islands, were distributed among the different stations. The tender plants were once or twice removed, and all perished, excepting those in my garden at Huahine, which I was happy to succeed in preserving. The climate was favourable to their growth, and they appeared to thrive well. After four years, each tree bore about forty berries, which when perfectly ripe were gathered, and sent to the several stations. They were planted, and have since flourished, so that in every island the coffee plant is now growing, and may be cultivated to almost any extent. The chiefs are fond of coffee as a beverage, and, with the people, will doubtless raise it for their own use; and as it requires but comparatively little attention, probably it may be furnished in greater abundance than either the sugarcane or cotton.

The tobacco plant is another exotic, common now in all the islands: it was introduced by Captain Cook, and has since been cultivated to a small extent by the natives, merely for their own use. Mr. Williams encouraged its culture to a considerable extent in the island of Raiatea, and the natives were taught to prepare it for the market of New South Wales, in a manner that rendered the Raiatean tobacco equal to any brought into Sydney. A lucrative branch of industry and commerce now appeared open to the enterprising and industrious inhabitants, when a heavy duty, which, according to report, in order to favour its growth in New Holland, was laid upon all taken into the port of Sydney, prevented their continuing its culture with the least expectation of profit. It was therefore in a great degree abandoned. The information, however, which the inhabitants received from the individual whom Mr. Williams employed to instruct them, not only in its growth, but in the methods of preparation, by which it assumes the different forms under which it is offered in the markets, was valuable; and though no very advantageous results have hitherto followed, it may hereafter be productive of good.

CHAP. XVII.

Renewed endeavours to promote industry among the people—Arrival of Messrs. Blossom and Armitage—Establishment of the cotton factory—First cloth made in Eimeo—Prospects of success—Death of Mrs. Orsmond—Voyage to Raiatea—Sudden approach of a storm—Conduct of the natives—Violence of the tempest—Appearance of the waterspouts—Emotions awakened by the surrounding phenomena—Influence of waterspouts on the minds of the natives—Conduct of a party overtaken by one at sea—Deliverance during a voyage from the Sandwich Islands—Abatement of the storm—Appearance of the evening—Arrival at Raiatea—Kindness of the inhabitants—District of Opoa—Visit to the settlement—Importance of education—Methods of instruction—Sabbath schools—Annual examination of the scholars—Public procession—Contrast between the present and former circumstances of the children.

ALTHOUGH the expensive and commendable measures adopted by the Directors of the Missionary Society, for encouraging industry among the South Sea Islanders, and furnishing them with a source of productive labour by introducing the manufacture of sugar, had not accomplished all that was designed, and Mr. Gyles had returned to England before the expiration of the period for which he had been engaged, the Directors still considered that it was their duty to endeavour to promote the temporal prosperity of the people—that the introduction of useful mechanic arts, and other means of advancing their civilization, though objects of only secondary importance, were not to be overlooked. Some stimulus to more regular employment than that to which the natives had been accustomed, during the unsettled and indolent state of society from which they were just emerging, was still necessary for their contentment and individual happiness, as well as their national prosperity.

The Directors of the Missionary Society were not influenced by their own choice, but by the necessities of the people, in making these and other secular arrangements, which were not contemplated in the original constitution and object of their association, but have resulted from the changes effected by their agents in the circumstances of those communities among which they have resided; and have sometimes involved an expense which could not always be met without difficulty. These collateral exertions often occasion embarrassment, and it would be highly gratifying, if other institutions were able to prosecute those departments of effort, which are rather appendages than proper parts of Missionary labour. Were the resources of those societies formed for the universal diffusion of education, and the means of the British and Foreign Bible Society such as to enable them to undertake entirely the instruction of the heathen, and the translation and circulation of the Scriptures, it would greatly facilitate the extension of Christianity. If, in addition to those already in existence, there was also an institution for the promotion of agriculture, mechanic arts, social order, and the general civilization of rude and barbarous tribes, such a society would exert a most beneficial and commanding influence, and furnish an able and important agency, in conjunction with those now engaged. It would enable Missionary institutions to follow more energetically their simple and primary labours, in sending forth messengers to preach the gospel to the heathen.

Such a society, however, did not exist. The promotion of industry and civil improvement were important objects, and, in order to accomplish them, especially in reference to the rising generation, two artisans, Messrs. Blossom and Armitage, were sent out with the deputation who visited the South Seas in 1821. The former was a carpenter, acquainted with the construction of machinery and wood-work in general; a department of labour highly advantageous to a rude, or but partially civilized people, and at this time in great estimation among the Tahitians. Mr. Blossom has been engaged in teaching native youth, and others, these arts; and though not altogether so successful as he desired, has nevertheless had the pleasure of beholding two or three excellent workmen trained under his care.

The introduction among an indolent people, of any art that requires constant, and sometimes heavy labour, must be gradual; but as building, and the use of household furniture, &c., increases among the people, skill in these departments will be held in higher esteem, and the number of workmen will necessarily increase with the demand for their labour, and the remuneration it receives.

It was known, that with but slight attention the cotton-plant might be cultivated in the islands to almost any extent; and it was supposed, that although the smallness of the returns it had brought, when offered for sale in the raw state, together with the difficulties attending their first attempt, had deterred the people from persevering in its culture; yet that they might be induced to resume it, if taught on the spot to manufacture cotton cloth. This was an article in great and constant demand throughout the islands. Mr. Armitage was therefore appointed to attempt to teach the natives to spin and weave the cotton grown in their own gardens. He was a native of Manchester, where the members of his family still reside. He was well qualified for the undertaking, possessing an intimate acquaintance with the various processes by which raw cotton is made into cloth, and having been overseer or foreman of an extensive manufactory.

In acceding to the proposal of the Directors, and engaging in this enterprise, he manifested a degree of generous devotedness seldom excelled. He exchanged inviting prospects of wealth, comfort, and usefulness at home, for the toil and self-denial inseparable from such an attempt. He was on the eve of entering into a matrimonial engagement. The gentleman who had hitherto been his employer had proposed to make him his partner, had arranged for the advance of a very considerable sum of money; part of the materials for commencing the new establishment were procured, and the results in that line of business have since been such, as to warrant the inference, that every advantage the parties anticipated might have been realized. This, however, he relinquished, and cheerfully engaged in an attempt to improve the temporal condition of the islanders, with no other remuneration than the Missionaries receive—a bare supply of the necessaries of life.

It may, perhaps, be thought that I am trespassing the bounds of propriety in giving these particulars to the public; but, in this instance, and there are others that might also be adduced, I feel it due, not more to the individual than to the cause in which he is embarked; to the friends by whom it is supported; and even to those who, in consequence of mistaken views and misrepresentation, may sometimes be induced to suppose mercenary motives influence those who engage in Missionary undertakings.

In the month of September, 1821, they reached Tahiti. The carding machine, looms, &c. were landed, and placed under the care of Paiti, a chief residing near the harbour of Taone; and in the adjacent village of Pirae, Messrs. Armitage and Blossom took up their abode.

Like every other undertaking that has yet been made to promote the true interests of the people, the cotton factory had to contend with great difficulty. At first the king and chiefs, under the recollection of the reported design and tendency of the sugar manufactory, expressed their wishes that the establishment should be formed near their principal residence, that all proceedings connected with it might be under their inspection. Subsequently, when they entered into its design, and began to consider that it would become a source of pecuniary advantage, although it was thought that Eimeo would be most eligible for its establishment, the chiefs of Pare and the adjoining districts refused to allow the machinery to be removed. In this state matters remained some time, and several of the finer parts of the iron-work were destroyed by the rust, and the whole greatly injured.

The deputation and the Missionaries, however, considering that the island of Eimeo afforded the greatest facilities for carrying on the work, removed it thither, and with great expense and labour Messrs. Armitage and Blossom erected the machinery and commenced their work. Shortly after this was completed, Mr. Blossom removed to the opposite side of the island, to take charge of the secular concerns of the South Sea Academy, and the work has since been carried on by Mr. Armitage alone.

The machinery, &c. were considered as belonging to the Missionary Society, but at a public meeting held in Eimeo, in May 1824, for the purpose of arranging the principles upon which its future operations should be conducted, it was distinctly stated by the deputation, and recognized by the Missionaries, "That the Society contemplates no other advantage in promoting the manufacture of cloth by this machinery, than the good of the inhabitants of these islands;" "That no charges by way of profit shall be made upon the cloth manufactured and sold to the inhabitants, more than is merely necessary to defray the expenses attending it," and "That all the inhabitants of the islands connected with both the Windward and Leeward Missions, shall be allowed to share alike in the advantages of this manufactory." At the same time it was recommended, that two young men and two young women from each island, should be sent, to learn the art of making looms, spinning, weaving, &c.

The work commenced with cotton belonging to the native Missionary Societies. Mr. Armitage taught them to card the cotton, and Mrs. Armitage instructed them in spinning. Their first attempts, as might be expected, were exceedingly awkward, and the warp they furnished difficult to weave. One piece of cloth, however, fifty yards in length, was finished, and presented to the king. Its appearance was coarse, and inferior to the imported calicoes of British manufacture; it was nevertheless grateful to the chiefs, from the fact of its being the first ever manufactured in their native islands.

Cotton for another piece was prepared, and the natives commenced spinning; but the confinement required being irksome, and their expectations rather lowered, as to the quality of the cloth they were to receive as wages for their labour,—before the warp was ready for the loom, they simultaneously discontinued their attendance at the factory. When interrogated as to their reasons for this sudden change in their conduct, it was found that they had not indeed struck for higher wages, but had left off to think about it, and that, until their minds were made up, they could not return. The spinning-wheels and the loom now stood still, excepting that Mrs. Armitage and Mrs. Blossom, with the assistance of their own servants, spun the cotton, which Mr. Armitage wove into about fifty yards of cloth, for the use of the academy.

Notwithstanding the inferior appearance of the cloth manufactured in Eimeo, it was soon found to be more durable than that procured from the ships. Yet the disappointment which the natives had experienced prevented their cultivating the cotton; and but little was available for the establishment, excepting that subscribed by the members of the native Missionary Societies: the people declined coming to learn, and prospects were most unpromising. This, however, was not the only source of discouragement.

Traders, influenced by the limited views and interested motives which too frequently regulate the proceedings of those who traffic with uncivilized nations, employed a variety of inducements to prevent the natives affording any encouragement to the establishment. At one time they assured them that it would be injurious to their interest, and, if successful, prevent their being visited by shipping, &c., offering at the same time, to give them for their raw cotton twice as much cloth as they could procure at the factory. At other times they threatened Mr. Armitage with ruin, and announced their determination to oppose him. Sometimes they endeavoured to persuade him to abandon so hopeless a project, as that of attempting to train the people to habits of industry.

Their threatenings to seek his ruin, by opposing his efforts, are rather amusing. They doubtless supposed the attempt was on his part a speculation for the accumulation of wealth; the only end which most propose, who visit those islands; and which, when pursued on fair upright principles, is not to be condemned. These proceedings, however, must have originated in the most contracted views of the influence of such an establishment, which, while it may induce and encourage habits of more regular employment, can never diminish, excepting in a very small degree, the demand for British calicoes, which will be superior in texture, pattern, &c. to any that can be made in the islands. It will also tend to encourage the more extensive culture of the cotton, and in the raw state the natives will never decline disposing of it to him who offers the best price.

Notwithstanding these and various other discouragements, Mr. Armitage was able to persevere; and as there was little prospect of the females he had taught to spin making up their minds to return, another party was selected. Nearly twenty girls, and eight or ten boys, engaged to learn to spin and weave. The conditions on which they were instructed were almost such as they or their friends chose to propose, both as to the time they should continue, and the hours they should labour; and instead of receiving a premium for teaching them, Mr. Armitage agreed to pay them for every ounce of cotton they should spin.

In every undertaking of this kind, the greatest embarrassments attend it at the outset, and the same difficulties that had suspended the instruction of the two former parties, were again to be overcome. The indolent habits of these young persons generally, their impatience of control, and the fugitive and unsettled mode of life to which many have been accustomed, were not to be at once overcome. Recent accounts, however, convey the pleasing intelligence, that the prospect of ultimately introducing this branch of labour very extensively among the people, is more encouraging than ever. The females were able to spin strong and regular thread, or yarn; one or two of the boys had been taught to make, all things considered, very good cloth. Mr. Armitage has also succeeded in dying the cloth, and thus furnishing different patterns and colours, which has greatly increased its value in their estimation. While the hands of the parties spinning or weaving are employed, the improvement of their minds is not neglected. Reading-lessons and passages of scripture are affixed to the walls and different parts of the factory.

The carding engine, and some of the other parts of the machinery, were turned by a large water-wheel, and the work has often been retarded by the repairs that the wheel or its appendages have required.

Several of the best native carpenters have, however, readily come forward to repair the wheel, and have received their payment in cloth made at the factory. The derangement of the machinery suspending the work of the spinners, some of them requested to take the cotton home, to prepare and spin at their own houses. The experiment has succeeded far beyond what was anticipated, and the natives now bring to the factory for sale the cotton yarn spun at their own dwellings, and ready at once for the loom.

This circumstance, though insignificant, is one of the most interesting that has yet transpired in connexion with this enterprise. The natives are now convinced that they can make cloth; others, besides those taught in the factory, will desire to learn; and as they can prepare and spin the cotton at their own dwellings, this employment, which is certainly adapted to their climate and habits, as they can take it up and lay it down at their convenience, will probably be very extensively followed through the islands. The native carpenters will be able to make looms, as they have made turning-lathes, which, though rude, will be such as will answer their purpose. The spinning-wheel will also become an article of furniture in most of their houses; and the father, the brother, and the son, will have the satisfaction of wearing native or home-spun garments, made with cotton grown in their own gardens or plantations, and spun by their wives' or sisters' or daughters' hands. The Tahitian, like the Indian weaver, may, perhaps, be seen fixing his rude and simple loom under the shadow of the cocoa-nut, or the banana tree, whilst the objects that often give such a charm to rural village scenery, and awaken so many ideas of contentment and happy simplicity in connexion with the peasantry of England, may be witnessed throughout the South Sea Islands.

In the month of December, 1818, when the Haweis sailed from Huahine, on her first voyage to New South Wales, Mr. and Mrs. Orsmond left us, as we mutually supposed, on a visit of a few months to the island of Raiatea, for the purpose of receiving Mr. Threlkeld's attentions at a season of unusual domestic anxiety. For two or three months contrary winds prevented any intercourse between us, when at length Mr. Orsmond's boat arrived, with the unexpected and melancholy tidings of the death of Mrs. Orsmond, which had taken place on the 6th of January, 1819. She had survived but a few hours the birth of an infant daughter, by whom, in the space of five short days, she was followed to the eternal world, and, we believe, to the abodes of holy and unending rest. The disconsolate partner of her days was thus left a widower and childless, far from all the alleviation which the sympathies and attentions of kindred and friends in such seasons impart,—a lone wanderer, amid a rude untutored race, in a solitary island of the sea. The kindness and the sympathy of his fellow-labourers mitigated, however, in a great degree, the poignancy of his distress; and the promises of inspired truth, with the consolations of religion, supported his mind under a bereavement which he had sustained in circumstances unusually distressing. The people around were touched with a feeling of compassion; but although their commiseration was fully appreciated by their teacher, there was not that reciprocity of feeling which could lessen, in any considerable degree, the burden of his grief. In the family of Mr. Williams he spent the greater part of his time, when not engaged in public duties, and experienced from its members every attention which kindness and attachment could prompt or bestow.

Early in 1819, circumstances rendered it desirable for us to visit Raiatea. We were anxious, also, to mingle our sympathies with those of our companions there, in that bereavement by which all were so deeply affected. We had been acquainted with Mr. and Mrs. Orsmond before leaving England. We had all left our native land about the same period, and had spent the greater part of our time, since arriving in the islands, either at the same station or under one roof, and felt very deeply the first breach now made by death, in the little circle with which we were more immediately connected. We therefore availed ourselves of the return of Mr. Orsmond's boat to visit the station.

About nine o'clock in the morning, Mr. Barff and myself, accompanied by five natives, and an English sailor who had charge of the boat, embarked from Huahine. Though the settlements were about thirty miles apart, yet, as the width of the channel was not much more than twenty miles, the mountains, and coast of the opposite island, were distinctly seen. The wind being fair, we expected to reach the Raiatean shore in three or four hours, and to arrive at the residence of our friends long before the close of day. We had not, however, been an hour at sea, when the heavens began to gather blackness, and dense lowering clouds intercepted our view of the shore we had left, and that to which we were bound. The wind became unsteady and boisterous, the sea rose, not in long heavy billows, but in short, cross, and broken waves. We had no compass on board. The dark and heavy atmosphere obscuring the sun, prevented our discerning the land, and rendered us unconscious of the direction in which the gathering storm was driving us. We took down our large sails, leaving only a small one in the forepart of the boat, merely to keep it steady.

The tempest increasing, the natives were alarmed, and during the occasional intervals in which the wind abated its violence, the rain came down in such torrents, as if the windows of heaven had been opened, and a deluge was descending. The rain calmed in a degree the broken and agitated surface of the dark blue ocean, that raged in fearful and threatening violence. Our boat being but small, not above eighteen feet long, and her edge, when the sea had been smooth, not more than a foot or eighteen inches above its surface; every wave that broke near, threw its spray over us, and each billow, in striking our little bark, forced part of its foaming waters over the bow or the sides. Happily, we had a bucket on board, by means of which we were able to bale out the water.

In this state we continued, I suppose, about two hours, hoping the clouds would disperse, and the winds abate; but, instead of this, the storm seemed to increase, and with it our danger. Most of the natives sat down in the bottom of the boat; and, under the influence of fear, either shut their eyes, or covered them with their hands, expecting every moment that the waves would close over us. We were not unconscious of our perilous situation; and, as a last resort, took down our little sail and our mast, tied the masts, bowsprit, and oars together in a bundle, with one end of a strong rope, and, fastening the other end to the bow of our boat, threw them into the sea. The bundle of masts, oars, &c., acted as a kind of buoy, or floating anchor; and not only broke the force of the billows that were rolling towards the boat, but kept it tolerably steady, while we were dashed on the broken wave, or wafted we knew not whither by the raging tempest and the pelting storm.

The rain soon abated, and the northern horizon became somewhat clear, but the joyful anticipation with which we viewed this change was soon superseded by a new train of feelings. *Huri, huri, tia moana,* exclaimed one of the natives; and, looking in the direction to which he pointed, we saw a large cylindrical waterspout, extending, like a massive column, from the ocean to the dark and impending clouds. It was evidently at no very remote distance, and seemed moving towards our apparently devoted boat.

The roughness of the sea forbade our attempting to hoist a sail in order to avoid it; and as we had no other means of safety at command, we endeavoured calmly to wait its approach. The natives abandoned themselves to despair, and either threw themselves along in the bottom of the boat, or sat crouching on the keel, with their faces downwards, and their eyes covered with their hands. The sailor kept at the helm, Mr. Barff sat on one side of the stern, and I on the other, watching the alarming object before us! While thus employed, we saw two other waterspouts, and subsequently a third, if not more, so that we seemed almost surrounded with them. Some were well defined, extending in an unbroken line from the sea to the sky, like pillars resting on the ocean as their basis, and supporting the black and overhanging clouds; others assuming the shape of a funnel or inverted cone, attached to the clouds, and extending towards the waters beneath. From the distinctness with which we saw them, notwithstanding the density of the atmosphere, the farthest could not have been many miles distant. In some, we imagined we could trace the spiral motion of the water as it was drawn to the clouds, which were every moment augmenting their portentous darkness. The sense, however, of personal danger, and perhaps almost immediate destruction, if brought within the vortex of their influence, restrained in a great degree all curious, and what, in other circumstances, would have been interesting observation, on the wonderful phenomena around us, the mighty agitation of the elements, and the terrific sublimity of these wonders of the deep.

The hoarse roaring of the tempest, and the hollow sounds that murmured on the ear, as the heavy billow rolled in foam, or broke in contact with opposing billows, seemed as if deep called unto deep; and the noise of waterspouts might almost be heard, while we were momentarily expecting that the mighty waves would sweep over us.

I had once before, when seized with the cramp while bathing at a distance from my companions, been, as I supposed, on the verge of eternity. The danger then came upon me suddenly, and my thoughts, while in peril, were but few. The danger now appeared more imminent, and a watery grave every moment more probable; yet there was leisure afforded for reflection, and the sensibilities and powers of the mind were roused to an unusual state of excitement by the mighty conflict of the elements on every side.

A retrospect of life, now perhaps about to close, presented all the scenes through which I had passed, in rapid succession and in varied colours, each exhibiting the lights and shades by which it had been distinguished. Present circumstances and connexions claimed a thought. The sorrow of the people—the dearest objects of earthly attachment, left but a few hours before in health and comfort on the receding shore—those unconscious infants that would soon, perhaps, be left fatherless, and dependent on their widowed mother, who, in cheerless loneliness, far from friends, and home, and country, might remain an exile among a strange, untutored race, emerging from the rudest barbarism;—these reflections awakened a train of feelings not to be described. But the most impressive exercise of mind was that referring to the awful change approaching. The struggle and the gasp, as the wearied arm should attempt to resist the impetuous waves, the straining vision that should linger on the last ray of retiring light, as the deepening veil of water would gradually conceal it for ever, and the rolling billows heaving over the sinking and dying body, which, perhaps ere life should be extinct, might become the prey of voracious inhabitants of the deep, caused scarcely a thought, compared with the appearance of the disembodied spirit in the presence of its Maker, the account to be rendered, and the awful and unalterable destiny that would await it there. These momentous objects absorbed all the powers of the mind, and produced an intensity of feeling, which for a long time rendered me almost insensible to the raging storm, or the liquid columns which threatened our destruction.

The hours that followed were some of the most solemn I have ever passed in my life. Although much recurred to memory that demanded deep regret and most sincere repentance, yet I could look back upon that mercy that had first brought me to a knowledge of the Saviour, with a gratitude never perhaps exceeded. Him, and Him alone, I found to be a refuge, a rock in the storm of contending feelings, on which my soul could cast the anchor of its hope for pardon and acceptance before God: and although not visibly present, as with his disciples on the sea of Tiberias, we could not but hope that He was spiritually present, and that, should our bodies rest till the morning of the resurrection in the unfathomed caverns of the ocean, our souls would be by Him admitted to the abodes of blessedness and rest. I could not but think how awful would my state have been, had I in that hour been ignorant of Christ, or had I neglected and despised the offers of his mercy; and while this reflection induced thankfulness to Him through whom alone we had been made to share a hope of immortality, it awakened a tender sympathy for our fellow-voyagers, who sat in mournful silence at the helm or in the bottom of the boat, and seemed averse to conversation. Our prayers were offered to Him who is a present help in every time of danger—for ourselves—and those who sailed with us; and under these, or similar exercises, several hours passed away. The storm continued during the day. At intervals we beheld, through the clouds and rain. one or other of the waterspouts, the whole of which appeared almost stationary, until at length we lost sight of them altogether, when the spirits of our native voyagers evidently revived.

The natives of the South Sea Islands, although scarcely alarmed at thunder and lightning, are at sea greatly terrified by the appearance of waterspouts. They occur much more frequently in the South than in the North Pacific, and although often seen among the Society Islands, are more rarely met with in the Sandwich group. But throughout the Pacific, waterspouts of varied form and size are among the most frequent of the splendid phenomena, and mighty works of the Lord, which those behold who go down to the sea in ships, and do business upon the great waters. They are sublime objects of unusual interest, when viewed from the shore; but when beheld at sea, especially if near, and from a small and fragile bark, as we have seen them, it is almost impossible so to divest the mind of a sense of personal danger, as to contemplate with composure or with satisfaction their stately movement, or the rapid internal circular eddy of the waters.

Nor is it easy for an individual, who has never beheld them in such a situation, to realize the sensation produced, when the solitary voyagers, from their light canoe, or their deckless boat, dancing on every undulating wave, descry these "liquid columns," towering from the surface of the water, uniting the ocean and the heavens, while the powerful agitation of the former indicates the mighty process by which they are sustained. By the natives of the Society Islands they are called *huri huri tia moana,* the meaning of which probably is, "turning, turning perpendicularly the ocean," or the deep: from *huri*, to turn over or up, the reduplication of the word donoting a repetition or continuance of the action; *tia*, to erect or to stand upright; and *maona*, the deep or ocean. Sometimes they have approached the shore, and although I do not recollect any instance of their actually destroying persons at sea, I am inclined to presume such a calamity must have occurred, or they would not be such objects of terror to the people.

During our abode in Huahine, a number of natives were on a voyage from the Leeward to the Windward Islands, in a boat belonging to Mr. Williams, when a waterspout approached them. They had heard that, when seen by navigators, they sometimes averted the threatened danger by discharging their artillery at the waterspout. Having a loaded musket in the boat, they at first thought of firing at the advancing column; but as it approached, the agitation of the water was so great, and the phenomenon so imposing and appalling, that their hearts failed; and when it was, according to their own account, within a hundred yards of their boat, and advancing directly upon them, they laid the musket down. The man at the helm now shut his eyes, and his companions threw themselves flat on their faces in the bottom of the boat. This is the exact position in which a captive, doomed to death, awaited the fatal stroke of a victor by whom he had been overcome in battle. After waiting in fearful suspense several minutes, the helmsman, hearing a rushing noise, involuntarily opened his eyes, and saw the column passing, with great velocity, a short distance from the stern of the boat. He immediately called his companions, who joined not only in watching its receding progress, but in acknowledging the protection of the Almighty in their preservation.

When returning from the Sandwich Islands on board the ship Russell, in 1825, we experienced a happy deliverance from one of these wonderful and alarming objects. Our Sabbath afternoon worship on the quarter deck had just terminated; Mrs. Ellis was lying on a sofa, and, observing unusual indications of terror in the countenance of the boy at the helm, she said, "What is it that alarms you?" He answered, in hurried accents, "I see a whirlwind coming," pointing to a cloud a little to the windward of the ship. His actions attracted the notice of the officer on deck, who instantly sent an able seaman to the helm, and called the captain. I had taken the books down into the cabin, and was putting them by, when I heard the officer, in a tone of unusual earnestness, ask the captain to come on deck. I hastily followed, and my attention was instantly directed to the waterspout.

The breeze was fresh, and as the object of alarm was still at some distance, it was possible we might avoid coming in contact with it. The captain, therefore, took in none of the sails, but called all hands on deck, ordered them to *stand by the halyards*, or ropes by which the sails are pulled up, so that, if necessary, they might let them go in an instant, and thus lower down the sails. We all marked its approach with great anxiety. The column was well defined, extending in an unbroken line from the sea to the clouds, which were neither dense nor lowering. Around the outside of the liquid cylinder was a kind of thick mist, and within, a substance resembling steam, ascending apparently with a spiral motion We could not perceive that much effect was produced on the cloud attached to the upper part of the column, but the water at its base was considerably agitated with a whirling motion; while the spray, which was thrown off from the circle formed by the lower part of the column, rose apparently twenty feet above the level of the sea. After watching in breathless suspense for some time its advance in a direct line towards our ship, we had the satisfaction to see it incline in its course towards the starboard quarter, and ultimately pass by about a mile distant from the stern. The ropes of the sails were again fastened, the men repaired to their respective stations, and we pursued our way under the influence of increased thankfulness for the deliverance we had experienced.

The storm, which had raged with violence ever since an hour after our departure from Huahine, began to abate towards the close of the day: we did not, however, see the land, and knew not whither we had drifted; but soon after the setting of the sun, the clouds dispersed, and a streak of light lingering in the western sky, indicated the direction in which we ought to proceed. The rain now ceased, the wind subsided; and although the surface of the sea was considerably agitated, it was no longer that quick dashing conflict of the waves to which we had been exposed while "a war of mountains raged upon its surface," but a long and heavy sluggish sort of motion. We pulled in our bundle of masts and oars, and placing the masts along the seats of the boat, the natives manned the oars, and rowed towards the west.

The moon rose soon after the light of the sun had departed, and although she shone not at first in cloudless majesty through an untroubled sky, yet the night was a perfect contrast to the day. The light fleecy clouds that passed over the surface of the sky, fringed with the moon's silver light, gave a pleasing animation to the scene, and

> "With scarce inferior lustre gleamed the sea,
> Whose waves were spangled with phosphoric fire,
> As though the lightnings there had spent their shafts,
> And left the fragments glittering on the field."

After rowing some time, we heard the hoarse roaring of the surf, as it broke in foam upon the coral reef surrounding the shore. To us this was a most welcome sound, indicating our approach to the land. Shortly afterwards we saw a small island, with two or three cocoa-nut trees upon it, and subsequently the coral reef appeared in view. We now found ourselves near the *Ava Moa*, Sacred Passage, leading to *Opoa*, the southernmost harbour in the island of Raiatea; and after rowing two or three miles, landed about midnight. Weary and famished, drenched with the rain, and suffering much from the cold occasioned by the wetness of our clothes, we were truly thankful, after the incidents of the day, to find ourselves once again on shore. The hospitable inhabitants of the dwelling which we entered soon rose from their beds, kindled a large fire in the centre of the floor, cooked us some provisions, and furnished us with warm and clean native cloth, to wear while our own clothes were hung up to dry. Having refreshed ourselves, and united in grateful thanksgiving to the

Preserver of our lives, we lay down upon our mats, and enjoyed several hours of comfortable and refreshing repose. I have often been overtaken with storms when at sea in European vessels, boats, and native canoes, but, to whatever real danger I may have been exposed, I never was surrounded by so much that was apparent, as during this voyage.

After a few hours of unbroken rest, we arose recruited the next morning, found our dried clothes comfortable, united with our host and his family in the morning devotions, and then, while they were preparing refreshments, took a view of the district. We found it not very extensive, though the land is rich and good. The gardens were large, and, at this time, well stocked with indigenous roots and vegetables. *Opoa* has long been a place of celebrity, not only in Raiatea, but throughout the whole of the Society Islands. It was the hereditary land of the reigning family, and the usual residence of the king and his household. But the most remarkable object connected with Opoa, was the large marae, or temple, where the national idol was worshipped, and human victims sacrificed. These offerings were not only brought from the districts of Raiatea and the adjacent islands, but also from the windward group, and even from the more distant islands to the south and south-east.

The worship of Oro, in the marae here, appears to have been of the most sanguinary kind; human immolation was frequent, and, in addition to the bones and other relics of the former sacrifices, now scattered among the ruins of the temple, there is still a large enclosure, the walls of which are formed entirely of human skulls. The horrid piles of sculls, in their various stages of decay, exhibit a most ghastly and affecting spectacle. They are principally, if not entirely, the sculls of those who have been slain in battle. A number of beautiful trees grow around, especially the tamanu, *callophyllum inophyllum*, and the aoa, *ficus prolixa*, resembling, in its growth and appearance, one of the varieties of the banian in India.

In the inland part of the district there is a celebrated *pare*, or natural fortress, frequently resorted to by the inhabitants in seasons of war; and with a little attention it might easily be made impregnable, at least to such forces or machines as the natives could bring against it.

A fine quay, or causeway, of coral rock had been raised along the edge of the southern side of the bay, on which the natives had erected the frame of a large and substantial place of worship. It appeared to have remained in the state in which we saw it for some months past. The king and chiefs, with their numerous attendants, had removed to the vicinity of the Missionary station on the other side of the island, and the district appeared comparatively deserted. The frame of the building had been prepared with great care, several of the pillars being of highly polished *aito*, or casuarina.

Early in the afternoon we left our kind friends, and enjoyed a pleasant sail within the reef, along the eastern shore of the island; which was remarkably broken, and beautiful in mountain scenery, as well as rich and verdant in the foliage with which the woody parts of the country were clothed. We passed between Tahaa and Raiatea, and arrived at the new Missionary settlement, on the north-west side of the latter, about noon. Here we received a cordial welcome from our friends Messrs. Orsmond, Williams, and Threlkeld, who were comfortable in their new sphere of action, and greatly attached to the people; by whom they were highly respected, and among whom they had reason to believe they were usefully employed.

Mr. Orsmond appeared to sustain his bereavement with Christian fortitude. We visited the grave of the first labourer that had been called from our little band, and (with mingled feelings of regret at her early departure from the field we had unitedly cultivated, and sympathy with him whom she had left behind,) beheld the humble mound under which her mortal remains were reposing, and around which a number of indigenous and exotic flowers had been planted. Mr. Williams and Mr. Orsmond had for some time past preached in the native language. They were not only anxious to instruct the people in religion, but to improve their present condition by encouraging them to build comfortable houses after our example, and to bring under cultivation a larger portion of the soil than they had hitherto been accustomed to enclose. While we remained, we visited the different parts of the district, and called upon the king,—whom we were delighted to find in a neat plastered house,—and, after spending two or three days with them at Vaoaara, we returned to Huahine.

No circumstances connected with the interesting station at Raiatea afforded us more satisfaction, than the favourable appearance under which the education of the inhabitants had been commenced.

Next to the direct communication of the gospel by the living voice, the schools have been considered as the most important department of regular instruction. We have always superintended the schools, and generally taught the higher classes. In some stations, the boys and the men have been educated in one school, and the women and girls in another; in others, the different sexes have been taught at different times; and in some, they have assembled in the same schools. This, however, has not been general. We have been highly favoured, in most of the stations, with valuable native teachers, in both the male and female schools. To this method of instruction we have looked for the perpetuity of the work, of which we had been privileged to witness the commencement; and from its influence on the rising generation, we have derived great encouragement in reference to the stability and increase of the Christian church.

In the island of Huahine, we had, during the latter part of our residence there, two district schools, one for the males and the other for the females, which we found more conducive to their improvement, than the method of instructing both sexes in the same school. After the departure of Mr. Davies in 1820, the superintendence of the schools had devolved entirely on Mr. Barff. The female school in Huahine was under the management of Mrs. Barff and Mrs. Ellis; and those at several of the other stations were also superintended by the wives of the Missionaries.

The habits of the people did not allow of their attending school with that regularity which scholars are accustomed to observe in England. Many of the pupils being adults, had other needful engagements. In order, however, to ensure as regular and punctual an attendance as possible, the principal instruction was given at an early hour every morning, that the people might attend the school before engaging in their ordinary avocations. The natives, therefore, assembled soon after sunrise: Mr. Barff usually repaired to the school for the men and boys about half past six o'clock in the morning, and, during the latter part of our residence in Huahine, Mrs. Barff and Mrs. Ellis, either unitedly or alternately, visited the female school at the same hour. It closed in general about eight, after which the people repaired to their daily employments. The boys' school was open at two o'clock in the afternoon, but it was principally for the instruction of children. Many of the adults received instruction more readily than the children, and acquired a knowledge of reading with much greater facility than persons of the same age would do in England. With many, however, more advanced in life, it was a most difficult task; and some, after two or three years' application, were still unable to advance beyond the alphabet, or the first syllables of the spelling-book. Another source of perplexity resulted from the injudicious methods of the native teachers, who at first, in their zeal to encourage and assist their scholars, repeated to them every word in the columns of spelling, and their lessons, so frequently, that many of their pupils could repeat from memory, perhaps, the whole of the book, without being able to read a single line. When they took the book, it was only necessary for them to be told the first word or sentence in a chapter, in order to their repeating the whole correctly, even though the book should be open at some other part, or the page be placed bottom upwards. Such individuals did not always like to go back to the lowest classes, and begin to learn the simplest words; yet it was necessary. In order to convince them of the propriety of this, they were told we could not distribute copies of the Scriptures to any but those who could read any part on looking at it, without pronouncing the words merely from memory. The native teachers had fallen into this practice, from the influence of former habits. All their knowledge, traditions, songs, &c. were preserved by memory; and the preceptor recited them to his pupil, till the latter could repeat them correctly. The matter of the lessons, they also thought was the great thing to be remembered; and this, together with a desire to facilitate the advancement of those under their care, led them to adopt the method of teaching the scholars to repeat lessons without due attention to the words of the book. It has been, however, discontinued.

After the conclusion of the usual school exercises, Mr. Barff appropriated half an hour to the instruction of the natives in the art of singing. The islanders in general are remarkably fond of singing, and always ready to learn. They have not such sweet melodious voices as the natives of Africa have, yet learn to sing, considering their circumstances, remarkably well. Many of the female voices are clear and soft, without being weak; and they usually perform parts appropriated to the female voice better than the men do theirs.

Translations of the most approved psalms and hymns, with a number that are original, have been prepared in the native language, in almost every variety of metre. To these the most popular English tunes are affixed; and with most of those sung by ordinary congregations in England, the natives are acquainted. Mr. Davies, I believe, first taught them to sing, and a tune usually called "George's" was the first they learned. On our arrival in the islands in 1817, it was in general use; and whenever we walked among the habitations of the people, some parts of it broke upon the ear. It is now, however, very seldom heard. The "Old Hundredth Psalm," "Denmark," "Sicilian Mariners," and others of a more moderate date, are among their greatest favourites.

The Bible has been the basis of the greater part of the instruction given in the schools, but not to the exclusion of other departments of knowledge. In addition to the various portions of Scripture, and numerous tracts that have been printed, a system of arithmetic has been prepared by Mr. Davies, and a table of chronology, which is extensively used; and, so soon as the entire volume of Scripture shall be completed, other useful works will be translated. Although a work on geography has not yet been printed, many of the natives have a tolerably correct idea of the extent, population, and relative positions of the most important countries of the world. They are certainly fond of calculations, and make themselves familiar with figures, so far as their books enable them to proceed. The schools are important appendages of every Missionary station, and are considered such by the most intelligent and influential of the people.

As it respects the spiritual improvement of the rising generation, the understanding of the Scriptures, and the extension of genuine Christianity, Sabbath-schools are the most interesting and encouraging sections of this department. The scholars are the same as in the day-schools, but the mode of instruction pursued is different. Writing, reading, and spelling are not taught, but the time is devoted to the religious instruction of the children. Each class is under the care of a native instructor, and we have in several of the stations been highly favoured in the co-operation of most valuable Sabbath-school teachers. In Huahine we found some most able assistants among them, especially the teachers in the girls' school. They were not satisfied with attending during the hours of school, and merely imparting the ordinary instruction, or hearing the usual recitals, but identified themselves with the advancement of the children, and exercised an affectionate care over them during the intervals between the Sabbaths.

By this means they gained the confidence and love of many of their pupils, and were resorted to for guidance and counsel in every engagement of importance, or season of difficulty. Frequently one of these teachers, in order to greater quietude, and more unreserved converse with the children, would take her little class to some retired spot in one of the valleys behind the settlement, for the purpose of talking in the most affectionate manner to each individually, and then uniting with them in prayer to the Most High. I cannot imagine a more cheering and affecting scene, than must often have been presented, when a native Sabbath-school teacher has seated herself on the grass, under the shade of a spreading tree, or by the side of a winding stream, and has there gathered her little class around her, for the purpose of unfolding, and impressing on their tender minds, the pure and sacred precepts of inspired truth; or has, under these circumstances, engaged with them in prayer to that God, who is not confined to temples made with hands, and who regards the sincerity of those who call upon him, rather than the circumstances under which their petitions are offered. Their delightful labours in this department of instruction have not been in vain. Several children and young persons, who have died, have left behind them the most consoling and satisfactory evidence, that they had departed to be with Christ; and others have been at an early age admitted members of the christian church.

The annual examinations of these schools are among the most exhilarating and interesting festivities now observed in the islands. They are usually held in the chapel of the station, in order to afford accommodation to a greater number of persons than could gain admittance to the schools. Sometimes the adults are examined as well as the children, but, in general, only the latter. Their parents attend, to witness the procedure, with great satisfaction. An entertainment and a procession usually terminate the exercises of the day.

One of these anniversaries, held at Burder's Point, the Missionary station, the district of Atehuru, in the year 1824, was unusually interesting. This district had formerly been distinguished, even among the districts of Tahiti, for the turbulent and warlike dispositions of its inhabitants, and the ardour of their zeal in the service of their idols—the magnitude of the idol temples—the sanguinary character of their worship—and the presence of Oro, the great war-god of the South Sea Islanders. Within the precincts of the Missionary station, not far from the place of worship, one of the great national maraes formerly stood,—where the image of Oro had often been kept, where human sacrifices had frequently been offered, where the inauguration of the last heathen king who reigned in Tahiti took place, and where every cruelty and every abomination connected with paganism had been practised for ages. After the subversion of idolatry, this marae was divested of its glory, stripped of all its idolatrous appendages, and robbed of its gods, while the houses they occupied were committed to the flames. Still the massy pile of solid stonework, constituting one end of the area which the marae included, remained in a state of partial dilapidation—an imposing monument of the reign of terror, as they denominated idolatry. The natives were, however, determined to remove even this vestige of the system of which they so long had been the vassals, and therefore levelled, for this occasion, the mighty pile, and with the materials formed a spacious solid platform, three feet high, one hundred and ninety-four feet long, and one hundred and fifty-seven feet wide; the whole surrounded with a stone wall cemented with lime. Here a festival was held on the 11th of June, 1824. Upon this platform ninety tables were prepared, after the manner of preparation for a feast in England. Seats, usually native-made sofas or chairs, were arranged along the sides of the tables, and all the children in the school, about two hundred and forty, dined together.

The Missionaries, and many of the parents of the children, were present—delighted to witness the cheerfulness of the boys and the girls, as they sat together, and unitedly partook of the bounties of Providence. Mr. Darling, the indefatigable Missionary of the station, remarks, "This was on the very spot where Satan's throne stood, and where, a few years ago, if a female had eaten but a mouthful, so sacred was the place considered, that she would have been put to death." What a spectacle of loveliness and peace must the platform have on this day exhibited, when compared with the scenes of abomination, absurdity, and cruelty, that had often been presented, when the very materials of which it was composed had formed part of an idolatrous temple. The children afterwards walked in procession through the settlement, halted at each of the extremities, sung a hymn, and then repaired to the chapel, where a suitable address was delivered to them by the pastor. These annual examinations and festivals are not peculiar to Bunaauia, but are instituted in other stations of the Georgian group.

In the Leeward or Society Islands the remembrance of these exercises are among the most pleasing recollections I retain of my intercourse with the people. In Huahine they are usually held at the close of the public services connected with the Missionary anniversaries.

On the 11th of May 1821, a large chapel was nearly filled with spectators. The school contained between four and five hundred children. Several from each class were examined, and manifested that they had been neither indolent nor careless. I beheld, with no common interest, a number of fine, healthy, and sprightly-looking children on that occasion assembled together, and saw a little boy, seven or eight years of age, with a little fringed mat wound round his waist, and a light scarf thrown over the shoulder, stand up on a form, and repeat aloud two or three chapters of one of the Gospels, and answer a variety of questions; and pass through the whole of his examination with scarcely a single mistake. This was the case with several on that occasion. At the close of the examination, the children were rewarded by Mr. Barff, who, on delivering the presents, which were different books in the native language, accompanied each by a suitable remark to the favoured proprietor. Often, as the little boy has walked back to the seat with his prize—perhaps a copy of one of the Gospels—I have seen the mother's eye follow the child with all a parent's fond emotion beaming in her eye, while the tear of pleasure has sparkled there; and, in striking contrast with this, the childless mother might be seen weeping at the recollection of the dear babes, which, under the influence of idolatry, she had destroyed—and which, but for her own murderous hands, might have mingled in the throng she then beheld before her. On the occasion above alluded to, when the examinations in the place of worship had terminated, the children walked, in the same order in which they were accustomed to proceed from the school to the chapel, to a rising ground in the vicinity of the governor's house. Here an entertainment had been provided for them by the chiefs. We followed, amid the multitude of their parents and friends; and, on reaching the place of the assemblage, beheld a most delightful scene. About three hundred boys sat in classes on the grass on the right-hand side of the rising ground, each teacher presiding at the head of his class. On the left-hand, about two hundred girls were arranged in the same manner. A plentiful repast had been prepared, which was carved, and handed to them as they sat upon the green turf. In the centre, tables were spread for the chiefs, and the parents and friends of the children: we sat down with them, gratified with their hospitality, but deriving far more pleasure from gazing on the spectacle on either side, than in partaking of the provision. Before the assembly departed, I gave a short address to the parents, teachers, and children. When I concluded, they all stood up; the boys formed a circle on one side, and the girls on the other, and sang alternately the verses of a hymn in the native language; after which, one of the teachers offered a short prayer,—and we retired, under the influence of those emotions of satisfaction which appeared to pervade the bosom of every individual; and it was not easy to say whether it was most powerfully exhibited in the countenances of the children or their parents.

Towards the evening of the day, the children walked two and two, hand in hand, from one end of the settlement to the other, preceded by the flag belonging to the schools. The best boy in the school carried the flag; which was not of silk emblazoned with letters of gold, but of less costly materials. The banners of the schools attached to the different stations were various; some of white native cloth, with the word "Hosanna" impressed upon it in scarlet dye; another was of light, but woven cloth, with the following sentiment inscribed upon it, *Ia ora te hui arii e ia maoro teienei hau,* "Life and blessing to the Reigning Family, and long be this peaceful reign!" The one at Huahine was of fine blue cloth, with a white dove and olive branch in the centre, beneath which was inscribed the Angels' Song, as the motto of the school. Sometimes the children, as they passed along, would sing, "Long be this peaceful reign," or any other motto that might be inscribed upon the banner. And when they walked through the district, a father or mother, or both, came out of the door of their little cottages, and gazed with highest pleasure on them as they passed by, walked beside them, or followed them with their eye until some clump of trees, or winding in the road, hid them from their view.

The meeting at Raiatea in the year 1824 was deeply affecting. It was held on a kind of pier or quay built in the sea. Six hundred children assembled to partake of the feast their parents had provided. The boys afterwards delivered public addresses. A religious service in the chapel closed the exercises of the day, and all retired to their respective homes, apparently delighted. Mr. Williams, in reference to this interesting spectacle, questions whether, but for the influence of Christianity, one-fourth of the children would have been in existence, and states his opinion, that they would not, and that *"the hands of their mothers would have been imbrued in their blood."* This was not groundless opinion, but an inference authorized by the most melancholy but unquestionable facts. At a former meeting held on the spot where the chapel stood, in which the children were examined, he was present. A venerable chief rose, and addressed the assembly, with impressive action, and strongly excited feeling. Comparing the past with the present state of the people, he said, "I was a mighty chief; the spot on which we are now assembled was by me made sacred for myself and family; large was my family, but I alone remain; all have died in the service of Satan—they knew not this good word which I am spared to see; my heart is longing for them, and often says within me, Oh! that they had not died so soon: great are my crimes; I am the father of nineteen children; *all of them I have murdered*—now my heart longs for them.—Had they been spared, they would have been men and women—learning and knowing the word of the true God. But while I was thus destroying them, no one, not even my own cousin, (pointing to Tamatoa the king, who presided at the meeting,) stayed my hand, or said, Spare them. No one said, The good word, the true word is coming, spare your children; and now my heart is repenting—is weeping for them!"

CHAP. XVIII.

Account of Taaroarii—Encouraging circumstances connected with his early life—His marriage—Profligate associates—Fatal effects of bad example—Disorderly conduct—His illness—Attention of the chiefs and people—Visits to his encampment—Last interview—Death of Taaroarii—Funeral procession—Impressive and affecting circumstances connected with his decease and interment—His monument and epitaph—Notice of his father—His widow and daughter—General ideas of the people relative to death and a future state—Death the consequence of Divine displeasure—State of spirits—Miru, or heaven—Religious ceremonies for ascertaining the causes of death—Embalming—The burying of the sins of the departed—Singular religious ceremony—Offerings to the dead—Occupation of the spirits of the deceased—Superstitions of the people—Otohaa, or lamentation—Wailing—Outrages committed under the paroxysms of grief—Use of sharks' teeth—Elegiac ballads singularly beautiful—The heva—Absurdity and barbarism of the practice—Institution of Christian burial—Dying expressions of native converts.

Although many of the parents had the satisfaction to behold their children growing up under the influence of those instructions which they were so anxious to impart, and realizing, as they arrived at years of maturity, all they had desired, there were instances of an opposite kind. The circumstances preceding the death of Taaroarii, the only son of the king, the chief of Sir Charles Sanders' Island, and the heir to the government of Huahine, were peculiarly distressing.

The young chieftain was in his nineteenth year; his high rank, and extensive influence, led us to indulge cheering anticipations; and, during his juvenile years, he was greatly beloved by the people. He had also, when it was supposed he could scarcely have arrived at years of discretion, shewn his contempt for the idols of his country, his desire to be instructed concerning the true God, and had prohibited the licentious and idolatrous ceremonies of the Areois, when there were very few in any degree favourable to Christianity. Subsequently, Taaroarii had become a diligent pupil of the Missionaries. We could not but hope that Divine Providence was raising him up to succeed his father, and to govern the islands under his authority, for the stability of the Christian faith, and the advancement of the people's true interests.

These hopes, however, were disappointed. He treated Christianity and the worship of God with respect, was a steady enemy to the introduction or use of ardent spirits by chiefs or people, and was not a profligate man. But, soon after our establishment in Huahine, a number of the most abandoned young men, of that and other islands, attached themselves to his retinue, which was always numerous, became his companions, flattered his pride, and, in many respects ministering to his wishes, they infused their own spirit into his mind.

Being naturally cheerful and good-natured, he was induced by his companions, first to neglect instruction, then the public worship of God, and, subsequently, to patronize and support his unprincipled followers. His venerable father beheld the change with the most poignant grief, and used all the affection, influence, and authority of a parent, to lead him from those evil courses; but his efforts, and those of other friends, failed.

In order to draw him from this influence, a matrimonial connexion was arranged, and he was united in marriage with the daughter of Hautia, who, next to Mahine, was the highest chief, and deputy-governor of the island. His daughter was near the age of the king's son; and though rather inferior in rank, she was in every other respect a suitable partner, and proved a faithful and affectionate wife.

A house was built for him near the dwelling of his wife's family, and subsequently a domestic establishment formed for the youthful couple, adjacent to his father's residence. It was, however, soon manifest that the baneful influence of his former associates was not destroyed. They gathered around him again, and he gave himself up to their guidance and counsel.

His wife was treated with cruelty, but still continued attentive to his comfort. A number of the most profligate of the young men attached to his establishment, having tataued themselves, he was induced to submit to the same, it is supposed, with a view to skreen them from punishment. They imagined the magistrates would not bring him to public trial; and if *he* was exempted, they knew *they* should escape. In this, however, they were mistaken.

When it was found that the young chief had actually violated the laws, the magistrates came to the king, to ask him whether he should be tried. The struggle was severe; but, under the influence of a patriotism worthy of his station, he said he wished the laws to be regarded, rather than those feelings which would lead him to spare his son the disgrace to which he had subjected himself.

To convince the people that the government would act according to the laws, and to deter others from their violation, he directed that his son should be tried. Taaroarii received his sentence with apparent indifference, but was so exasperated with his father, that he more than once threatened to murder him, or to cause his destruction.

Some months after this, he broke a blood-vessel, it is supposed, with over-exertion at the public work appointed as a penalty for his crime. After this, he laid aside his labour; his people would have performed the work for him at once, but he would not allow it, and appeared to identify himself with them, in the humiliating situation to which they had reduced themselves. In the conversations we sometimes had with him, he seemed to regret having connected himself with the party who now considered him as their head and leader.

Shortly after this event, symptoms of rapid consumption appeared, and assumed a most alarming character. Every available means were promptly employed, but without effect. His father frequently visited him, and his wife was his constant attendant. In order to try the effect of change of air, he was laid upon a litter, and brought on men's shoulders into the valley, where a temporary encampment had been erected near our dwelling. The chiefs of the island, with their guards, attended, and, when they reached the valley, fired three volleys of musketry, indicative of their sympathy.

While he remained here, we often saw him; he was generally communicative, and sometimes cheerful, excepting when the topic of religion was introduced, and then an evident change of feeling took place; he would attend to our observations, but seldom utter a syllable in reply, and seemed unwilling to have the subject brought under consideration. This was the most distressing circumstance attending his illness, and to none more painfully affecting than to his aged father.

On the last day of his life, Mrs. Ellis and our two elder children, to whom he had always been partial, went to see him: he appeared comparatively cheerful, listened to all that was said, and shook the children by the hand very affectionately, when they said *Ia ora na*, or Farewell. I spent some time with him during the same afternoon, and it was the most affecting intercourse I ever had with a dying fellow-creature.

The encampment was fixed on an elevated part of the plain, near which the river, that flowed from the interior mountain to the sea, formed a considerable curvature. The adjacent parts of the valley were covered with shrubs, but the margin of the river was overgrown with slender branching purau, and ancient chestnut-trees, that reared their stately heads far above the rest, and shed their grateful shade on the waters, and on the shore. Near the edge of the cool stream that rippled among the pebbles, and at the root of one of these stately trees, I found the young chieftain, lying on a portable couch, surrounded by his sorrowing friends and attendants.

I asked why they had brought him there: they said that he complained of heat or want of air, and they had brought him to that spot that he might enjoy the refreshing coolness of the stream and the shade. I could not but admire their choice as I sat beside him, and felt, after leaving the portions of the valley exposed to the sun's rays, as if I had entered another climate. The gentle but elastic current of air swept along the course of the river, beneath the foliage that often formed beautiful natural arches over the water, and through which a straggling sun-beam was here and there seen sparkling in the ripple of the stream.

After mingling my sympathy with the friends around, I spoke at some length to the young man, whose visage had considerably altered since the preceding day. I endeavoured to direct his mind to God, for mercy through Christ, and affectionately urged a personal and immediate application, by faith, to him who is able to save even to the uttermost, and willing to receive even at the eleventh hour, &c.

All prospect of his recovery had ceased; our solicitude was therefore especially directed to his preparation for that state on which he was so soon to enter. This indeed had been our principal aim in all our intercourse with him. On this occasion he made no reply, (indeed I suppose he was unable, had he been disposed,) but he raised his head after I had done speaking, and gazed stedfastly upon me, with an expression of anguish in his whole countenance, which I never shall forget, and which is altogether indescribable. Whether it arose from bodily or mental agony, I am not able to say, but I never beheld so affecting a spectacle.

Before I left his couch, I attempted to direct his mind to the compassionate Redeemer, and, I think, engaged in prayer with him. The evening was advancing when I took leave, and the conviction was strongly impressed on my mind, that it was the last day he would spend on earth. My eye lingered on him with intense and mingled interest, as I stood at his feet, and watched his short heaving and laborious respiration; his restless and feverish head had been long pillowed on the lap of his affectionate wife, whose face, with that of every other friend, was suffused with tears. His eye rolling its keen fitful glance on every object, but resting on none, spoke a state of feeling very remote indeed from tranquillity and ease. I could not help supposing that his agitated soul was, through this her natural window, looking wishfully on all she was then leaving; and as I saw his eye rest on his wife, his father, his friends around, and then glancing to the green boughs that waved gently in the passing breeze, the bright and clear blue sky that appeared at intervals through the foliage, and the distant hills whose summits were burnished with the splendour of the retiring sun—I almost imagined the intensity and rapidity of his glance indicated an impression that he would never gaze on them again. Such was the conviction of my own mind; and I reluctantly retired, more deeply than ever impressed with the necessity of early and habitual preparation for death.

O! how different would the scene have been, had this interesting youth, as earth with all its associations receded from his view, experienced the support and consolations of the gospel, with the hopes of immortality. I presume not to say that in his last hours, in those emotions of the soul, which nature was too much exhausted to allow him to declare, and which were known only to God and to himself, he was not cheered by these anticipations. I would try to hope it was so: for indications of such feelings his dear sorrowing and surviving friends anxiously waited.

How striking the contrast between his last day on earth, and that of Teivaiva, another youth of Huahine, and, like Taaroarii, an only son and an only child, who, when he saw his sorrowing parents weeping by the side of the couch on which he lay, collected his remaining strength, and rousing himself, said—"I am in pain, but I am not unhappy; Jesus Christ is with me, and he supports me; we must part, but we shall not be parted long; in heaven we shall meet, and never die. Father, don't weep for me. Mother, don't weep for me. We shall never die in heaven." But the latter of these, while in health and comfort, had been happy in the ways of religion, seeking the favour of God: the former had neglected and departed from those ways, and had lived in the practice of sin.

About nine o'clock in the evening, Mahine sent word that his son was worse. Mr. Barff and myself hastened to the encampment, and found him apparently dying, but quite sensible. We remained with them some time, endeavoured to administer a small portion of medicine, and then returned. A short time before midnight, on the 25th of October, 1821, he breathed his last.

When the messenger brought us the tidings of his death, we repaired to the tent, found his parents, his wife, and an aunt who was exceedingly fond of him, sobbing and weeping bitterly by the side of the corpse. The attendants joined in the lamentation; it was not the wild and frantic grief of paganism, so universal formerly on such occasions, but the expression of deepest anguish, chastened and subdued by submission to the Divine will. We mingled our sympathies with the mourners, spent a considerable time with them endeavouring to impart consolation to their minds, and then returned to rest, but not to sleep.

The sudden departure of the young chieftain, and the circumstances connected with it, powerfully affected our minds. We had been intimate with him ever since our arrival in the islands, had received many tokens of kindness from him, had watched his progress with no ordinary interest, especially since his removal to Huahine in 1818. We had considered him as the future sovereign of the island in which we should probably spend our days, but he was now for ever removed. We hoped we had been faithful to him. But at times such as this, when one and another was removed from the people amongst whom we laboured, we were led to ponder on the state into which they had entered; and when their prospects had been dark, and their character doubtful, we could not but fear that we perhaps had not manifested all the solicitude we ought to have done, nor used means available for the purpose of leading them to Him, who alone could deliver from the fear of death, and all the consequences of conscious guilt. Reflections of this kind were now solemn and intense, and I trust profitable.

The funeral was conducted in the christian manner: a coffin was made for the body, and a new substantial stone vault was built in the south-west angle of the chapel-yard; on account of which, his interment was deferred until five days after his decease.

About three o'clock in the afternoon of the 30th of October, we repaired to the encampment of the king, and found most of the people of the island assembled. About four the procession left the tent. Mr. Barff and myself walked in front, followed by a few of the favourite attendants of the young chief. The coffin was borne by six of his own men; it was covered with a rich yellow pall, of thick native cloth, with a deep black border. Six young chiefs, in European suits of mourning, bore the pall; amongst them was the son of the king of Raiatea. His wife, his father, and near relations, followed, wearing also deep European mourning. Mrs. Barff and Mrs. Ellis, with our children, walked after these; the tenantry of his own district, and servants of his household, came next; and after them the greater part of the population of the island.

When we reached the place of sepulture, I turned and looked towards the valley, and beheld, I think, a scene of the most solemn interest that ever I witnessed. Before us stood the bier, on which was laid the corpse of the individual of highest hopes among all I beheld, destined for the highest distinction the nation knew, whose tall, and, for his years, gigantic form, open and manly brow, had promised fair for many years of most commanding influence, an influence which we once had hoped would have advanced his country's welfare. Beside that bier stood his youthful widow, weeping, we have reason to believe, tears of unfeigned sorrow; and who, in addition to the loss she had sustained, was on the eve of becoming a mother. Near her stood his venerable sire, gray with age, and bending with infirmities, taking a last sad look of all that now remained of what had once been the stay of his declining years, his hope and joy; towards whom, in all his wayward courses, he had exercised the affection of a father.

Around them stood the friends, and, along the margin of the placid ocean, and emerging from the shadowy paths that wound along the distant valley, the mourning tribes, the father, and the mother, with their children, were seen advancing slowly to the spot. Each individual in the whole procession, which, as they walked only two abreast, extended from the sepulchre to the valley, wore some badge of mourning; frequently it was a white *tiputa* or mantle, with a wide black fringe.

When the greater part had reached the chapel yard, Mr. Barff addressed the spectators, and I offered a prayer to the Almighty, that the mournful event might be made a blessing to the survivors. The body was then deposited in the tomb, the pall left on the coffin. The father, the widow, and several other friends, entered, took a last glance, and retired in silence, under strong and painful emotion. When we withdrew, the servants placed a large stone against the entrance, and left it till the following day, when it was walled up. The tomb was whitewashed, and a small coral stone placed perpendicularly, at the end towards the sea, on which was inscribed in the native language, this simple epitaph, "Taaroarii, died October 25th, 1821." On the following Sabbath, the solemn event was improved, in a discourse from 2 Kings xx. 1.

I never saw persons more deeply affected than the friends of the deceased had been during his illness, especially his excellent father, and his wife. For many days prior to his death, the latter sat by his couch, supporting his aching head in her lap, wiping the cold perspiration from his brow, or refreshing him with her fan, watching with fondest solicitude his look, and aiming, if possible, to anticipate his wishes. It ended not with his decease. She scarcely left his body until it was interred, sitting on one side, while his aunt, or some other relative, sat on the other, through the day; and when overcome with fatigue and watching, falling asleep in the same station at night; yet I never heard the least murmur or repining word against the dealings of God. It was but the excess of sorrow, on account of the bereavement. Two months afterwards she became a mother, and, during our continuance on the island, Mrs. Ellis was considered as the guardian of her infant daughter. Since our departure the child has been trained, by its fond mother, according to the direction of Mrs. Barff, and will probably succeed to the government of the island at its grandfather's death.

Mahine, the pious and venerable chief, still lives to be an ornament to the Christian religion, a nursing father to the infant churches established in his country, and the greatest blessing to the people whom he governs. His daughter-in-law in some degree supplies to him the place of his departed son, and is, indeed, the comfort and solace of his declining years. Her behaviour to him and his family has been uniformly affectionate and respectful—the whole of her public and domestic conduct, such as to deserve the imitation of her own sex, among whom she has maintained her elevated station with becoming propriety, and to confirm those pleasing anticipations we have often indulged respecting her religious character. We have reason to hope that she is not a Christian merely in sentiment and opinion, but that her mind is under the decisive influence of the principles inculcated in the sacred volume.*

Having had occasion to speak of the decease and interment of Taaroarii, an account of the views of death and a future state, and the rites of burial, which formerly

* Among a number of letters which have recently come to hand from the Society Islands, Mrs. Ellis had the satisfaction to receive one from the widow of Taaroarii, of an extract from which, the following is a literal translation. "Peace to you from the true God, from Jehovah, and from Jesus Christ. My word to you is, that my affection for you and your children is unabated. Through the goodness of God your breath has been lengthened out. We did not know whether you were living or not, and, behold, your little presents arrived, and we knew that you were still living. On account of the goodness of God, our breath is lengthened;

prevailed in the islands, will not be inappropriately introduced here. Some of their usages and opinions on these subjects were remarkably curious. Every disease to which they were subject, was supposed to be the effect of direct supernatural agency, and to be inflicted by the gods for some crime against the tabu, of which the sufferers had been guilty, or in consequence of some offering made by an enemy to procure their destruction. Hence, it is probable, in a great measure resulted their neglect and cruel treatment of their sick. The same ideas prevailed with regard to death, every instance of which they imagined was caused by the direct influence of the gods.

The natives acknowledged that they possessed articles of poison, which, when taken in the food, would produce convulsions and death, but those effects they considered more the result of the god's displeasure, operating by means of these substances, than the effects of the poisons themselves. Those who died of eating fish, of which several kinds found on their coasts are at certain seasons unsuitable for food, were supposed to die by the influence of the gods; who, they imagined, had entered the fish, or rendered it poisonous. Several Europeans have been affected by these fish, though only in a slight degree, usually causing swelling of the body, a red colour diffused on the skin, and a distressing head-ache. Those who were killed in battle were also supposed to die from the

and our dwelling prolonged in this land; but we know not that we shall see each other's faces again. You know that frail and feeble is the body of man. Tamarii (her infant daughter) is learning the word of God. Come back to Huahine. Peace be to you all, from Jesus Christ."—The letter from which the above is taken, is dated October 6, 1827, and is signed, "Taaroarii-vahine."

influence of the gods, who, they fancied, had actually entered the weapons of their murderers. Hence, those who died suddenly were said to be seized by the god.

Their ideas of a future state were vague and indefinite. They generally spoke of the place to which departed spirits repaired on leaving the body, as the *po*, state of night. This also was the abode or resort of the gods, and those deified spirits that had not been destroyed. What their precise ideas of a spirit were, it is not easy to ascertain. They appear, however, to have imagined the shape or form resembled that of the human body, in which they sometimes appeared in dreams to the survivors.

When the spirit left the body, which they called *unuhi te varua e te atua*, the spirit drawn out by the god, (the same term, *unuhi*, is applied by them to the drawing of a sword out of its scabbard,) it was supposed to be fetched, or sent for, by the god. They imagined that *oramatuas*, or demons, were often waiting near the body, to seize the human spirit as it should be drawn out (they supposed) from the head; and, under the influence of strong impressions from such superstitions, or the effects of a disordered imagination, when dying, the poor creatures have sometimes pointed to the foot of the mat or the couch on which they were lying, and have exclaimed, "There the *varua*, spirits, are waiting for my spirit; guard its escape, preserve it from them," &c.

On leaving the body, they imagined it was seized by other spirits, conducted to the *po*, or state of night, where it was eaten by the gods; not at once, but by degrees. They imagined that different parts of the human spirit were scraped with a kind of serrated shell, at different times; that the ancestors or relatives of the deceased performed this operation; that the spirit thus passed through the god, and if it underwent this process of being eaten, &c. three different times, it became a deified or imperishable spirit, might visit the world, and inspire others.

They had a kind of heaven, which they called *Miru.* The heaven most familiar, especially in the Leeward Islands, is *Rohutu noanoa,* sweet-scented Rohutu. This was situated near *Tamahani unauna,* glorious Tamahani, the resort of departed spirits, a celebrated mountain on the north-west side of Raiatea. The perfumed Rohutu, though invisible but to spirits, was somewhere between the former settlement and the district of Tipaehapa on the north side of Raiatea. It was described as a beautiful place, quite an elysium, where the air was remarkably salubrious, flowers abundant, highly odoriferous, and in perpetual blossom. Here the Areois, and others raised to this state, followed all the amusements and pursuits to which they had been accustomed in the world, without intermission or end. Here was food in abundance, and every indulgence. It is worthy of remark, that the misery of the one, and enjoyments of the other, debasing as they were, were the destiny of individuals, altogether irrespective of their moral character and virtuous conduct. The only crimes that were visited by the displeasure of their deities were the neglect of some rite or ceremony, or the failing to furnish required offerings. I have often, in conversations with the people and sometimes with the priests, endeavoured to ascertain whether they had any idea of a person's condition in a future state being connected with his disposition and general conduct in this; but I never could learn that they expected, in the world of spirits, any difference in the treatment of a kind, generous, peaceful man, and that of a cruel, parsimonious, quarrelsome one. I am, however, inclined to think, from the great anxiety about a future state, which some have evinced when near death, that natural conscience, which I believe pronounced a verdict on the moral character of every action throughout their lives, is not always inactive in the solemn hour of dissolution, although its salutary effects were neutralized by the strength of superstition.

As soon as an individual was dead, the tahua tutera was employed, for the purpose of discovering the cause of the deceased person's death. In order to effect this, the priest took his canoe, and paddled slowly along on the sea, near the house in which the body was lying, to watch the passage of the spirit; which they supposed would fly upon him, with the emblem of the cause for which the person had died. If he had been cursed by the gods, the spirit would appear with a flame, fire being the agent employed in the incantation of the sorcerers; if *pifaod*, or killed, by the bribe of some enemy, given to the gods, the spirit would appear with a red feather, the emblem or sign of evil spirits having entered his food. After a short time, the tahua, or priest, returned to the house of the deceased, and told the survivors the cause of his death, and received his fee, the amount of which was regulated by the circumstances of the parties.

The taata faatere, or faatubua, was then employed, to avert the destruction of the surviving members of the family. A number of ceremonies were performed and prayers offered, according to the cause of the death that had taken place; and when these were concluded, the priest, informing the family that he had been successful, and that the remaining members were now safe, received another fee, and departed.

The disposal of the corpse was the next concern. The bodies of the chiefs, and persons of rank and affluence, were preserved; those of the middle and lower orders buried: when interred, the body was not laid out straight or horizontal, but placed in a sitting posture, with the knees elevated, the face pressed down between the knees, the hands fastened under the legs, and the whole body tied with cord or cinet wound repeatedly round. It was then covered over, and deposited not very deeply in the earth.

However great the attachment between the deceased and the survivors might have been, and however they might desire to prolong the melancholy satisfaction resulting from the presence of the lifeless body, on which they still felt it some alleviation to gaze, the heat of the climate was such, as to require that it should be speedily removed, unless methods were employed for its preservation, and these were generally too expensive for the poor and middle ranks. They were therefore usually obliged to inter the corpse sometimes on the first, and seldom later than the second day after death. During the short period that they could indulge the painful sympathies connected with the retention of the body, it was placed in a sort of bier covered with the best white native cloth they possessed, and decorated with wreaths and garlands of the most odoriferous flowers. The body was also placed on a kind of bed of green fragrant leaves, which were also strewed over the floor of the dwelling. During the period which elapsed between the death and interment of the body, the relatives and surviving friends sat round the corpse, indulging in melancholy sadness, giving vent to their grief in loud and continued lamentations, often accompanied with the use of the shark's tooth; which they employed in cutting their temples, faces, and breasts, till they were covered with the blood from their self-inflicted wounds. The bodies were frequently committed to the grave in deep silence, unbroken excepting by occasional lamentations of those who attended. But on some occasions, the father delivered an affecting and pathetic oration at the funeral of his son.

The bodies of the dead, among the chiefs, were, however, in general preserved above ground: a temporary house or shed was erected for them, and they were placed on a kind of bier. The practice of embalming appears to have been long familiar to them; and the length of time which the body was thus preserved, depended altogether upon the costliness and care with which the process was performed. The methods employed were at all times remarkably simple: sometimes the moisture of the body was removed by pressing the different parts, drying it in the sun, and anointing it with fragrant oils. At other times, the intestines, brain, &c. were removed, all moisture extracted from the body, which was fixed in a sitting position during the day, and exposed to the sun, and, when placed horizontally, at night was frequently turned over, that it might not remain long on the same side. The inside was then filled with cloth saturated with perfumed oils, which were also injected into other parts of the body, and carefully rubbed over the outside every day. This, together with the heat of the sun, and the dryness of the atmosphere, favoured the preservation of the body.

Under the influence of these causes, in the course of a few weeks the muscles dried up, and the whole body appeared as if covered with a kind of parchment. It was then clothed, and fixed in a sitting posture; a small altar was erected before it, and offerings of fruit, food, and flowers, daily presented by the relatives, or the priest appointed to attend the body. In this state it was preserved many months, and when it decayed, the skull was carefully kept by the family, while the other bones, &c. were buried within the precincts of the family temple.

It is singular that the practice of preserving the bodies of their dead by the process of embalming, which has been thought to indicate a high degree of civilization, and which was carried to such perfection by one of the most celebrated nations of antiquity, some thousand years ago, should be found to prevail among this people. It is also practised by other distant nations of the Pacific, and on some of the coasts washed by its waters.

In commencing the process of embalming, and placing the body on the bier, another priest was employed, who was called the *tahua bure tiapapau,* literally "corpse-praying priest." His office was singular: when the house for the dead had been erected, and the corpse placed upon the platform or bier, the priest ordered a hole to be dug in the earth or floor, near the foot of the platform. Over this he prayed to the god, by whom it was supposed the spirit of the deceased had been required. The purport of his prayer was, that all the dead man's sins, and especially that for which his soul had been called to the *po,* might be deposited there, that they might not attach in any degree to the survivors, and that the anger of the god might be appeased.

The priest next addressed the corpse, usually saying, *Ei ia oe na te hara e vai ai*, "With you let the guilt now remain." The pillar or post of the corpse, as it was called, was then planted in the hole, perhaps designed as a personification of the deceased, to exist after his body should have decayed—the earth was thrown over, as they supposed, the guilt of the departed—and the hole filled up.

At the conclusion of this part of the curious rite, the priest proceeded to the side of the corpse, and, taking a number of small slips of the *fa maia*, plantain leaf-stalk, fixed two or three pieces under each arm, placed a few on the breast, and then addressing the dead body, said, There are your family, there is your child, there is your wife, there is your father, and there is your mother. Be satisfied yonder, (that is, in the world of spirits.) Look not towards those who are left in this world.—The concluding parts of the ceremony were designed to impart contentment to the departed, and to prevent the spirit from repairing to the places of his former resort, and so distressing the survivors.

This was considered a most important ceremony, being a kind of mass for the dead, and necessary for the peace of the living, as well as the quiet of the deceased. It was seldom omitted by any who could procure the accustomed fees for the priest, which for this service were generally furnished in pigs and cloth, in proportion to the rank or possessions of the family.

All who were employed in embalming, which they called *muri*, were, during the process, carefully avoided by every person, as the guilt of the crime for which the deceased had died, was supposed in some degree to attach to such as touched the body. They did not feed themselves, lest the food defiled by the touch of their polluted hands, should cause their own death, but were fed by others.

As soon as the ceremony of depositing the sins in the hole was over, all who had touched the body or the garments of the deceased, which were buried or destroyed, fled precipitately into the sea, to cleanse themselves from the pollution, called *mahuruhuru*, which they had contracted by touching the corpse; casting also into the sea, the clothes they had worn while employed in the work. Having finished their ablutions, they gathered a few pieces of coral from the bottom of the sea, and, returning with them to the house, addressed the dead body by saying, "With you may the *mahuruhuru*, or pollution, be," and threw down the pieces of coral on the top of the hole that had been dug for the purpose of receiving every thing contaminating connected with the deceased.

The ceremonies in general were now finished, but if the property of the family was abundant, their attachment to the deceased great, and they wished his spirit to be conveyed to *Rohutu noanoa*, the Tahitian paradise, the fifth priest was employed. Costly offerings were presented, and valuable articles given to the priest of Romatane, the keeper of this happy place; Urutaetae was the guide of such as went thither, and the duty of the priest now employed was to engage him to conduct the spirit of the departed to this region of fancied enjoyment.

The houses erected as depositories for the dead, were small and temporary buildings, though often remarkably neat. The pillars supporting the roof were planted in the ground, and were seldom more than six feet high.

The bier or platform on which the body was laid, was about three feet from the ground, and was moveable, for the purpose of being drawn out, and exposing the body to the rays of the sun. The corpse was usually clothed, except when visited by the relatives or friends of the deceased. It was, however, for a long time carefully rubbed with aromatic oils once a day.

A light kind of altar was erected near it, on which articles of food, fruits, and garlands of flowers were daily deposited; and if the deceased were a chief of rank or fame, a priest or other person was appointed to attend the corpse, and present food to its mouth at different periods during the day. When asked their reason for this senseless practice, they have said they supposed there was a spiritual as well as a material part of food, a part which they could smell; and that if the spirit of the deceased returned, the spirit or scent of the offering would be grateful. Connected with the depositories of the dead, there was what they called the *aumiha*, a kind of contagious influence, of which they appeared to be afraid; and hence, at night especially, they avoided the place of sepulture. The family, district, or royal maraes were the general depositories of the bones of the departed, whose bodies had been embalmed, and whose skulls were sometimes preserved in the dwelling of the survivors. The marae or temple being sacred, and the bodies being under the guardianship of the gods, were in general considered secure when deposited there. This was not, however, always the case; and in times of war, the victors sometimes, not only despoiled the temples of the vanquished, and bore away their idol, but robbed the sacred enclosure of the bones of celebrated individuals. These spoils were appropriated to what the nation considered the lowest degradation, by being converted into chisels or borers, for the builders of canoes and houses, or transformed into fishing-hooks. In order to avoid this, they carried the bones of their chiefs, and even the recently deceased corpse, and deposited them in the caverns of some of the most inaccessible rocks in the lofty and fearful precipices of the mountain defiles.

Notwithstanding the labour and care bestowed on the bodies of the dead, they did not last very long; probably the most carefully preserved could not be kept more than twelve months. When they began to decay, the bones, &c. were buried, but the skull was preserved in the family sometimes for several generations, wrapt carefully in native cloth, and often suspended from some part of the roof of their habitations. In some of the islands they dried the bodies, and, wrapping them in numerous folds of cloth, suspended them also from the roofs of their dwelling houses.

The tribes inhabiting the islands of the Pacific were remarkably superstitious, and among them none more so than the inhabitants of the Georgian and Society Islands. They imagined they lived in a world of spirits, which surrounded them night and day, watching every action of their lives, and ready to avenge the slightest neglect, or the least disobedience to their injunctions, as proclaimed by their priests.

These dreaded beings were seldom thought to resort to the habitations of men on errands of benevolence. They were supposed to haunt the places of their former abode, to arouse the survivors from their slumbers by making a squeaking noise, which when the natives heard they would sometimes reply to, asking what they were, what they wanted, &c. Sometimes the spirits upbraided the living with former wickedness, or the neglect of some ceremonious enactment, for which they were unhappy.

When a person was seized with convulsions or hysterics, it was said to be from seizure by the spirits, who sometimes scratched their faces, tore their hair, or otherwise maltreated them. For some time after the death of Taaroarii we could seldom induce any of our servants to go out of the house after it was dark, under an apprehension that they should see, or be seized by, his spirit. They were, however, very ignorant young persons. The natives in general laugh at their former credulity. The whole system of their superstition seems to have been, in every respect, wonderfully adapted to debase the mind, and keep the people in the most abject subjection to the priests, who, in order to maintain their influence, had recourse to this extensive and imposing machinery of supernatural agency; and it must be confessed that, considering their isolated situation, their entire ignorance of science, of natural and experimental philosophy, their ardent temperament, the romantic nature of the country, and the adventurous character of many of their achievements, there was something remarkably imposing to an uncultivated mind in the system here inculcated.

Almost every native custom connected with the death of relations or friends, was singular, and none perhaps more so than the *otohaa*, which, though not confined to instances of death, was then most violent. It consisted in the most frantic expressions of grief, under which individuals acted as if bereft of reason. It commenced when the sick person appeared to be dying; the wailing then was often most distressing, but as soon as the spirit had departed, the individuals became quite ungovernable.

They not only wailed in the loudest and most affecting tone, but tore their hair, rent their garments, and cut themselves with shark's teeth or knives in a most shocking manner. The instrument usually employed was a small cane, about four inches long, with five or six shark's teeth fixed in, on opposite sides. With one of these instruments every female provided herself after marriage, and on occasions of death it was unsparingly used.

With some this was not sufficient; they prepared a short instrument, something like a plumber's mallet, about five or six inches long, rounded at one end for a handle, and armed with two or three rows of shark's teeth fixed in the wood, at the other. With this, on the death of a relative or a friend, they cut themselves unmercifully, striking the head, temples, cheek, and breast, till the blood flowed profusely from the wounds. At the same time they uttered the most deafening and agonizing cries; and the distortion of their countenances, their torn and dishevelled hair, the mingled tears and blood that covered their bodies, their wild gestures and unruly conduct, often gave them a frightful and almost inhuman appearance. This cruelty was principally performed by the females, but not by them only; the men committed on these occasions the same enormities, and not only cut themselves, but came armed with clubs and other deadly weapons.

The otohaa commenced with the nearest relations of the deceased, but it was not confined to them; so soon as the tidings spread, and the sound of the lamentations was heard through the neighbourhood, the friends and relatives repaired to the spot, and joined in the tragic performance.

I am not prepared to say that the same enormities were practised here as in the Sandwich Islands at these times, but on the death of a king or principal chief the scenes exhibited in and around the house were in appearance demoniacal. The relatives and members of the household began; the other chiefs of the island and their relatives came to sympathize with the survivors, and, on reaching the place, joined in the infuriated conduct of the bereaved; the tenantry of the chiefs also came, and, giving themselves up to all the savage infatuation which the conduct of their associates or the influence of their superstitions inspired, they not only tore their hair, and lacerated their bodies till they were covered with blood, but often fought with clubs and stones till murder followed.

Auna has now some dreadful indentations on his skull from blows he received by stones on one of these occasions at Huahine; and in almost one of the last *otohaa* observed in the same island, a man was killed by the contents of the musket of another. Since the introduction of fire-arms, they have been used in these seasons; and the smoke and report of the guns must have added to the din and terrible confusion of the scene. I cannot conceive of a spectacle more appalling, than that which the infuriated rabble, smeared with their own blood, presenting every frightful distortion in feature, and frantic madness in action, must often have exhibited. This scene was sometimes continued for two or three successive days, or longer, on the death of a person of distinction.

I have often conversed with the people on their reasons for this strange procedure, and have asked them if it was not exceedingly painful to them to cut themselves as they were accustomed to do. They have always answered that it was very painful in some parts of the face—that the upper-lip, or the space between the upper-lip and the nostril, was the most tender, and a stroke there was always attended with the greatest pain—that it was their custom, and therefore considered indispensable, as it was designed to express the depth of their sorrow—that any one who should not do so, would be considered deficient in respect for the deceased, and also as insulting to his family. The acts of violence committed, they added, were the effects of the paroxysms of their sorrow, which made them *neneva*, or insensible. They continued till their grief was *ua maha*, or satisfied, which often was not the case till they had received several severe blows upon the tender part above mentioned.

The females on these occasions sometimes put on a kind of short apron of a particular sort of cloth, which they held up with one hand, while they cut themselves with the other. In this apron they caught the blood that flowed from these grief-inflicted wounds, until it was almost saturated. It was then dried in the sun, and given to the nearest surviving relations as a proof of the affection of the donor, and was preserved by the bereaved family as a token of the estimation in which the departed had been held.

Had the otohaa been confined to instances of death, or seasons of great calamity, it would not have appeared so strange, as it does in connexion with the fact, that it was practised on other occasions, when feelings, the most opposite to those of calamity, were induced. In its milder form, it was an expression of joy, as well as grief; and when a husband or a son returned to his family, after a season of absence, or exposure to danger, his arrival was greeted, not only with the cordial welcome, and the warm embrace, but loud wailing was uttered, and the instrument armed with shark's teeth applied, in proportion to the joy experienced.

The early visitors, and the first Missionaries, were much surprised at this strange and contradictory usage; and, in answer to their inquiries, were informed, that it was the custom of Tahiti. The wailing was not so excessive, or the duration so long, nor were the enormities committed so great, as in the event of a death. The otohaa appears to have been adopted by the people to express the violence or excess of the passion with which they were exercised, whether joy or grief.

There was another custom associated with their bereavements by death, of an opposite character, and more agreeable to contemplate. This was their elegiac ballads, prepared by the bards, and recited for the consolation of the family. They generally followed the otohaa, and were often treasured up in the memory of the survivors, and eventually became a part of the ballads of the nation. Though highly figurative and beautiful in sentiment, breathing a pathetic spirit of sympathy and consolation, they were often historical, or rather biographical, recounting, under all the imagery of song, the leading events in the life of the individuals, and were remarkably interesting, when that life had been one of enterprise, adventure, or incident.

"In every nation it has been found that poetry is of much earlier date than any other production of the human mind," and I am disposed to ascribe the highest antiquity to these ballads. Much of their mythology is probably to be ascribed to this source, and many of their legends were originally funeral or elegiac songs, in honour of departed kings or heroes. I have heard them recited, and have often been struck with their pathos and beauty; two lines of one, which Mr. Nott heard recited for the consolation of a mother and family, on the death of an only son, have always appeared exceedingly beautiful. The grief generally felt was described in affecting strains, and then, in reference to sympathy of a higher order, it was added—

To rii rii te ua ite iriatae:
Eere ra te ua, e roimata ïa no Oro.

The literal rendering of which would be—

"Thickly falls the small rain on the face of the sea,
They are not drops of rain, but they are tears of Oro."

The sentiment of the second line is weakened by the introduction of the plural pronoun and the conjunction; but, preserving the idiom, as well as the sense, the line would be—

Not rain, but the *weeping* it is of Oro.

In the Tahitian, the word for tears, *roimata*, is the same in the singular and plural, and accords with the singular pronoun.

Scarcely had Taaroarii, the young chieftain of Huahine, been consigned to the tomb, when a ballad was prepared, after the ancient usage of his country. I heard it once or twice, and intended to have committed it to paper, but my voyage to the Sandwich Islands, shortly afterwards, prevented. It commenced in a truly pathetic manner; the first lines were—

Ua moe te teoteo o Atiapii i roto te ana
Ua rava e adu tona uuuuna.

"The pride of Atiapii* sleeps in the cavern;
Departed has its glory, or its brightness," &c.

* One of the names of the island of Huahine.

It was throughout, adapted to awaken tenderness, and feelings of regret at the event, and sympathy with the survivors.

Several weeks after the decease of a chief or person of distinction, another singular ceremony, called a *heva,* was performed by the relatives or dependants. The principal actor in this procession was a priest, or relative, who wore a curious dress, the most imposing part of which was the head-ornament, or parae. A cap of thick native cloth was fitted close to the head; in front were two large broad mother-of-pearl shells, covering the face like a mask, with one small aperture through which the wearer could look. Above the mask a number of beautiful, long, white, red-tipped, tail feathers of the tropic bird, were fixed, diverging like rays; beneath the mask was a curved piece of thin yet strong board, six or nine inches wide in the centre, but narrow at the ends, which were turned upwards, and gave it the appearance of a crescent.

Attached to this was a beautiful kind of net-work of small pieces of brilliant mother-of-pearl shell, each piece being about an inch, or an inch and a half long, and less than a quarter of an inch wide. Every piece was finely polished, and reduced to the thinness of a card; a small perforation was made at each corner, and the pieces fastened together by five threads passed through these perforations. They were fixed perpendicularly to the board, and extended nearly from one end to the other. The depth varied according to the taste or means of the family, but it was generally nine inches or a foot.

The labour in making this part of the parae must have been excessive. The many hundred pieces of mother-of-pearl shell that must have been cut, ground down to the required thickness, polished, and perforated, without iron tools, before a single line could be fixed upon the head-dress, required a degree of patience that is surprising.

This part covered the breast of the wearer; a succession of pieces of black and yellow cloth fastened to the pearl-shell netting, surrounded the body, and reached sometimes to the loins, to the knees, or even to the ankles. The beautiful mother-of-pearl shell net-work was fringed with feathers; a large bunch of man-of-war-bird's plumage was fixed at each end of the board, and two elegantly shaped feather tassels, hanging from each end, were attached to the light board by cords, also covered with feathers.

In one hand the heva carried a paeho, a terrific weapon about five feet long, one end rounded for a handle, the other broad and flat, and in shape not unlike a short scythe. The point was ornamented with a tuft of feathers, and the inner or concave side armed with a line of large, strong, sharks' teeth, fixed in the wood by the fibres of the tough *ieie*. In the other hand he held a *tete* or kind of clapper, formed with a large and a smaller pearl-oyster shell, beautifully polished.

The man thus arrayed led the procession, and continued, as he walked along, to strike or jingle the shells against each other, to give notice of his approach. He was attended by a number of men and boys, painted with charcoal and red and white clay, as if they had endeavoured to render themselves as hideous as possible. They wore only a maro or girdle, and were covered with these coloured earths. Sometimes the body was painted red, with black and white stripes; at other times the face painted red or black, and the rest of the body red and white. They were armed with a club or cudgel, and proceeded through the district, seizing and beating every person they met with out of doors. All who saw their approach instantly fled, or hid themselves.

They did not enter any of the dwellings, but often struck them as they passed by, to the great terror of those within. They appeared and acted as if they were deranged, and were supposed to be inspired by the spirit of the deceased, to revenge any injury he might have received, or to punish those who had not shewn due respect to his remains.

Since the introduction of Christianity, these and other barbarous and heathen customs, in connexion with the death and burial of the natives, have ceased; the rites and usages of Christian burial, as far as it seemed desirable, or the circumstances of the people would admit, have been introduced, and are generally observed. At each of the Missionary stations, a piece of ground near the sea-shore, and at some distance from the houses, has been devoted by the government to the purposes of interment, and all who die near are buried there.

Those who die in the remote districts are buried by their friends near the place; sometimes in the vicinity of their little rustic chapel, at others in the garden near their dwelling. They are not always deposited in a coffin, as the survivors are often destitute of boards and nails; they are, however, decently interred, usually wrapped in native cloth and matting, and placed in the keel or lower part of a canoe.

If there be a native Missionary or teacher near, he is called to officiate at the interment; if not, a male branch of the family usually offers up a prayer when the body is committed to the earth. Some inconvenience was sustained when the natives first embraced Christianity, with regard to the burial of those who died at a distance from the Missionary station. The heat of the climate was such as often to render it necessary to inter them on the day of their decease, or on that which followed, and they had not time to send for a native teacher. To obviate this, a prayer suitable to be offered up at the time of interment was written, and distributed among the natives, for the use of those who resided at a distance. This appeared not only according to christian propriety, but necessary, from any latent influence of the former superstitions, which might lurk in the minds of those who, though they renounced idolatry, were but very partially instructed in many points of Christian doctrine.

At the Missionary stations, the corpse has seldom been brought to the place of worship. We in general repair to the house, and, offering up a prayer with the family, accompany the procession to the place of interment; our practice, however, in this respect is not uniform, but is regulated by circumstances.

On reaching the burying-ground, we stand by the side of the grave, which is usually about six feet deep, and when the coffin is lowered down, address the friends of the deceased, and the spectators, and conclude the service with a short prayer.

At first they believed that the deceased must be in some degree benefited by this service; and that such should occasionally have been their ideas, is by no means surprising, when we consider the mass of delusion from which they had been so recently delivered. This, however, rendered it necessary for us to be more explicit in impressing upon their minds, that the state of the dead was unalterably fixed, and that our own benefit alone could be advanced by attending it.—But the views and ceremonies connected with death, and with the disposal of the body, either in the pagan or Christian manner, are unimportant in comparison with the change in the individuals who have died, and the views and anticipations which, under these systems, different individuals have entertained. "One thing, of all I have read or heard," said the aged and venerable Matahira, "now supports my mind: Christ has said, I am the way."

> "He the beloved Son,
> The Son beloved, Jesus Christ,
> The Father gave,
> That we through him might live,"

was sung by another in the native language, with the last breath she drew. "I am happy, I am happy," were among the last words of the late distinguished regent of the Sandwich Islands. These are expressions no pagan ever used, in looking forward to his dissolution. They result alone from the effects which the mercy of God in Christ is adapted to kindle in our hearts, augmented by gratitude to Him who hath brought life and immortality to light.

END OF VOL. I.

London: Henry Fisher, Son, and P. Jackson, Printers.

www.ingramcontent.com/pod-product-compliance
Ingram Content Group UK Ltd.
Pitfield, Milton Keynes, MK11 3LW, UK
UKHW040700180125
453697UK00010B/309